P9-CJZ-834

RESOURCES AND INDUSTRY

RESOURCES AND INDUSTRY

GENERAL EDITOR

Dr Ian Hamilton

New York
OXFORD UNIVERSITY PRESS
1992

CONSULTANT EDITOR
Professor Peter Haggett, University of Bristol

Dr Claes Alvstam, Gothenburg University, Sweden
The Nordic Countries
Dr Kathy Baker, School of Oriental and African Studies, London, UK
Central Africa
Professor Graham P Chapman, School of Oriental and African Studies, London, UK
The Indian Subcontinent
Dr Susan Cunningham, London, UK
South America
Professor Maurice T Daly, University of Sidney, Australia
Australasia, Oceania and Antarctica
Dr Evangelia Dokopoulou, London, UK
Italy and Greece, Central Europe, Spain and Portugal
Dr David Fox, University of Manchester, UK
Central America and the Caribbean
Dr Charles Gurdon, London, UK
Northern Africa, The Middle East
Dr Ian Hamilton, London School of Economics, UK
Raw Materials and Industrialization, Northern Eurasia, Japan and Korea, Eastern Europe
Dr Roger Hayter, Simon Fraser University, British Columbia, Canada
Canada and the Arctic
Dr Rupert Hodder, London, UK
China and its Neighbors
Graham Keats, Johannesburg College of Education, South Africa
Southern Africa
Dr E Willard Miller, The Pennsylvania State University, USA
The United States
Dr Stewart Richards, University of British Columbia, Canada
Southeast Asia
Professor Suzane Savey, Université Paul Valéry, France
France and its Neighbors
Dr Kenneth Warren, Jesus College, Oxford, UK
British Isles
Professor Egbert Wever, Buck Consultants International, Netherlands
The Low Countries

AN EQUINOX BOOK

Planned and produced by
Andromeda Oxford Limited
9-15 The Vineyard, Abingdon
Oxfordshire, England OX14 3PX

Published in the United States of America by
Oxford University Press, Inc.,
200 Madison Avenue,
New York, N.Y. 10016

Oxford is a registered trademark of
Oxford University Press

Library of Congress
Cataloging-in-Publication Data

Resources and industry / general editor, Ian Hamilton
p. cm.
Includes index.
ISBN 0-19-520943-5
1. Economic geography. 2. Raw materials. 3. Natural resources. 4. Industry. 5. Resource-based communities.
I. Hamilton, F. E. Ian.
HF1025.R385 1992
330.9--dc20 92-19727
CIP

Volume Editor Fiona Mullan
Editor Victoria Egan
Designers Chris Munday, Frankie Wood
Cartographic Manager Olive Pearson
Cartographic Editors Katrina Ellor, Sarah Rhodes
Picture Research Manager Thérèse Maitland
Picture Researcher David Pratt

Project Editor Susan Kennedy
Art Editor Steve McCurdy

ISBN 0-19-520943-5

Printing (last digit): 9 8 7 6 5 4 3 2 1

Printed in Singapore by C.S. Graphics

INTRODUCTORY PHOTOGRAPHS
Half title: *Copper worker, Zaire (Gamma, P. Maitre)*
Half title verso: *Uranium enrichment factory, Brazil (Explorer, Gerard Boutin)*
Title page: *Hand-crafted violins, Cremona, Italy (Frank Spooner, Marc Deville)*
This page: *Kodak packaging line, Rochester, New York, USA (Frank Spooner, Matt/Liaison)*

Contents

VOITH

PREFACE

SINCE THE DAWN OF CIVILIZATION, PEOPLE HAVE ACQUIRED AND APPLIED technical knowledge and skills to harness the Earth's resources for making the materials and tools they require. At the most basic level these satisfy their needs or wants in food, shelter, clothing, defence, transport and leisure. The ways people meet their needs have become more complex, diverse and sophisticated in the past 250 years since the advent of the first Industrial Revolution. The rise of modern industrial society is associated with the discovery and exploitation of new resources on a vast scale; reappraisal of traditional resources; far-reaching changes in the products, processes and services people use; and the spread of mining, manufacturing and services to more regions of the world.

At the beginning of the 20th century, the introduction of electricity, the internal combustion engine, chemicals manufacturing and mass-production assembly lines brought about startling transformations in the ways people live, work, and travel. Today, once again, the world is experiencing rapid industrial change. Even wider use of computers and robotics, development of biotechnology, solar power, and vastly improved communications are among innovations that promise to revolutionize how and where we live, work and interact with each other on Earth – and probably also in Space – in the 21st century.

Industry, though, in providing for our needs and desires can deplete the Earth's resources, scar landscapes, and impair people's health through noise, dirt, odors and effluent. Millions of families in many regions of the world, depend directly or indirectly on industry for employment and income, though few may derive wealth or job satisfaction from it. Such dilemmas are a recurrent theme in this book.

The starting point in each region is the quality and location of its natural resources and how they can be exploited both now and in the future. The second stage examines the process of change as old manufacturing sectors and methods decline, and new ones emerge. How have industries across the world reacted to the changes forced upon them by world events, changing governments and fluctuating supplies of raw materials? In several regions service industries are replacing manufacturers as key employers. The third major topic in every region is the workforce. How are they – as producers and consumers, entrepreneurs and workers, government policymakers and public watchdogs – shaping contemporary and future resource use and industrial development? You the reader are offered information and comment to assess these issues for yourself.

Dr. F.E. Ian Hamilton

London School of Economics and School of Slavonic & East European Studies

Paper manufacturing, Union Camp Paper, North Carolina, USA

Molten tin being refined at Oruro, Bolivia (overleaf)

RAW MATERIALS AND INDUSTRIALIZATION

RESOURCES FOR INDUSTRY

FROM WHEEL TO MICROCHIP

THE ECONOMIC CONTEXT

What is a Resource?

A RESOURCE IS ANYTHING PEOPLE USE OR might want to use to achieve an end. More specifically, it is a natural element, organic substance or human skill used to produce commodities, manufacture goods or offer services that people need. No material, however much of it exists, becomes a resource until people find a use for it and value that use. What amounts to a resource, therefore, changes in response to developments in technology, the economic climate, and people's values.

Petroleum is a good example of a natural substance that has altered radically in commercial value over the centuries. Although present in the Earth for millions of years, it was not exploited by primitive peoples in the way that precious metals were. The Persians considered seeping oil to be a nuisance, but later found that its water repellent qualities made it suitable for caulking ships. Only centuries later did new technology make oil a resource useful enough for men to search it out and drill for it. It was not until the invention of the kerosene lamp in 1853 that oil rose to prominence as a potential fuel valued for its lighting and heating properties. In the 20th century, the development of the internal combustion engine made oil and related products the world's leading fuel for vehicles, planes and ships as well as providing energy for industry and domestic use. Nicknamed "black gold", oil became an economic and political asset for world governments, essential to the armed forces, and a trading commodity for states and entrepreneurs.

Renewable and nonrenewable resources

The wealth and variety of the Earth's resources are unparalleled – to our present knowledge – in the universe. Fossil fuels, metals and minerals are its "stock" resources. Formed during the Earth's long geological evolution, their supply is finite. Fossil fuels are burnt for energy and once used, are lost for ever. Metals, however, are largely recyclable, and with care, could become one of the world's group of renewable resources.

The Earth is also abundant in largely untapped "flow" resources: solar energy, air, wind, running water, waves and tides. Farming, forestry and fishing yield varied "biotic" resources that we use for food, shelter and clothing; for example, wheat for flour, hops for beer, wool for clothing and wood for furniture and paper. Like flow resources, biotic resources are renewable with efficient management. However, massive and insensitive exploitation of forests and fish mean that neither can regenerate, converting them from flow to stock status, and threatening their future supply.

Prospecting and exploring

New tools, such as satellite imagery, aid the discovery of new reserves, but knowledge of where resources are and how to manage them remains imperfect, subjective and also debatable. The most secure predictions are estimates of proven reserves; the known deposits of sufficient quality and accessibility to be commercially workable with current technologies. Just how valuable these resources are depends on market demand and prices, extraction and processing costs and whether they can compete with artificial or naturally occurring alternatives.

Crucial for future generations are the Earth's conditional reserves – fossil fuels or mineral deposits that are presently uneconomic to extract. These known deposits may become workable if economic

MAIN ELEMENTS OF THE EARTH'S CRUST

Of 92 natural elements, just nine account for 99 percent of the crust's weight. Oxygen and silicon are widespread and are very frequently found combined with other elements as oxides or silicates. Of the remainder, some, such as iron, occur in many regions. Key industrial metals such as nickel, tin and copper occur sporadically, localized in few places. Platinum, gold and silver are very rare.

% of main elements	
Oxygen	46.6
Silicon	27.7
Aluminum	8.1
Iron	5.0
Calcium	3.6
Sodium	2.8
Potassium	2.6
Magnesium	2.1
Titanium	0.5
*other elements	1.0

*other elements/parts per million	
Zirconium	170
Chromium	100
Nickel	70
Zinc	70
Copper	50
Gallium	15
Lead	13
Tin	3
Uranium	2
Silver	0.05
Platinum	0.01
Gold	0.005

Pooling nuclear power (*above*) Uranium, the key ingredient in nuclear fission reactors and some military weapons, is a highly radioactive element. It must be securely stored to prevent any leakage. Main producers are the United States and Australia.

The cost of mining (*left*) Open-cast mining is a cost-effective way of obtaining ore close to the surface. However, the damage done to the land is enormous, as this iron ore mine in western Australia demonstrates.

and political conditions change and better techniques emerge to mine, process and conserve them. Beyond the resources we already know about are hypothetical resources, which experts predict on the basis of geological evidence and past experience. Typically these might include future oil discoveries in the North Sea, Arctic North America, the Middle East or the South China Sea. Finally, speculative resources describe deposits yet to be found in hitherto unexplored regions, for example, in South America, Africa, Australia, Antarctica, Siberia in the former Soviet Union or under the oceans.

There is a continuing debate about whether world resources are adequate to withstand the way we manage them. Commentators in the 1970s saw combined economic and population growth as unsustainable across the globe. Increased materials and energy consumption, together with environmental damage, would exhaust resources, causing a massive decline in factory and farm output in the early 21st century. Today, new technologies are reducing consumption of energy and minerals. At the same time, paper, glass, metal and plastic are being recycled, and fossil fuels and renewable energy souces are being used more efficiently. With careful use of resources, we can begin to face the future with cautious optimism.

Fuels for Energy

Among the most important of the Earth's resources are those that provide fuel – for industry, for transportation, for electric light, for domestic cooking and heating. By far the greatest part of our energy needs comes from conventional fossil fuels – coal, oil and natural gas – so-called because they have been formed in the ground over millions of years from plant and animal remains, providing 86 percent of fuel and power worldwide. They are nonrenewable. Estimates put nature's original endowment at 10.75 trillion tonnes of coal, 2.3 trillion barrels of oil and 2 trillion barrels oil-equivalent of natural gas.

Coal is still a desirable fuel. At the end of the 1980s, it was the source of 28 percent of all fuel and power. Improved surveys raised proven world reserves in 1990 to 1.6 trillion tonnes (3 percent of which has been used). This is still only 15 percent of the Earth's total stocks, but is sufficient at current production levels for another 340 years. Coal occurs in several different forms, some more valuable than others as fuels. The geologically oldest coals are the hardest, and have high heating (calorific) value. Sold as anthracite, bituminous and coking coals, they are effective sources of fuel for energy.

Softer, younger brown coal (lignite) has much poorer heating properties. Its only advantage is that it lies closer to the surface, and can be more easily and cheaply mined, especially in regions that lack alternative fuels. Following steep petroleum price rises in the 1970s, coal was substituted for oil to meet energy demands in coal-rich but oil-deficient countries such as the United States, South Africa and Australia, leading to an 80 percent expansion of production since 1960.

In the older industrialized parts of the world, hard coal has been exploited for so long that the most accessible supplies have been exhausted. Continued output of hard coal often involves sinking deep mineshafts, both costly and hazardous. In western Europe most remaining hard coal is in thin and geologically disturbed deposits, which are expensive to mine. This has led to pit closures, as mines must compete against cheaper imports of American, Australian and Polish coal. By contrast, huge brown coal basins in remote areas from markets – in the Great Plains of the United States and in Siberia in Russia – may long remain untouched.

The future for oil

Petroleum now supplies three-fifths of world energy – 39 percent from oil, and 21 percent from natural gas. First drilled in 1859 at Titusville, Pennsylvania in the United States, and Krosno, in present-day Poland, oil rapidly became the 20th century's most important energy source. It still fuels most forms of transportation, as well as being a versatile raw material in a range of manufacturing industries, including chemicals, detergents and plastics. Between 1960 and 1992 output of oil had trebled and gas quadrupled. The advantages they have over coal are greater heating value, more efficient combustion, almost no waste, easier and cheaper extraction by drilling, transportation by pipeline or supertanker, and a much wider range of applications.

Oil under water (*below*) The enormous consumption of world oil has made exploration and drilling more difficult as land reserves are used up. Oil rigs in the North Sea are expensive and dangerous, but provide substantial fuel supplies.

OPEC – CONTROLLING THE WORLD'S OIL

In 1960 countries from the Middle East, Northern Africa, western Africa and Asia came together to form the Organization of Petroleum Exporting Countries (OPEC). This has attempted, as a cartel, to exploit oil's strategic importance by coordinating production and raising prices. OPEC controls 77 percent of proven oil reserves. The Middle Eastern and North African members alone have 29 of the world's 37 "supergiant" oilfields, with stocks of more than 5 billion barrels each, together with 41 percent of the world's proven reserves of natural gas. Most of the major oil-consuming areas such as the United States, western Europe and Japan must import OPEC oil, but at present drilling rates, oil could run out by 2035. Already 32 percent of proven oil reserves and 11 percent of proven natural gas reserves have been used.

It seems unlikely that OPEC will continue to control large energy reserves for much longer. More of the world's underutilized natural gas reserves are being tapped to substitute for oil. The gas is brought by pipeline from the remote districts of Northern Eurasia, or by tanker as liquefied natural gas (LNG) from Alaska, Australia, Brunei and Malaysia. In the long term, improved technology can reduce oil usage, but alternative forms of energy will also have to be developed for use by future generations.

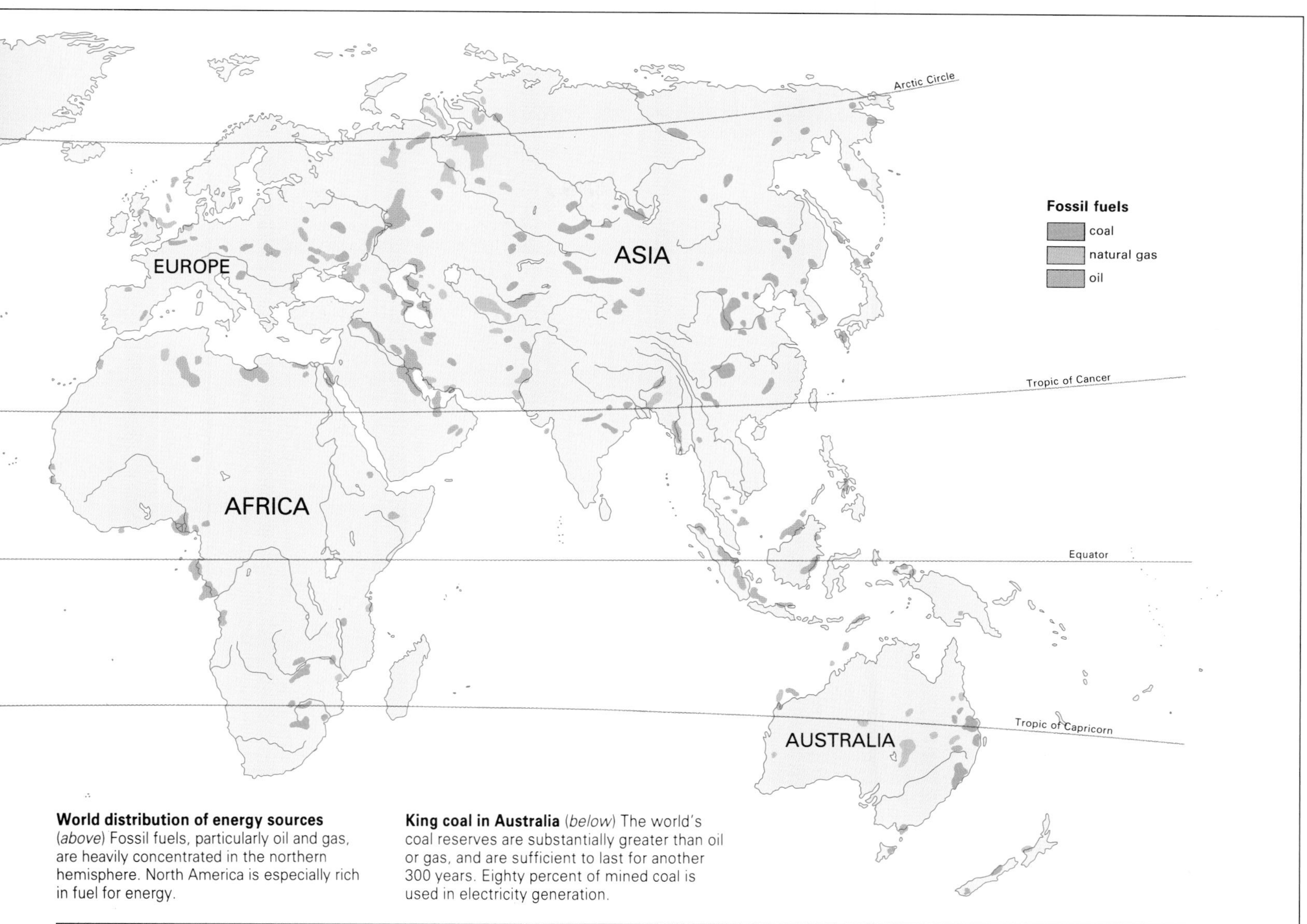

World distribution of energy sources (*above*) Fossil fuels, particularly oil and gas, are heavily concentrated in the northern hemisphere. North America is especially rich in fuel for energy.

King coal in Australia (*below*) The world's coal reserves are substantially greater than oil or gas, and are sufficient to last for another 300 years. Eighty percent of mined coal is used in electricity generation.

Electrifying the World

ELECTRICITY IS A MANUFACTURED FORM OF energy that can be produced from any source – a fossil fuel, water, wind or even biowaste. Commercial output began in 1882 with the first hydroelectric power (HEP) station, in Wisconsin in the United States. This was followed in 1884 by the introduction of coal-fired steam turbines in England.

Electricity production grew massively during the second half of the 20th century. The volume produced in the United States alone in 1988 far exceeded the total world figure for 1960. This growth is accounted for by a significant increase in population, widespread industrialization, and the extension of power supplies to rural areas of the Third World, the countries of the former Soviet Union and Eastern Europe.

Spreading the burden

In the 1970s, following price rises in oil, coal became more competitive for large-scale power generation. Larger coal-fired power stations were commissioned, some exceeding outputs of 4,000 megawatts, mostly in inland coalfields such as Britain's East Midlands or close to brown-coal mines in central and Eastern Europe.

Oil remains competitive for smaller-scale plants, especially in remote, less-developed or thinly populated regions. Natural gas may become the generating fuel of the future, being more environmentally friendly than coal or oil as well as more efficient. "Combined cycle" gas turbines achieve 45 percent higher thermal efficiency than new coal stations. In future, it is likely that piped natural gas from the North Sea, Northern Eurasia and the Middle East, or gas carried in liquid

TOP TEN ELECTRICITY PRODUCERS

Country	Billion kilowatt-hours 1960	1988
USA	844	2,854
Soviet Union	292	1,698
Japan	116	754
China	58	538
Canada	115	504
West Germany	119	429
France	72	392
United Kingdom	137	308
India	18	238
Brazil	23	214

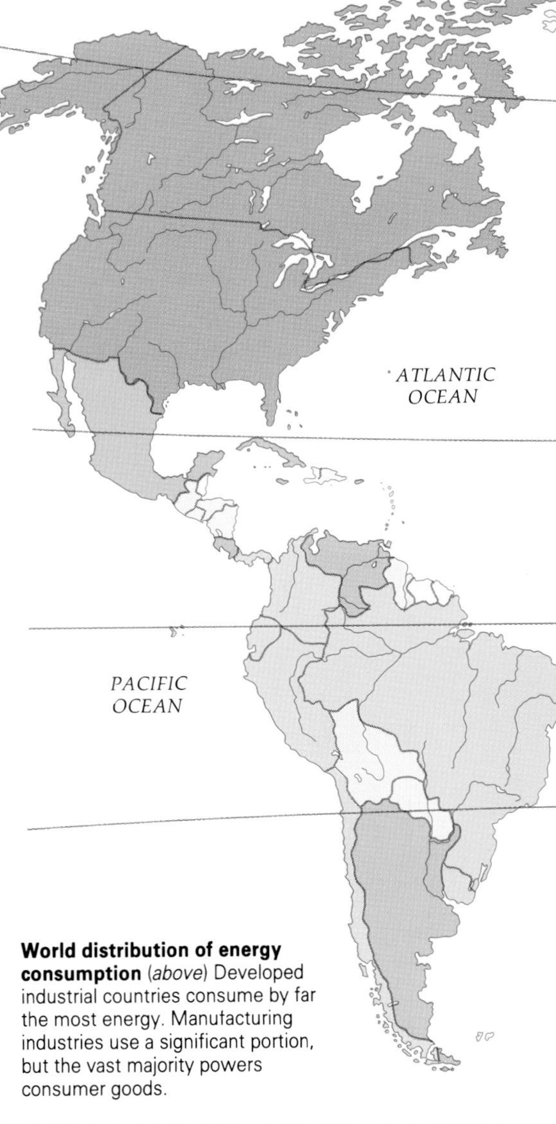

World distribution of energy consumption *(above)* Developed industrial countries consume by far the most energy. Manufacturing industries use a significant portion, but the vast majority powers consumer goods.

Transforming power *(below)* For all the advantages of electric power, electricity generation and transmission are very wasteful. Up to half of the potential energy is lost during conversion from a primary fuel (such as coal or oil) to electricity itself.

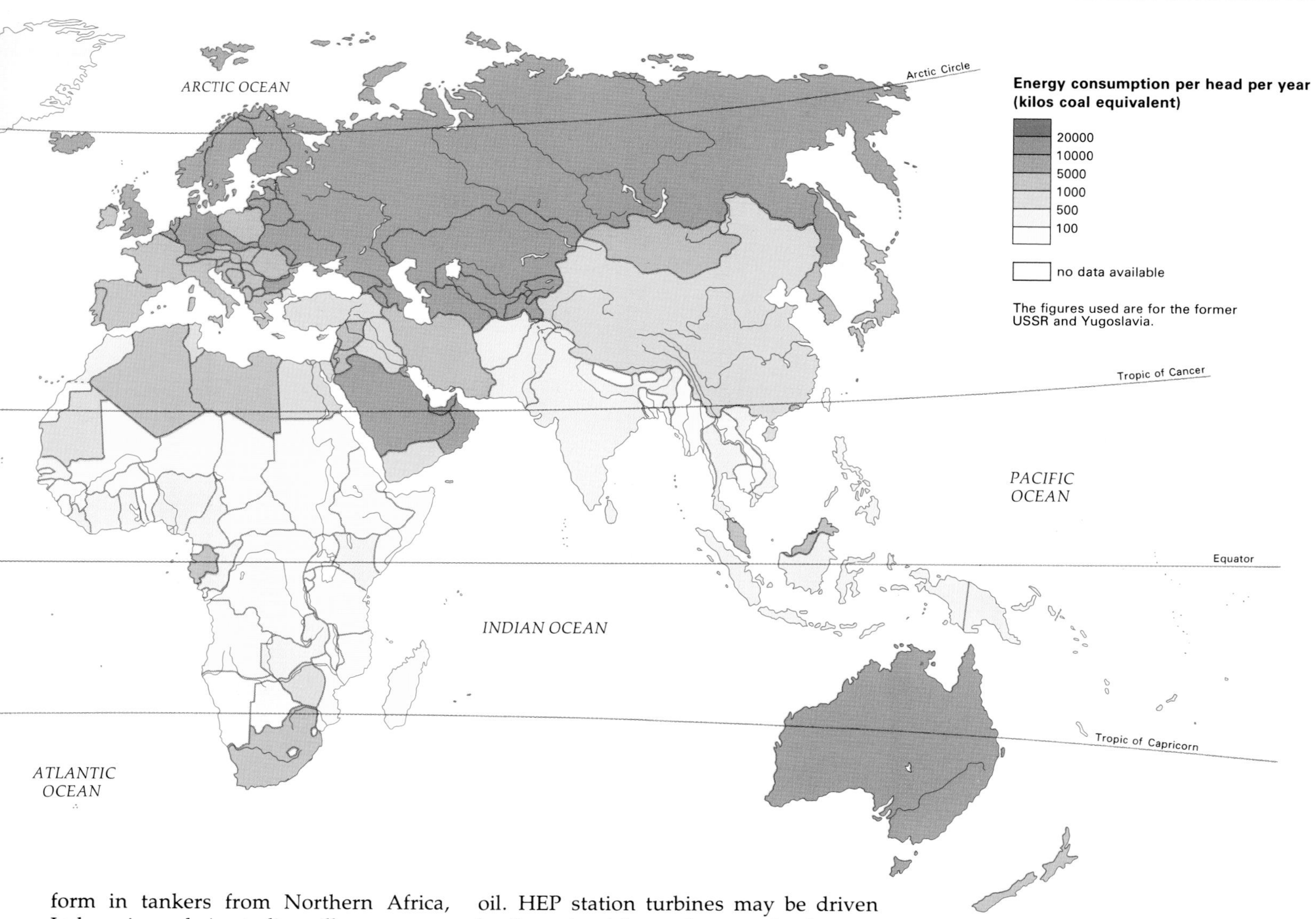

form in tankers from Northern Africa, Indonesia and Australia will serve new gas-fired power stations located on coastal sites in Europe, Africa and Asia.

The alternating current

Water – harnessed in HEP stations – generates almost one-fifth of world electricity, equivalent to 3.85 billion barrels of oil. HEP station turbines may be driven by the natural force of mountain rivers as they fall from high plateaus to the plain. Niagara Falls is probably the best example. In the absence of a natural waterfall, dams placed across deep or low-lying broad valleys create lakes and powerful artificial heads of water. A number of major new projects have made HEP the dominant electricity supply source in many countries in Africa, South America and Southeast Asia, where it guarantees cheap energy without reliance on expensive oil imports.

At the present time, all other renewable energy sources yield only half of one percent of world electricity. Economic and technical inefficiencies restrict large-scale modern windmill projects to local use. Suitable sites are flat coastlands and islands; and countries that have launched successful schemes include Denmark and the Low Countries. Biofuels, which harness methane gas from enormous volumes of organic urban waste, have also had local success, but have not yet been developed on a larger scale.

Of the various alternative energy sources, solar energy is potentially the most abundant, but is the hardest to collect and store. The Sun could supply 12,000 times the present world energy demands. It is in the equatorial countries and arid areas of the world, which mainly lack the money and expertise to develop it, that solar energy has the greatest potential use.

THE NUCLEAR POWER DEBATE

Nuclear power – the production of electricity by harnessing energy released when radioactive uranium atoms are split – is probably the most controversial means of generating electricity. Production has grown rapidly since the construction of the first laboratory nuclear reactor in Chicago in the United States in 1942. In 1956 power was generated commercially from the 20 megawatt Sellafield plant in Cumbria, Britain and today, 300 atomic power stations worldwide yield 17 percent of all electricity, almost as much as that produced by HEP.

Nuclear power is attractive in many ways. Virtually limitless power output from uranium is possible for centuries. Annually it saves the equivalent of 3.3 billion barrels of oil or 675 million tonnes of coal. It supplies two-thirds of all power in France and Belgium, between 10 and 50 percent in many other European states, the former Soviet Union, Japan, South Korea, Canada, the United States and Argentina. The major producers are the United States (31 percent of world nuclear power), France (13.3), Japan (10.3), the former Soviet Union (10.1) and Germany (7.5).

The explosion at the Chernobyl plant, in Ukraine, in the Soviet Union, in 1986 and its global fallout drew world attention to the enormous costs to life and the environment when nuclear power goes wrong. Incidents such as this raise safety standards, but they are unlikely to stop nuclear power programs until other safer, renewable energy sources are commercially viable.

Industrial Metals

Purification by fire Iron is refined and processed with other metals to make steel at this foundry in Qatar in the Middle East. Like many nations, Qatar has to import iron ore and other raw materials to make this industrial necessity.

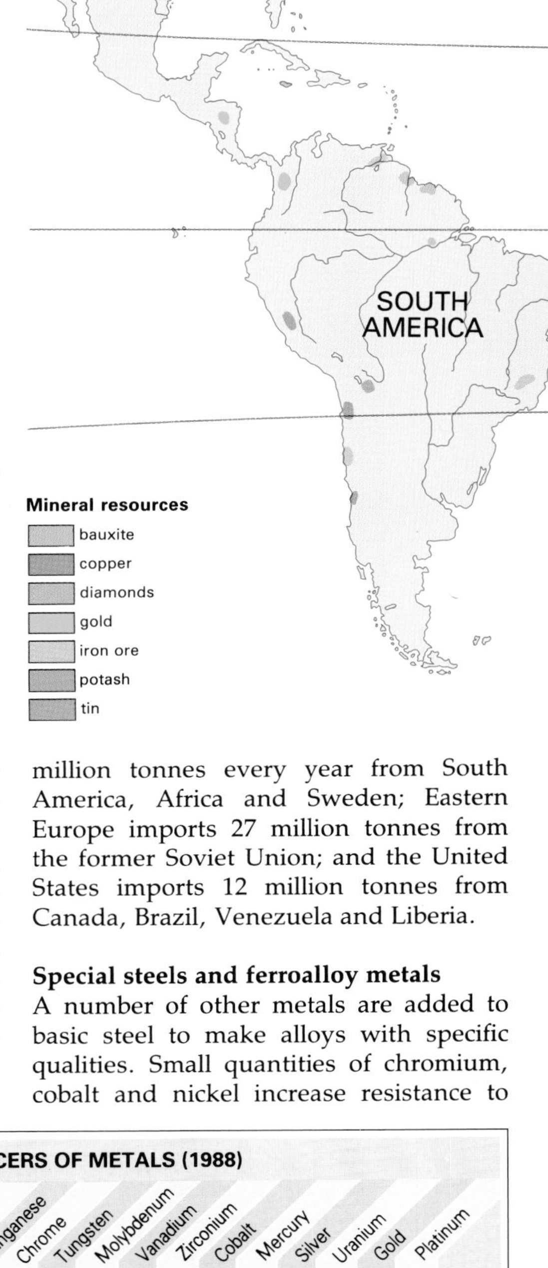

ALL METALS OCCUR IN NATURE AS ORES; mixtures of metal with waste rock and unwanted minerals. Copper, gold, silver and platinum do exist as pure metals within the ores, but most occur as compounds of the metal with sulfur (sulfides) or oxygen (oxides). Whether or not it is cost-effective to mine specific deposits depends on their size, quality, the presence of other valuable minerals and how they are processed. Market demand, trading prices and the ease and cost of transportation to customers are also taken into consideration.

Iron, abundant in many areas, is by far the most widely used metal. Most of it is made into steel, an alloy of iron. The iron and steel industry uses 565 million tonnes of refined iron annually.

Compared with other ores, the metal content of iron ore is high. The best deposits are rich in iron but low in phosphorus and sulfur impurities. These high-quality ores occur in major iron-producing countries such as Brazil, India, Australia and Sweden. Large, long-established steel industries have exhausted the richest deposits in Europe, which now imports iron ore. The United States and former Soviet Union rely increasingly on poorer ores. These must be concentrated near mines before it makes economic sense to transport them. Bolivian reserves, though high quality, are unworked because of their remoteness.

Iron ore is mined in 54 countries and exported by 27, yet two-thirds of known reserves are found in just two areas, the former Soviet Union (39 percent) and Brazil (24 percent). Japan and Korea are the biggest importers, receiving 100 million tonnes annually from Brazil, Australia and India. The European Community, Finland and Austria import 80 million tonnes every year from South America, Africa and Sweden; Eastern Europe imports 27 million tonnes from the former Soviet Union; and the United States imports 12 million tonnes from Canada, Brazil, Venezuela and Liberia.

Special steels and ferroalloy metals

A number of other metals are added to basic steel to make alloys with specific qualities. Small quantities of chromium, cobalt and nickel increase resistance to

WORLD TOP PRODUCERS OF METALS (1988)

Country / Rank in Output	Iron Ore	Copper	Nickel	Bauxite	Lead	Zinc	Tin	Manganese	Chrome	Tungsten	Molybdenum	Vanadium	Zirconium	Cobalt	Mercury	Silver	Uranium	Gold	Platinum
United States	5	2			3	7					1				4	3	2	3	4
Canada	7	4	1		4	1					4			4		5	1	5	3
Australia	4		3	1	1	3	7	5		5			1	5		6	5	4	
South Africa	8		7		10			2	1			1	2				4	1	1
Soviet Union	1	3	2		2	2	5	1	2	2	3	2		3	1	4	3	2	2
China	3		9	7	5	4	2	6		1		3	6					6	
India	6			5				7	5				5						
Brazil	2			3			1	4	7	7			3					7	

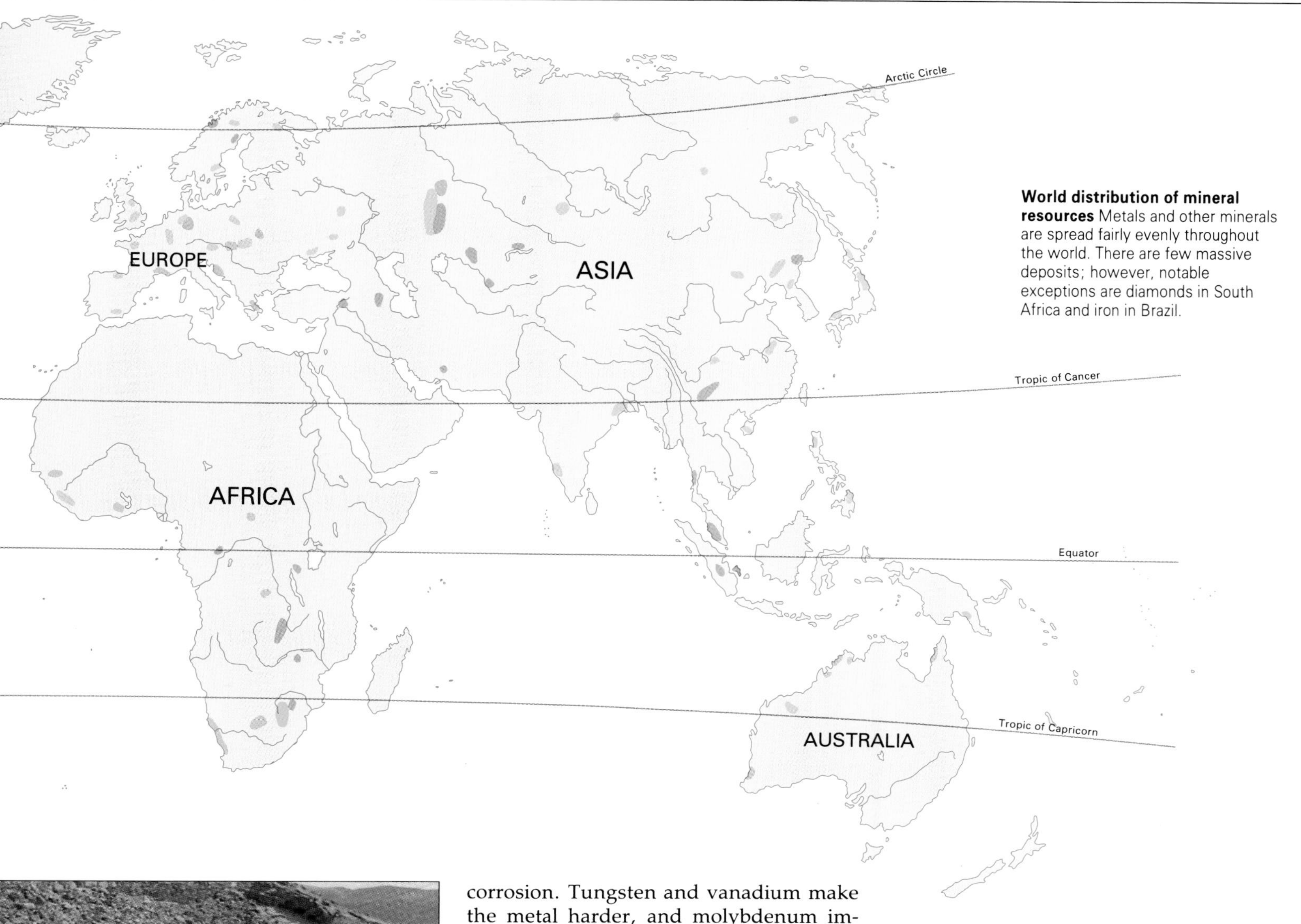

World distribution of mineral resources Metals and other minerals are spread fairly evenly throughout the world. There are few massive deposits; however, notable exceptions are diamonds in South Africa and iron in Brazil.

Small-scale zinc mining in Bolivia (*above*) Traditional methods of extraction and processing are still used in some of the world's developing countries. Once processed, this ore is most likely to be used in galvanizing steel to make it rustproof.

corrosion. Tungsten and vanadium make the metal harder, and molybdenum improves workability. In addition, some metals are used as part of the refining process. Approximately 8 million tonnes of manganese are used in steelmaking each year to absorb the impurities found in furnace materials, as well as being added to the final product to strengthen the resistance of steel rails.

Ores of these metals occur throughout the world, not necessarily associated with iron ore. While most can be refined to a pure metal, the process is often difficult and expensive. If the metal is intended for use in steel manufacture, the ore is normally refined with some iron to produce an alloy known as a ferroalloy (iron plus at least one other metal). The steel industry normally uses refined iron and ferroalloys to achieve the correct proportions in each particular kind of steel.

Rarer and precious metals

Much less plentiful than iron, deposits of copper, aluminum (from bauxite), lead, tin and zinc are major industrial resources and several million tonnes are mined annually across the globe. Bauxite mining has increased most in recent years, as aluminum replaces more expensive copper in kitchen utensils, heavier steel in automobiles and high-rise buildings, and is light enough to meet aerospace requirements. Many of these metals have wide-ranging uses in making alloys. Copper, for example, retards rust, and both zinc and tin are anticorrosive. Copper and aluminum are excellent conductors of electricity, and are critical to the electronics industry. Lead and zinc are used in the manufacture of batteries, and lead is valuable to the medical profession and in scientific laboratories because it cuts down radioactivity.

Gold and silver have been made into precious objects since the Bronze Age (4000–1400 BC) and many deposits have been worked out. More recently platinum and uranium have risen dramatically in value, as the result of gaining industrial prominence in new technologies. Platinum, for example, is used in catalytic converters to reduce harmful emissions from petrol engines, and for lining glass-making furnaces. Uranium commands high prices on the world market because of its use as a nuclear fuel.

The Larger Resource Pool

ENERGY AND METALS ARE THE KEY RESOURCES in modern manufacturing. Combined with other mineral resources, and also with biotic and flow resources (forests, crops, livestock, air and water), they form a still far greater pool of resources on which industries and services depend.

Nonmetallic minerals

With only few exceptions, nonmetallic minerals – including sand, gravel, limestone, clay, chalk and marble – are bulky, heavy, of low commercial value and costly to move. Unless cheap water transportation is available, they are used locally or processed near the minehead or quarry. Some have multiple and diverse uses. Gypsum is used to fill interior wall paneling in buildings and for plaster casts for broken limbs. Limestone is used for building wherever it can be quarried abundantly – for example in Mediterranean countries – and after processing it produces lime for fertilizers and for use in smelting and glassmaking. The majority of minerals, however, have only one dominant use.

Sand, gravel, clay, chalk and limestone are quarried all over the world to make cement, glass or bricks. Marble is much rarer, a luxury material in the construction trade found only in Greece, Italy, Iran, Spain and Mexico.

Other nonmetallic minerals have widespread use as chemical fertilizers. These include phosphates, potash and sulfur. Although abundant as Earth elements, high-grade, workable deposits are few. In the United States, for example, sulfur has to be transported from the Texas salt domes to the Midwest and to the northeast. Europe transports most of its phosphate from North Africa, its potash from Germany and its sulfur from Poland.

Diamonds are exceptional among the diverse range of nonmetallic minerals, in that they are relatively scarce, easy to

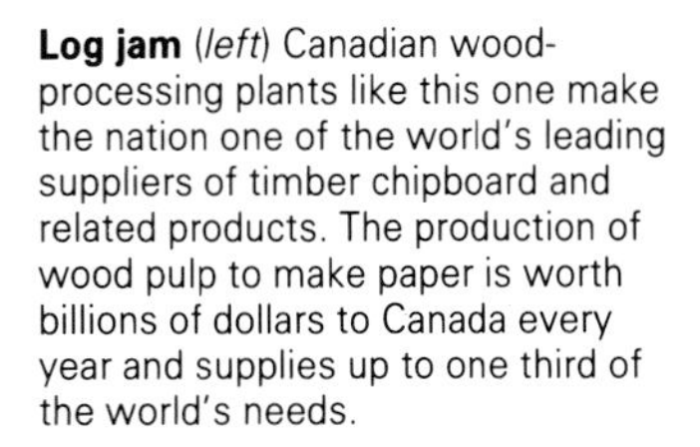

Log jam (*left*) Canadian wood-processing plants like this one make the nation one of the world's leading suppliers of timber chipboard and related products. The production of wood pulp to make paper is worth billions of dollars to Canada every year and supplies up to one third of the world's needs.

Brilliant natural yellow (*right*) Sulfur, common both in a natural, uncombined state and as a constituent of many minerals, is a soft, yellow substance used in the chemical industry. At this processing plant it is converted to sulfuric acid, a common ingredient in the manufacture of fertilizer and in petroleum refining.

LEADING WORLD PRODUCERS OF NONMETALLIC MINERALS

	Natural phosphates	Potash	Sulfur	Salt	Diamonds	Gem diamonds	Gypsum	Graphite
USA	13.8	1.5	3.2	35.5	-	-	14.9	-
Soviet Union	12.0	10.4	3.0	14.8	6.5	4.5	-	83.5
China	4.1	-	-	22.6	-	-	8.1	200
Canada	-	8.2	-	10.6	-	-	9.0	-
South Africa	-	-	-	-	4.6	3.7	-	-
Iraq	2.4	-	0.7	-	-	-	-	-
West Germany	-	2.9	-	7.2	-	-	2.3	-
Poland	-	-	5.0	6.2	-	-	-	-
Mexico	-	-	2.1	7.0	-	-	2.6	43.8
Botswana	-	-	-	-	4.2	11.0	-	-
Zaire	-	-	-	-	15.2	-	-	-

Figures for diamonds and gem diamonds are given in million carats. All other figures are given in million tonnes. All figures are for 1988.

transport, and very highly valued on the world market. Once dredged or mined, they are despatched by air from mining centers in central and southern Africa and eastern Siberia for cutting and polishing into jewelry and gemstones in Amsterdam or Brussels, or for industrial use in cutting tools and machines.

The environment as a resource

Fresh air is one of industry's most valuable materials. For decades, nitrogen has been extracted from the atmosphere and mixed with ammonia to produce fertilizers. Recently, however, the demand for industrial gases from air separation plants has soared. Steelmaking furnaces require oxygen, food manufacturers need liquid nitrogen for instant chilling and electronics firms require ultrapure gases in order to treat microchips and circuit boards.

The natural environment of land and water provides foods, fibers, hides and forest products, all of which are resources for industry. Crops and livestock are the backbone of the food-processing industries, and provide natural fibers such as wool, leather and silk for clothing and furnishing. Forests supply timber for building, construction and manufacturing processes, as well as paper for the printing industry, and they also yield a range of products including rubber for the chemical industries.

People are now beginning to use the environment as a resource in a different way – as a tourist attraction. Greater affluence, longer leisure time and the advances in transportation systems mean that more people are traveling farther afield in search of new cultural and leisure experiences. In some places, such as the South Pacific islands, this can be beneficial, bringing in foreign currency and creating much needed employment for the local population. However, all too often a comparatively sudden influx of foreigners can stretch natural resources to the limit and destroy the nature of the landscape without bringing lasting benefit to the people who live there; the profits instead go to multinational organizations and are not reinvested.

Machineless Manufacturing

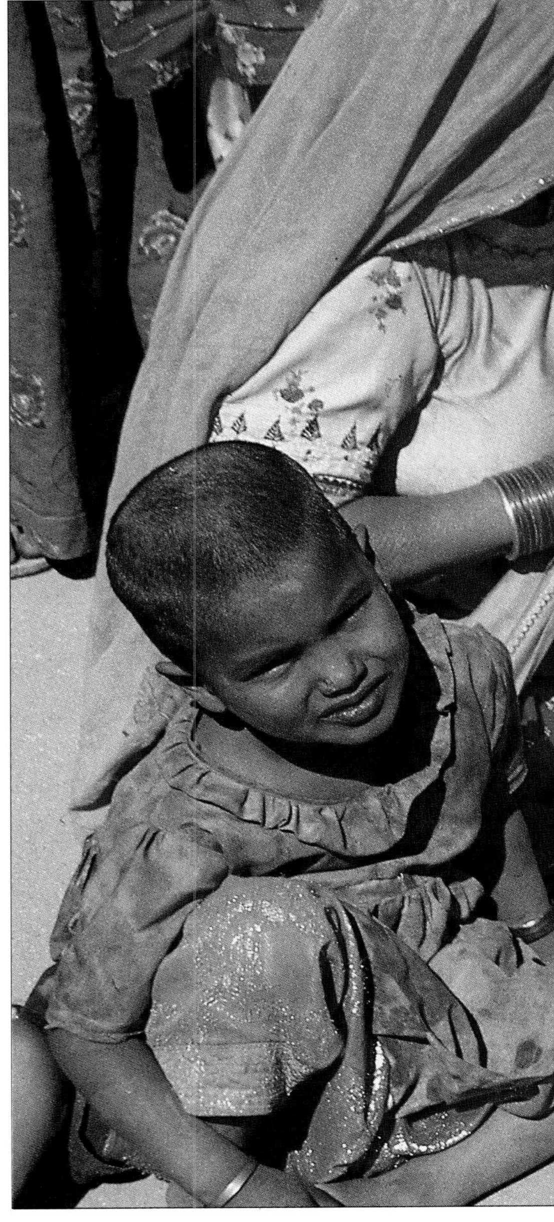

For most of human history, people have built their own shelters and made their own household utensils, furnishing, clothing, footwear, tools of trade, transport, storage and weaponry using simple manual skills and locally available materials. Metals and gemstones have been crafted to make jewelry for self-adornament, as the trappings of power or for religious purposes. The finished product was, and still is in parts of the world where handicrafts are widely practiced, determined by the materials close to hand and by the techniques that local craftsmen had acquired. In time, what were originally part-time household subsistance activities evolved into full-time trade occupations.

The pace of change varied enormously from society to society and at different periods in history. For example, Polynesians living on Pacific islands had no access to metals, so they used bones of fish and sharks or seashells and corals to make weapons, carved objects and jewelry. They also skillfully lashed together cord and fiber from coconut, hibiscus or other tropical trees to make wooden canoes – even fishing boats up to 30 or 40 metres long which could withstand the force of huge waves.

By contrast, the Chinese were casting bronze in the 3rd millennium BC for hunting knives, workmen's drills, weapons and even chariots. At the same time they were perfecting the firing of clay to make ceramics for pots and ornaments. In many areas of the world wood remained crucial to handicrafts for centuries. It was used as a fuel for firing clays and smelting metals and had multiple other uses including making cabins, furniture, household utensils, carts and sledges, tools and weapons.

Technical progress in using metals and the innovation of wheels for potters, spinners, millers and carters ushered in new crafts and revitalized old ones. Improved transport, especially by sea, enabled products to be moved longer distances. From very early times major trade centers emerged specializing in handicrafts brought from all over the world for a wide international market. They often developed in or near ports with extensive land links. European examples include Byzantine Constantinople (jewelry, silks, perfumes), Venice (glass) or Florence (silks).

The European Renaissance (1250–1527) brought to the western world crafts such as papermaking, printing and instrument-making all flourishing in China for some time before this. It fostered an environment in which arts and science could blossom in the West. In doing so it sowed the seeds of further inventions – particularly industrial machinery – which would later form a basis for larger-scale factory production and, in time, undermine the very existence of handicrafts.

Decline, survival and revival

The growth and spread of modern manufacturing techniques have led directly to the decline, even disappearance, of crafts from many areas of the world. Craftspeople could not compete in price, quality, volume, or speed of delivery with standardized mass factory output and aggressive marketing.

Yet handicrafts do survive widely. They persist throughout large areas of Africa, Asia, Central and South America and in many Pacific islands and Indian Ocean communities, where poverty makes it impossible for the majority to buy manufactured goods. Sharp social divisions often create a dualistic society where informal handicrafts provide for the needs of most people, while new commercial industries (supplying mainly export markets in developed countries) benefit those few with political, economic or social status.

Handicrafts also survive in more industrial nations. Indigenous peoples in countries such as North America and Australia preserve some traditional crafts, frequently with government aid. Local people in remoter, rural areas, bypassed by industrialization, have clung to crafts to preserve their communities, perhaps with the help of local guilds or cooperatives. Some crafts survived because the markets they served were ignored by state-planned factory production.

The period since the 1960s has seen a revival of handicrafts in many regions of the world. The phenomenal growth and global spread of tourism has been the major force reviving innumerable traditional crafts in many parts of the world. Greater affluence and changing consumer tastes have stimulated a shift away from mass-produced goods in search of unique, individual, unusual or custom-made articles, in traditional styles giving a fillip to crafts workshops some of which now employ modern machinery.

Italian craftsmanship (*left*) Before the Industrial Revolution replaced people with machines, most goods were made individually with the skilled worker having to rely on his or her own ability. These classic Italian shoes, made by hand using traditional methods, are very expensive – the price reflects their quality.

Religious customs, traditional practices (*right*) This metalsmith in Jerusalem prepares candlesticks to be used in Jewish religious ceremonies. At the crossroads between East and West, workshops in the Middle East combine manufacturing sophistication with traditional values and customs.

No break with the past (*below*) In some parts of the developing world, the Industrial Revolution has not taken place yet. People still use time-honored ways of making things because that is all they have available to them. These women making and decorating pottery near Jodhpur in India use dye from local plants to make pigments and dry their urns in the sunshine or in wood-fired kilns.

The Industrial Revolution

THE INDUSTRIAL REVOLUTION IS OFTEN POPUlarly believed to have begun in 18th- and 19th-century Britain and spread gradually to other regions of the world. More modern, if controversial, views suggest that, viewed globally, it has passed through four distinctive cycles of technical and economic change, roughly dated 1770–1825, 1825–1880, 1880–1930s, 1930–1980s, and is now in a fifth. Each cycle is associated with the rise of new industries using new technologies, and with each new cycle the focus moves to a new world region.

The first wave 1770–1825

Three radical innovations during this first phase laid the foundations for all further industrial development. The invention of the steam engine, using coal to generate power; metal smelting, leading to the invention of harder and stronger materials; and the development of machinery manufacture. Before this period, producing

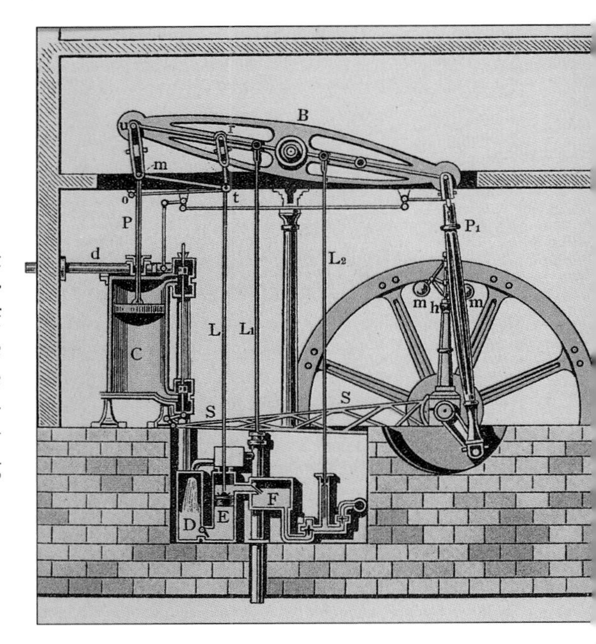

The driving force of change (*right*) James Watt's improvements to the steam engine helped to spark the Industrial Revolution. He was able to reduce the amount of steam that the Newcomen engine wasted and consequently increased the power of the engine.

textiles, metal goods, pottery and materials for construction were relatively dispersed cottage and handicraft activities. These new discoveries transformed them into largescale industries concentrated on or near coalfields.

It was a Scottish engineer, James Watt (1736–1819) who made practical the first of these innovations by improving the steam engine invented in 1705 by Thomas Newcomen (1663–1729). His alterations led to a drastic reduction in coal consumption and power costs and greatly improved efficiency. As a result, steam engines were used as a source of power in a broadening range of industrial activities from mining to distilling.

The textile industry

The textiles industry underwent a period of dramatic change and growth between 1770 and 1840, acquiring many features associated with modern factories. Before these changes, textile weaving had been a cottage industry. Then handlooms were worked by two weavers who were supplied with yarn by six to ten women working spinning wheels at home. The development of Kay's flying shuttle (1733), enabling one weaver to operate a handloom instead of two, created a "domino effect" as the quest began to create machines that would speed up the stages of preparing yarn for weaving by automating the combing, carding and spinning of wool or cotton. Once the Arkwright and Crompton powered spindles (developed in the 1760s and 1770s) had pioneered factory mass-production of yarns, the search turned back to the development of a power loom to increase production rates. This was only perfected for varied uses after 1800.

Although people resisted or destroyed mechanical innovations for fear of losing their jobs, employment in the English cotton industry alone rose from about 100,000 in 1770 to 350,000 by 1800. By 1812, 360 mills were operating 4.6 million Crompton spindles, but only 2,400 power looms, though this rose to 100,000 by 1830. As steam power became increasingly important the industry concentrated near coalfields in the north of England, and the traditional centers in the east and west of the country which depended on water power, declined.

After 1800 the impetus for further innovation began to shift abroad, primarily to New England in the United States. Although Britain still made 60 percent of the world's cotton goods in 1835, the textile and clothing industry was spreading to new centers around the world, especially to regions such as India and China with a local cotton or silk supply. Today it is one of the world's most widespread branches of manufacturing.

The Satanic mills *(above)* The textile industry in England during the Industrial Revolution imposed great hardship and suffering on its workforce. New machines took weaving out of people's homes and into factories. Unscrupulous mill owners and machine-wary workers made both industrial accidents and child exploitation commonplace.

Coalbrookdale by night *(left)* The first person to successfully smelt iron ore with coke – making a lighter and stronger iron cast – was Abraham Darby (1678–1717) at his Coalbrookdale works in Shropshire, England. Using local coal, which had a low sulfur content, Darby was able to develop a largescale iron foundry, and soon the area became the world's leading center in iron smelting.

MASS-PRODUCED POTTERY

Most European pottery was handcrafted and crudely glazed in 1770, though aristocrats bought fine tin-glazed Delftware, Meissen or Sevres porcelain. English potters in Staffordshire were the first to embrace the industrial revolution, probably influenced by changes taking place in Birmingham just 40 miles to the south. They brought superior china clay from the south by sea and canal, began using local coal in ovens and from 1776, replaced waterwheels or windmills with steam engines to grind and mix materials, turn potters' wheels and operate presses to squeeze prepared clay into molds.

These changes marked the beginning of the era of higher quality mass-produced chinaware. Josiah Wedgwood (1730–95) led the revolution, experimenting with clays, glazes, colors, and temperature control and introducing tools and machines driven by steam engines wherever possible. Aware of a range of different markets for pottery, Wedgwood divided production between "useful" mass-produced, inexpensive, yet high-quality chinaware and the renowned "ornamental", employing artists to design new patterns. He invented some of the processes of modern marketing, opening a London showroom, mailing illustrated advertisements and trading direct with customers all over the world.

The Second Wave 1825–1880

As the pace of change quickened, industrial growth itself created new challenges. How could the composition and design of metal parts be improved to cope with greater heat and friction in working machinery? How could increasing quantities of raw materials such as cotton, metal ores or coal for fuel be procured more quickly and cheaply?

Finding ways to solve these difficulties created more new industries. The application of steam power to land and water transportation was of central importance. Development of railroads and steamships interacted with innovations in mechanical engineering, coalmining and steelmaking. These in turn fostered new manufacturing techniques such as continuous processing used in steelmaking and conveyor-belt or batch methods used in food canning and packaging.

The transportation revolution

It was the harnessing of high-pressure steam for railroad locomotives and steamships that set in motion the second industrial revolution on a global scale. The British engineer Richard Trevithick (1771–1833) pioneered the first steam railroad locomotive, but it was the entrepreneur George Stephenson (1781–1848) who firmly established the railroad as a means of carrying passengers and freight between cities and even across continents by making major contributions to track, bridge and rolling-stock engineering. Railroad speeds and safety were enhanced by the introduction of steel rails after 1856 by Henry Bessemer (1813–98) who also invented the Bessemer converter for making steel, and of airbraking systems in 1869 by the American inventor George Westinghouse (1846–1914).

Steel from iron and hot air *(above)* The English engineer Henry Bessemer discovered that air blown through molten cast iron purified the metal and allowed the iron to be easily poured. His converter paved the way for the steel industry.

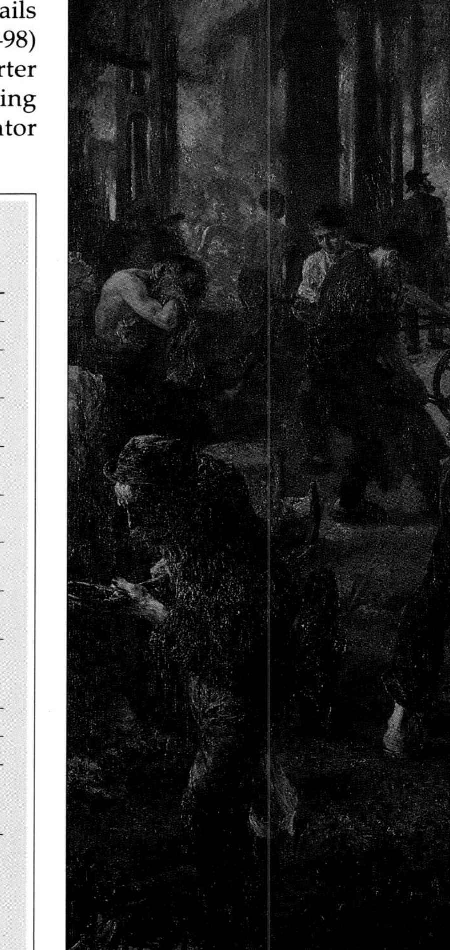

KEY INVENTIONS IN IRON AND STEEL PRODUCTION SINCE 1700

Date	Inventor	Key development	Place
1709	Abraham Darby (1678–1717)	First furnace to smelt iron using coke	Britain
1778	John Wilkinson (1728–1808)	First to use a steam engine to pump air into a blast furnace	Britain
1783	Henry Cort (1740–1800)	Invented the rolling process to make iron rods, plates and sheets	Britain
1784	Henry Cort (1740–1800)	Invented the puddling process to convert pig iron into wrought iron	Britain
1828	James Neilson (1792–1865)	Used hot air instead of cold in a blast furnace, reducing coal use by 66% per tonne of iron	Britain
1856	Henry Bessemer (1813–98)	Invented the steel converter to turn iron (not containing phosphorous) into steel	Britain
1861	William Siemens (1823–83)	Invented the open hearth furnace, adapting a process used in glassmaking to use waste gases from the blast furnace	Britain
1864	Pierre-Emile Martin (1824–1915)	Improved the open hearth furnace, increasing its efficiency and adapting it to use scrap metal	France
1876	Sidney Gilchrist Thomas (1850–85) Percy Gilchrist (1851–1935)	Invented the Thomas–Gilchrist process using limestone to line the furnace to convert phosphoric iron into steel	Britain
1878	William Siemens (1823–83)	Invented the electric arc process	Britain
1899	Paul L. T. Héroult (1863–1914)	Improved electric arc furnace for commercial use	France
1949	Austrian Steel	Pioneered the basic oxygen or Linz–Donawitz process, using pure oxygen rather than air in furnaces to refine pig iron into steel	Austria
c. 1970	various	Invented direct reduction, using enriched iron-ore pellets to make steel in an electric hearth furnace	USA & Italy

Storms at sea frequently disrupted cargo and passenger movement by sailing ship. Substitution of steamships from 1825–1880 revolutionized river and intercontinental water traffic. The American inventor Robert Fulton (1765–1815) and the Scot, Henry Bell (1767–1830) introduced the first commercial steamboats on the rivers Hudson (1807) and Clyde (1811) respectively. This use rapidly spread along rivers and across lakes in North America, and from 1830 transatlantic steamships began operating. The invention of the first ironclad steamship in 1837 by the British engineer Isambard K. Brunel (1806–59) reduced fuel consumption by halving the overall weight of each vessel compared with wooden ships. Brunel's first propeller-screw steamship (1843), did away with the need for paddles to propel the ship and reduced the amount of space needed to house the engine, increasing cargo capacity. The opening of the Suez Canal in 1869 brought further improvements and benefits to steam shipping and cut journey times. It shortened the sea route from London to India, Australia, and Japan by 3,400 miles (5,450 km) by opening a passage between the Indian Ocean and Mediterranean Sea.

The transport revolution had far-reaching effects on the location of industry, stimulating the emergence of new centers of innovation, especially in the north-eastern United States, France and Germany. Industries clustered beside railroads and in ports such as London, Glasgow, Hamburg or New York. The use of coal drew other industries such as steelmaking and textiles to the coalfields, though by the 1870s major concentrations had also appeared in the Ruhr region of Germany, in northern and eastern France, and in and around Pittsburgh in the United States.

German steelworking, 1875 (*above*) Making machines for the Industrial Revolution created huge demands in the iron industry. Improvements introduced by Bessemer, Siemens and others were quickly adopted in foundries all over Europe to boost production and to push forward the frontiers of manufacturing.

The birth of steam railways, 1830 (*left*) The Manchester to Liverpool line in the north of England was the first to realize the full commercial potential of the railroads. Pioneering locomotives like "The Fury" were able to move heavy, bulky goods over long distances quickly and cheaply.

THE DYNAMIC STEEL INDUSTRY

Until the 1920s iron and steel furnaces were located close to sources of coking coal since transporting fuel in bulk was wasteful of time and resources. Early steelmaking furnaces burned huge amounts of coal. In 1760 it took 8–11 tonnes of coking coal to produce one tonne of iron and in 1860 it took 7 tonnes of coal per tonne of steel. Since Neilson's revolutionary hot blast furnace (1828) innovations in steelmaking have cut fuel use steadily reducing the locational "pull" of coalfields.

Furnaces using metal scrap, especially the superior Siemens–Martin open hearth, widely adopted before 1955, led to plants being located close to markets – along the Great Lakes (Chicago–Hamilton) axis, for example. Electric-arc and mini furnaces have brought high-quality steel manufacture to areas deficient in coal, but with cheap hydroelectric or gas supplies, such as the western United States or Sweden.

The ability of new furnaces to operate on combinations of coal, oil, natural gas, liquid propane gas or electricity makes coastal sites particularly suitable for the large new integrated plants if materials have to be shipped in. Most integrated plants use the more effective oxygen steelmaking process that has largely displaced open hearths. New refining and concentrating techniques raise the iron content of ores beyond 90–95 percent, making them easily transportable and permitting a vast increase in blast furnace capacity.

The Age of Electricity

THE PERIOD 1880–1930 SAW RAPID GROWTH and consolidation of a cluster of new manufacturing sectors. They grew around new power sources (oil and electricity) and new forms of communication and transportation (the telegraph, telephone and automobile).

Power for the future

Major advances in the development of practical largescale electric power generation and in the design of electrical equipment were made after 1865, particularly where effective links existed between research scientists and commercial manufacturers. At the forefront of innovation were the German electrical engineer Werner von Siemens (1816–92) in Berlin and Alexander Graham Bell (1847–1922) working alongside others in the Boston–Baltimore–Chicago triangle of the northeastern United States.

Another area of major innovation was the emerging chemical industry. Manufacturing chemicals involves isolating substances in a variety of raw materials – coal, oil, gases, wood or nonmetallic minerals, and converting them into products for use in other industries. Research and innovation, especially in Germany, resulted in the rapid development of sulfuric acid, soda ash, caustic soda, chlorine, aniline dyes, artificial fertilizers and explosives after 1870. Production of most of these products was concentrated on riversides, lakes and coasts as the chemical processes often needed large supplies of water. Plants manufacturing chemicals also tended to consume large quantities of coal, coke, brine or limestone, and the waterways offered easy transportation for bulky materials, as well as for the finished product.

The emergence of large firms

The period between 1870 and 1900 is closely associated with the birth and rapid growth of large industrial firms, especially in the United States, Germany, France and Italy. Many remain leading names today: Siemens and Philips in the electrical industry; Hoechst and Dow in chemicals; Agfa and Kodak in photographic materials; Goodyear and Pirelli in rubber tires; Ford and Peugeot in automobiles are just a few. Control of technology enabled them to gain dominant positions in domestic and foreign markets, growing with the emergence of a relatively affluent urban population. Early success induced them to establish subsidiary factories and become multinational enterprises. Faster bulk transportation by rail and sea, and easier managerial control by telegraph and telephone, made this kind of expansion easy.

KEY INNOVATIONS IN THE EARLY ELECTRICAL INDUSTRY

Date	By	Development	Place
1800	Allessandro Volta (1745–1827)	Electric battery, first generation of electric current	Italy
1831–45	Michael Faraday (1791–1867)	Electric motor and dynamo Demonstrated electromagnetic induction	Britain
1837	Charles Wheatstone (1802–75)	First workable telegraph	Britain
1843	Charles Wheatstone (1802–75)	Wheatstone bridge, to measure electrical resistance	Britain
1847	Werner von Siemens (1816–92)	Commercial manufacture of improved telegraph	Germany
1866	Werner von Siemens (1816–92)	Dynamo for largescale electricity generation	Germany
1869–76	Elisha Gray (1835–1901)	Microphone transmitter and receiver incorporated by Bell in the telephone	USA
1876	Alexander G. Bell (1847–1922)	The telephone, enabling human speech to be transmitted along a wire	Canada
1877	Thomas Edison (1847–1931)	The phonograph, the first machine to record and play back sound	USA
1880	Thomas Edison (1847–1931)	Incandescent electric lamp	USA
1875-79	Charles Brush (1849–1929)	Electric arc lamps used in the first streetlighting system	USA
1883-1900	Elihu Thomson (1853–1937)	Alternating current motor, high-frequency generators and transformers	USA
1884	Giovanni Pirelli (1848–1932)	Manufactured the first electric cables	Italy
1885	George Westinghouse (1846–1914)	Constant voltage alternating current generator establishing the advantage of alternating current over direct current	USA

Philips' Incandescent Lamp Works, 1910 (*above*) Hundreds of women were employed at the first Philips' factory at Eindhoven in the Netherlands to mount fragile tungsten filaments in the new-style lamps.

Lighting the road ahead (*right*) Early training with Thomas Edison, inventor of the electric light bulb, influenced the German industrialist, Robert Bosch, to pioneer automobile headlights in the 1880s.

AUTOMOBILE CLUSTERS

The early history of the automobile industry is a good illustration of the way clusters of factories producing the same product tend to concentrate in particular areas. This is because a pool of technical expertise and labor is built up in one place. The first experiments in automobile production were in Paris, France, in southwest Germany, and in the midwestern United States. Étienne Lenoir (1822–1900) made the first automobiles in Paris, but the focus soon shifted to southwest Germany where Nikolaus Otto (1832–91) introduced the 4-stroke engine in Mannheim. German success was consolidated when Gottlieb Daimler (1834–1900) and Carl Benz (1844–1929) working respectively in Stuttgart and Mannheim, manufactured the first high-speed gasoline-driven engines for motor cycles and cars. After 1886 Robert Bosch (1861–1942) began making spark plugs, magnetos and other electrical devices for cars at his Stuttgart works. The German engineer Rudolf Diesel (1858–1913) demonstrated a more efficient 25hp diesel engine in 1897 and by 1922 the engine had been sufficiently reduced in size for use in truck production by Daimler–Benz.

Between 1890 and 1900 more than 50 companies began making automobiles in the midwestern United States. By the early 20th century the industry localized in the Detroit–Flint–Lansing triangle of eastern Michigan. One reason was that Henry Ford (1863–1947) began pioneering conveyor-belt assembly methods at his factory near Detroit. By 1930 Ford and General Motors had entered the European market, concentrating manufacturing and assembly in Britain and Germany.

Bosch-
Licht
BERN
HARD
· HOLLERBAUM & SCHMIDT · BERLIN · N · 65 ·

New-Age Materials

THE PERIOD BETWEEN 1930 AND 1980 SAW A disruptive world war and the growth of an unprecedented number of key industries. Electronics, nuclear energy, petrochemicals, pharmaceuticals, plastics and synthetic materials were the most significant. At the same time, older industries followed important new directions. The jet engine became airborne and the aerospace industry took off; and telegraph and telephone networks stimulated more innovations in communications, particularly in radio, audio equipment, cinema and television.

The importance of oil

Without doubt, the greatest stimulus to innovation was the growth of the oil industry. Cheap and plentiful supplies encouraged a fourth wave of industrialization in the United States and western Europe, which extended to Japan and the Soviet Union by the middle of the century. Being cheaper and easier to extract, transport and process, oil displaced coal as an energy source and replaced it altogether in rail and sea transportation. Catalytic cracking, a refining process that separates crude oil into several different products, was developed in 1936 and gave a spur to sustained research into industrial chemicals. It brought breakthroughs in bulk manufacturing of synthetic fibers, pharmaceuticals and petrochemicals. Catalytic cracking also yielded cheaper and superior raw materials to the textiles industries creating a revolution in synthetics manufacture after World War II.

Early synthetic fiber, rayon, was made from wood cellulose. The first plastics, phenols, were the byproducts of coal-processing. Bakelite, a formaldehyde named after Leo Baekeland (1863–1944), a Belgian–American chemist, was an early landmark that revolutionized electrical and radio insulation. In the 1930s, scientists produced three chemical compounds basic to modern plastics: polyvinylchloride (PVC), polystyrene and polyethylene, the latter by ICI for initial use in radar. Toward the end of the 20th century nylon, acrylic and polyester fibers had replaced natural fibers in many areas because of their strength and durability. Plastics, too, have been adapted to amazingly diverse uses, replacing metals in some automotive and aerospace components, wood in the building trade, glass for food

KEY INNOVATIONS IN COMMUNCATIONS

Date	Inventor	Innovation	Place
1887	Heinrich Hertz (1857–1894)	Demonstrated radio-wave transmission	Germany
1890	Herman Hollerith (1860–1929)	Punch card electric machine to count the census	USA
1894	Oliver Lodge (1851–1940)	Radio wave detector used in radio receivers	Britain
1897	Ferdinand Braun (1850–1918)	Cathode ray tube, the basis of the television set	Germany
1899	Guglielmo Marconi (1874–1937)	Long-distance broadcasting of radio messages	Britain
1904	John Fleming (1849–1945)	Thermionic valve (diode), made radio signals detectable by telephone receiver	Britain
1906	Lee de Forest (1873–1961)	Thermionic valve (triode) made live broadcasting possible, and was the key component in radio, telephone, radar and TV equipment until the invention of the transistor	USA
1925	John Logie Baird (1888–1946)	Mechanical television picture transmission. Demonstrated color television pictures in 1928	Britain
1931	Massachusetts Institute of Technology	Mechanical analog computer	USA
1933	Vladimir Zworykin (1889–1982)	Showed moving pictures on television using an all electric system and a cathode-ray camera	USA
1940	Chester Carlson (1906–1968)	Xerographic copying	USA
1945	University of Pennsylvania	Electronic numerical integrator and calculator	USA
1947	Bell Laboratories	Transistor, replaced the thermionic valve and was a fraction of its size	USA
1949	University of Cambridge	Electronic digital computer	Britain
1954	Texas Instruments	Silicon transistor, replaced conventional transistors, increasing capacity and reducing size	USA
1958	Texas Instruments	Integrated circuit, combined the functions of electric components onto a single slice of silicon	USA
1969	Intel Corporation	Microprocessor, combined all the circuits in a computer onto one silicon chip – miniaturization	USA
1977	Bell Laboratories	Fiber optic cables, improving telecommunications and medical technology	USA

Airbus assembly at Toulouse, France (*above*) Formed in 1970, Airbus Industrie is a joint European manufacturing venture, aiming to make the European aerospace industry more competitive with the American giants, Boeing and McDonnell Douglas. Parts for Airbus jets are made separately in France, Germany, Britain, Spain, the Netherlands and Belgium. Sections of the fuselage, wings and other components are then transported to the consortium's headquarters in Toulouse for assembly.

General Electric, New York (*right*) Radio sets using thermionic valves were mass produced in the United States (mostly by women) during the 1940s and 1950s. Originally cased in wood, they were large enough to be a piece of furniture. After the mid-40s bakelite, the newly developed forerunner of plastic, became a popular casing.

HIGH-FLYING INDUSTRY

The first practical airplanes were built in 1903 in the United States by the Wright brothers, Wilbur (1867–1912) and Orville (1871–1948). Subsequently, aluminum and alloys replaced wood in the bodywork; while substantial improvements in aerodynamics, engine design, fuel technology and structural engineering brought startling advances in speed and capability.

Aerospace technology developed very rapidly in the 20th century to serve military purposes and to expand the horizons of civilian travel. Throughout its development, aerospace has depended heavily on the electronics industry for its navigation equipment. Faster, longer-distance and accurately targeted flight requires better radio communication and quicker and more complex computer control. Defense aircraft in particular need sophisticated radar and satellite systems.

Military demands helped establish aircraft manufacturers such as Fokker and Messerschmitt in Germany; British Aerospace in Britain; Dassault in France and Mitsubishi in Japan. Since the 1950s there has been a spectacular expansion of civilian passenger and cargo-carrying aircraft production. Passenger aircraft such as the American Boeing 747 jet airliner and the Anglo-French supersonic Concorde have dramatically reduced travel times, making it possible to reach almost anywhere on earth within 24 hours of setting out from a major city.

storage and becoming the dominant manufacturing material for furniture, toys and household utensils.

Revolution in communications

During the mid 20th century electronics, aided by huge advances in physics, developed to improve people's abilities to communicate information in sound and pictures far beyond the facilities offered by telegraph and telephone. Each stage of innovation and development has made equipment more powerful and sophisticated, smaller, lighter, faster, and cheaper. In four decades computers have shrunk from the size of a room to the size of a laptop typewriter. Calculators are the size of credit cards, and pocket televisions the size calculators were 10 years ago.

Invented in 1945, the Electronic Numerical Integrator and Calculator (ENIAC) was an early prototype of an electronic computer. In fact it was more like a giant adding machine with no memory to store data or programs. A series of advances since then has transformed computers into tiny sophisticated machines with a huge capacity for memory. The most far-reaching developmental leap was the invention of integrated circuits (silicon chips) in 1958, incorporating the functions of hundreds of electric circuits on one slice of silicon. Further refinements led to the first microprocessor in 1969, with the capacity to concentrate on a single chip all the circuits that do the work of different parts of a computer. Now microprocessors are widely used in electronic equipment ranging from washing machines and wordprocessors to international telecommunication systems.

Information Technology

SINCE THE LATE 1960s HUGE TECHNICAL innovations in the way we store and the speed we can access information have led to a fifth wave of industrial development focused on information technology. In part, information technology has developed so fast because it was potentially valuable to the governments, military forces and businesses that funded its growth. Rapid access to much greater amounts of information helped them to make better-informed decisions about directing foreign policy, controlling unrest or staying ahead of the competition in an international field. In part, too, its success is self-fueled. Once the facility had been invented for receiving information and analysis about events as they happened, it created an insatiable demand for newer, faster and more user-friendly technology.

A vast range of service industries from the news media to the money markets depend on the speed and reliability of their information-gathering abilities to maintain their competitive edge, and are prepared to invest huge sums in updating their technology. The industry is in a constant state of flux – equipment is surpassed almost as soon as it is on the market, and specific user requirements create a huge volume of work for the

A helping hand (*left*) Computer-aided design (CAD) has rapidly grown into an indispensable tool for the engineering professions. Computers can develop a three-dimensional representation of a structure, analyze and draw it, saving enormous time and effort.

support industries – programming, software design, employee training and installation services.

Creating a new way of working

During this same period it became evident that a huge shift of emphasis was taking place in industrial activity across the globe. Employment in services was growing rapidly in the West, while employment in manufacturing stagnated, then decreased. The slump in growth was provoked by the world oil crisis and was part of a larger pattern of labor-intensive manufacturing relocating to newly industrializing and Third World countries with lower labor costs.

The most marked expansion in the service sector has been in a cluster of interdependent advanced industries that create, process, repackage and apply knowledge. They serve the layperson using computers as a tool as well as computer professionals. Diversity in this field is relatively new. In the 1970s, the industry was dominated by computer giants such as IBM running mainframes, huge centralized databases operated by computer experts. The effect was to restrict information processing to the headquarters or specialized divisions of firms, nationalized industries and government ministries, mainly in key cities.

Out of the hands of experts

Since then, progressively smaller, more powerful and versatile machines – which drastically cut information-processing costs – have brought computing within reach of small firms and individuals. Desk-top personal computers have transformed information technology into a labor-saving device used by all kinds of people in their everyday work, many of them based at home. Innovations in telecommunications mean that people no longer have to share the same building to work from the same database or remain in close contact. Modems allow computer users all over the world to access information from a central database, or from each other by telephone. Satellites and fiber-optic cables have vastly improved communication networks, setting up direct links between widely separated locations, through telefax and electronic mail.

The information technology revolution has also made possible faster and more efficient manufacturing systems for wider adoption. Computer-aided design, engineering and manufacturing (CAD, CAE, CAM) enable firms to alter the design and specifications of products more rapidly to meet the changes in markets or in fashion. Computer-controlled machinery in industry has brought robotics to the production lines of the automotive, electrical, furniture and textile industries. It has also introduced automated production systems into metal processing, glassmaking, and into the chemicals, pharmaceutical and food processing industries.

Ears of the universe (*left*) Radar is an acronym of radio detection and ranging. It is a way of registering distance and movement by the reflection of electromagnetic impulses. Developed during World War II, radar is now used in space research.

Light through glass (*below*) Research into fiber optics has had an enormous impact on communications. Images and information are carried quickly and cheaply along beams of light travelling through fine silica glass strands.

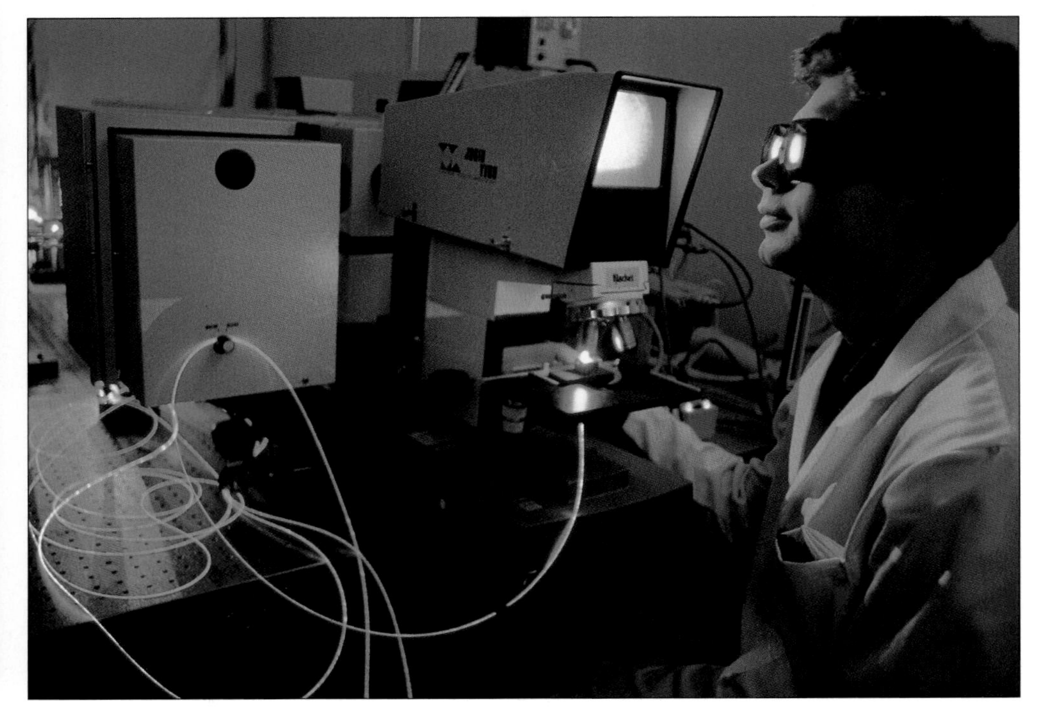

People Power

PEOPLE ARE THE WORLD'S MOST FLEXIBLE industrial resource. They have discovered countless ways to use their intellectual abilities, muscle power and manual dexterity to harness the Earth's extraordinary natural resources for a multiplicity of purposes. Paramount in this process is their ability to learn quickly from experimentation, have ideas, create new products and apply their knowledge and organizational skills to finding ways to mass produce and market them.

In industrial terms "human resources" generally refers to the workforce collectively, comprising all the creative, managerial and productive layers that generate products and services for consumption. Far from all the world's 3.2 billion people of working age (approximately 15–64) are active members of the workforce. Many are still in education and training, many more, especially women, work full-time looking after their families and in the home. The proportion of the 15–64 age group who are economically active ranges from 80 percent in Scandinavia to below 60 percent in Ireland, southern Europe, and Turkey; fluctuates about 55–60 percent in South America, but falls as low as 41 percent in parts of the Middle East and North Africa. By contrast, from 75 to 88 percent of 15–64

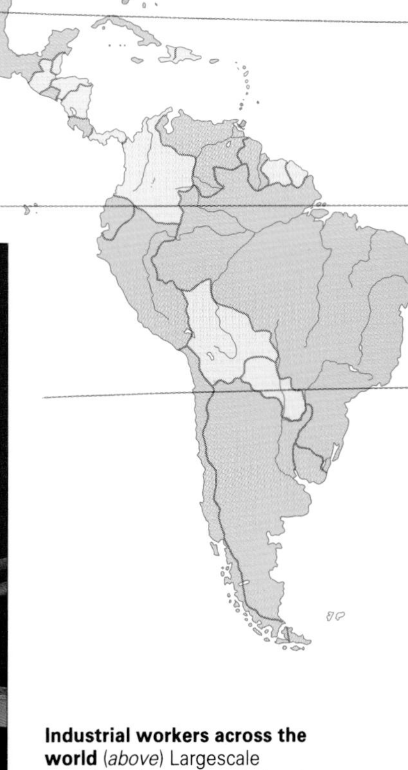

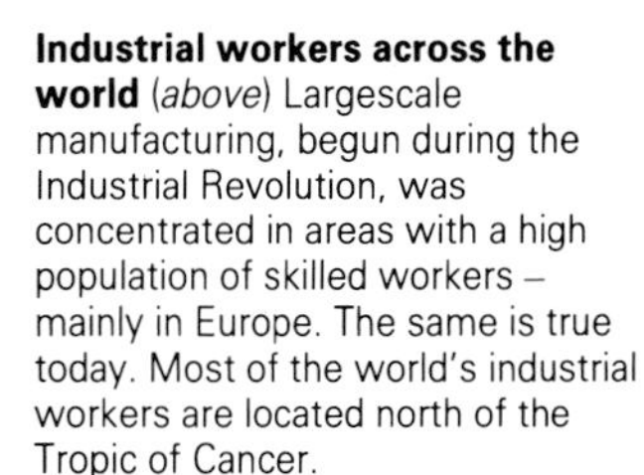

Industrial workers across the world (*above*) Largescale manufacturing, begun during the Industrial Revolution, was concentrated in areas with a high population of skilled workers – mainly in Europe. The same is true today. Most of the world's industrial workers are located north of the Tropic of Cancer.

Keeping many jobs afloat (*left*). This paint sprayer, working at the China Shipbuilding Corporation in Taiwan, is applying a special undercoat made from zinc to prevent corrosion. In most western shipbuilding countries this job would be done by machines as part of the continuous process to lower costs by reducing the number of employees. In Taiwan, however, like many Asian, African, and some South American countries, the workforce is often the cheapest industrial resource available, and automation is comparatively expensive. In fact, Taiwan's shipbuilding industry is so rich in people power that it profitably breaks up old ships to recycle scrap metal, an activity that would be completely uneconomic in other world regions.

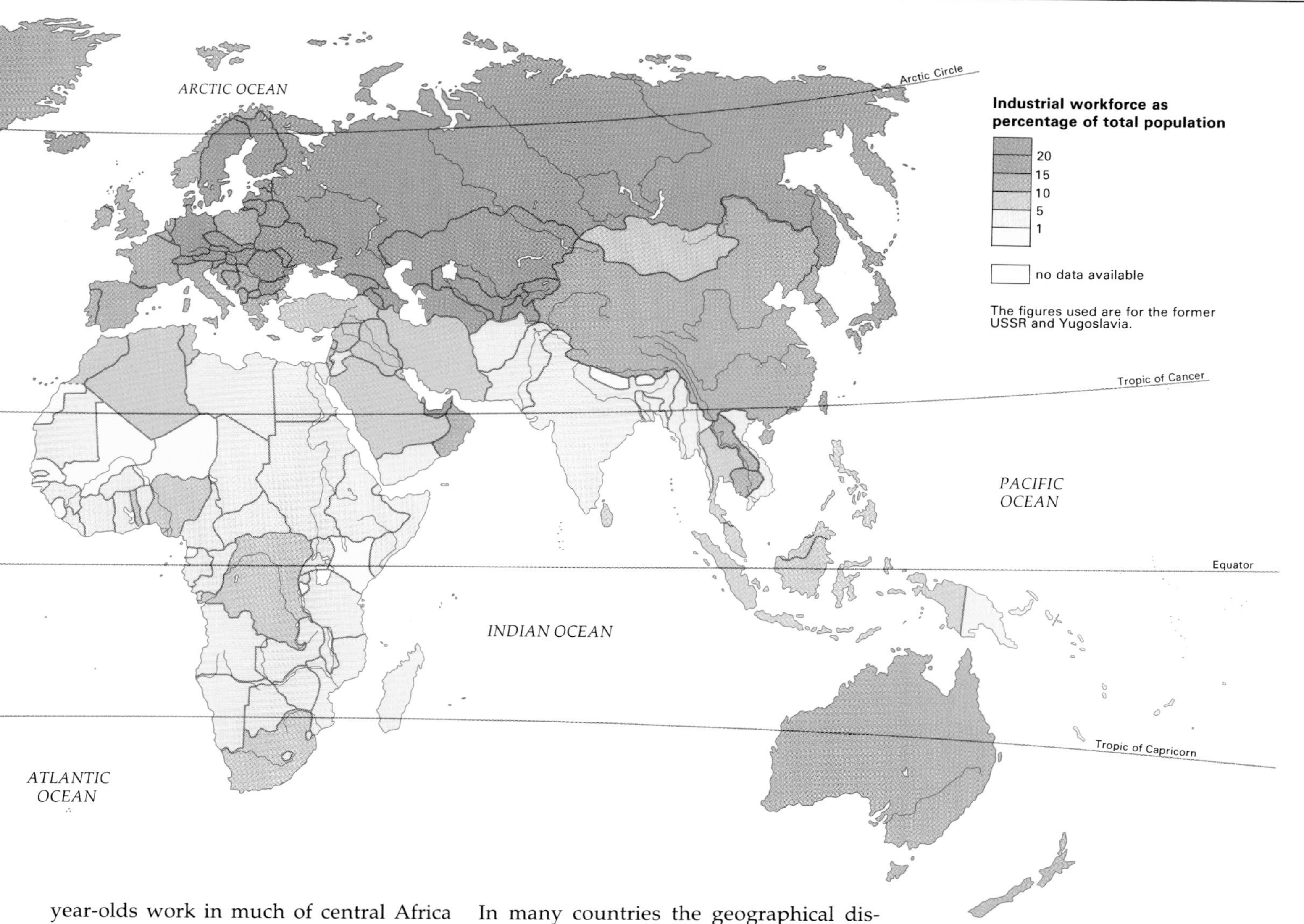

year-olds work in much of central Africa where the potential workforce is small and women are important producers in agriculture and handicrafts.

Labor and location

Mining and manufacturing require certain minimum supplies of labor in any part of the world to operate satisfactorily. In many countries the geographical distribution of human resources may be quite different from the location of natural resources. Historically, most people lived as subsistence farmers, supplemented by fishing or trading. Industrialization changed that way of life, encouraging migration from rural areas.

The development of largescale mass production methods during the third wave of the industrial revolution made supplies of unskilled and semiskilled workers an important resource. This was what attracted the automotive industry, electrical assembly plants and photographic industry to larger cities. Most other types of manufacturing, especially if it is research-led, have tended to follow this pattern, establishing plants in cities or in specially developed industrial areas. These have to be near research facilities (often universities), and able to attract the more mobile managerial and technically qualified members of the labor force. These more sophisticated industries are usually clustered in "rural" or "playground" areas close to metropolitan centers with colleges and laboratories.

At the same time, the earlier-established industries such as textiles, clothing and shoemaking have progressively moved to parts of the world where labor costs are lower. Asia has been particularly popular since high population densities, traditional handicraft skills or dexterity, and a history of hard work have led to high levels of productivity.

INVENTORS AND ENTREPRENEURS

The past 200 years of history is littered with names of individuals who had the talent, drive and flair to combine invention (having a good idea, or developing a new method of manufacturing) and entrepreneurship (turning it into a commercial success). Many of the industrial innovators of the 19th and early 20th centuries founded companies that are still household names today. Josiah Wedgwood (1730–95) in pottery, Alfred Krupp (1812–87) in steel, Werner von Siemens (1816–92) in telecommunications, Friedrich Bayer (1825–80) in chemicals, Robert Bosch (1861–1942) in electrical equipment, John Dunlop (1840–1921) and Giovanni Pirelli (1848–1932) in tires and rubber goods are just a few of the many examples.

Sometimes innovations are made quite independently by people in different places at the same time. Alexander Graham Bell (1847–1922) and Elisha Gray (1835–1901) both invented the telephone, working in direct competition a hundred miles apart in the northeastern United States. Earlier, in 1859 Edwin Drake (1819–1880) sank the first oil well at Titusville, Pennsylvania, United States about the same time as an oil well was drilled near Krosno (now southeast Poland).

Few individuals today can take personal credit for technical breakthroughs. Innovation has become the complex and costly result of teamwork by the top brains in university, government and private corporation laboratories, with many fields of research controlled by international consortiums.

Servicing Industry

As communities have become more industrialized and urbanized, people both need and can afford more services to meet business requirements, provide comforts and entertainment. In advanced countries, services employ more of the labor force than any other kind of industry. In most Third World countries, services usually employ fewer workers than farming, but more than industry.

Economists usually divide services broadly into two categories: tertiary and quaternary (sometimes called "advanced services"). Tertiary services tend to be supplied to the consumer, bringing goods from farmers or manufacturers, or services, to the point of consumer purchase. Advanced services tend to be supplied to the producer, often in the areas of finance, law, information, education, safety or quality regulation.

From road freight to education

Tertiary services include distribution, retail sales and consumer services. Distribution is preeminently concerned with moving things – food, raw materials, manufactured goods, parcels and messages or information – from their place of origin to where they will ultimately be used. Shops and supermarkets, hotels, restaurants, hairdressers and leisure centers all sell goods or services to individuals or companies and the distribution networks that support them are all employed in tertiary services.

Advanced services encompass business, financial, public-sector and government services including education, health and law enforcement. There are broadly speaking five types of advanced services, many of which are closely linked or interdependent. First are the administrative, regulatory and decision-making services. They include government ministries and agencies at national or regional level, corporate business headquarters or decision-making centers, legal and professional organizations.

Second are financial services – banking, insurance, stock markets and commodity trading together with property and land development (real estate), accounting and auditing services.

Many firms in this field also employ a third type of service: information-

THE EVOLUTION OF INTERNATIONAL NEWS

Reuters, the London-based international news agency, is a good illustration of the changing character of the service sector. Paul Julius Freiherr von Reuter (1816–1899) first became aware of the potential for transmitting information while working in publishing in Berlin, Germany in the 1840s. From 1849 he established a news service in Paris using electric telegraphy, and in 1850 set up a carrier-pigeon service for selected, edited messages between Aachen and Brussels as a way of linking up the German and Franco-Belgium telegraph terminuses.

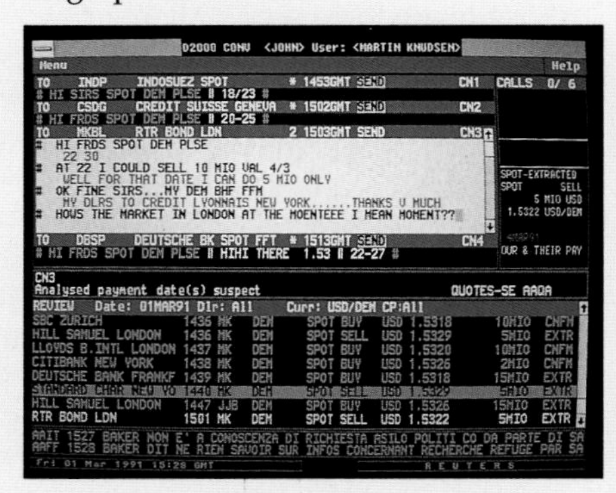

Bull market (*above*) Reuters brings vital, up-to-the-minute news of the world's stock markets to financial brokers everywhere.

Stimulated by his interest in financial affairs, Reuter moved to London in 1851, opening a telegraphic office near the Stock Exchange and Bank of England – a base from which he could supply news to publishers and newspapers all over the world. To expand business, he encouraged a growing number of national and daily newspaper publishers – then opening offices and printing works in London's Fleet Street – to take advantage of a worldwide distribution facility by subscribing to his news information service.

The volume of his business and number of customers grew as London became an important center in the expanding world network of telecommunications. Eventually Reuter was able to maintain a network of correspondents throughout the world who provided and exchanged regular information on political, technical and financial news. Today, the Reuters headquarters coordinates one of the world's most modern telecommunications networks for up-to-the-minute news transmission, maintaining a global network of 190,000 computers that transmit instantaneous data via satellites on currency-exchange and commodity price movements worldwide.

Eating out *(above)* The most immediate and personal involvement with a service industry for many people is going to a bar or restaurant. Successful restaurants, such as this one in Chicago, see their staff as more than a convenient way to deliver food. The way they relate to the customers can make or break their employer's reputation.

Trains, trucks and transportation *(left)* Railroads were one of the first industrial services to be offered to manufacturers and consumers alike. Although the railroad offered a fast and cheap way of moving goods from the factory warehouse to the customer, it was limited in its delivery areas. The advent of the automobile – and later the truck – gave manufacturers the power to reach every part of the country. Today transportation and distribution have become integrated services, and trains, trucks and even planes work together to deliver the goods.

processing. Information-based services collect data and analyze it for private and public sector decision-making in all kinds of spheres. Consultancy work, printing and publishing, the media, advertising and marketing also fall into this group.

The fourth category of advanced services is specialized technical skills. This group includes design and planning in computing, engineering, construction or architecture. Workers in these sectors often plan new ways to improve conditions in other parts of service and manufacturing industries – setting up a more efficient computer system within a banking company, for example, or building new highways to improve transportation.

The fifth group covers the vast area of health and education, government-supported services that have a direct effect on the capability, performance and potential of the workforce.

The Role of Government

GOVERNMENTS SEEK TO INFLUENCE THE management of resources and industrial development when they become anxious that existing conditions do not fulfill their economic or political needs or aspirations. Often intervention is the result of lobbying by vested interests outside the government – the military, business community, workers, consumer pressure groups or environmentalists. It can be passive, enforcing limited regulations on resource-use and industrial activity; indirect and incidental through broad economic, fiscal or social policies; or active and direct, deliberately managing resources and industries.

The most common tool used by governments to protect domestic industry is tariffs against imports. However, since the 1950s the General Agreement on Tariffs and Trade (GATT), an international set of agreements to which more than 90 countries are signatory, has had a good deal of success in lowering or removing tariffs on most raw materials and manufactured goods. This has led governments – on their own initiative or under pressure from specific industrial interests – to erect diverse nontariff barriers to trade, including strict quotas, voluntary restraint agreements, technical specifications, legal requirements, subsidies and overt or covert "buy national" campaigns. Inevitably, protectionism attempts to support a larger volume of industrial activity within a country than might be the case under free trade, while supporting high levels of exports with subsidies and promotion strategies.

Influencing new development

Government funding of innovative research often allocates resources to universities and laboratories in less industrialized areas. This rapidly attracts clusters of new high-technology firms, and in turn laboratory and university researchers may establish their own manufacturing or advanced service firms nearby. The indirect effect of this kind of government funding, when it is successful, is to create major new industrial zones. The most outstanding example is where United States' government funding has promoted the growth of high-technology manufacturing in the Santa Clara Valley south of San Francisco – giving it the popular name "Silicon Valley".

In Western Europe in particular, and to a lesser extent in Japan, governments have attempted to relocate industry from expanding industrial areas such as London, Paris, Brussels, Turin or Milan to depressed industrial areas of higher unemployment. In recent years, the policy has been modified to attract more foreign investment, especially in more advanced research-led manufacturing.

Chinese taskforce in Tanzania *(above)* Governments support each other's industries for all kinds of reasons. In the late 1970s China sent a labor force to the leftist Revolutionary Party in Tanzania to help maintain their transportation system.

Smoke screen *(right)* Governments play a major role in the armaments industry. Although some states buy their arms from private manufacturing companies, others develop their own weapons industries. These white phosphorous smoke bombs were made by a state-owned Israeli company.

Across the world, central governments are tending to reduce this regulatory role in industry, slimming down central bureaucracies and devolving more management responsibility to private enterprises and to local and regional state authorities. In part, this stems from the realization that bureaucracy is a major cost borne by industry. Furthermore, the inefficiencies of a centralized state apparatus may inhibit the international competitiveness of industry and limit national levels of productivity.

THE ARMAMENTS INDUSTRY

State and military cooperation have brought into existence increasingly sophisticated, expensive, destructive and deadly weapons. Thriving industries have grown around the manufacture of armored vehicles and tanks, bomber and fighter aircraft, automatic guns, chemical and biological gases and agents, plastic explosives, rockets and missiles, nuclear bombs, thermonuclear warheads, computer and laser-guided launch systems.

Governments of all countries rank arms expenditure high among their budget priorities – none more so, however, than where single-party dictatorships need and find ready military backing. In 1987 total world expenditure on armaments exceeded US$952 billion, approximately US$200 for every man, woman and child on Earth. At the beginning of the 1990s there were some 168,160 tanks, 36,136 aircraft and several thousand missiles with nuclear warheads in the hands of the military forces across the world.

Up until 1990 there was a marked geographic imbalance between the centers of armament production and the government markets for arms. This was largely because production was tied to several leading industrial nations with the material resources, innovative research, technological experience, skilled labor and huge financial reserves needed to fund manufacturing and development.

Historically, most production is in the United States, former Soviet Union and Western Europe, while the market is usually the developing countries. In fact some Middle Eastern countries import far more armaments than major industrialized European states export. World recession, causing Western and Soviet governments to question arms expenditure, the ending of the cold war and the dissolution of the former Soviet Union may change all this over the next few decades.

WORLD'S LEADING ARMAMENTS EXPORTERS AND IMPORTERS (1987)

Exporters	US$ billions
Soviet Union	21.2
United States	12.6
France	2.6
United Kingdom	2.1
West Germany	1.8
China	1.0
Czechoslovakia	0.975
Poland	0.8
Brazil	0.6
North Korea	0.41
Importers	
Iraq	5.6
Saudi Arabia	3.8
India	3.2
Syria	1.9
Vietnam	1.9
Cuba	1.8
Israel	1.6
Angola	1.6
Iran	1.5
Egypt	1.5

105 mm
WP SMOKE
8 XA 12/82
105-7T-M6
665 XA 11/82

Threatening the Quality of Life

NO MATTER HOW BENEFICIAL OR IMPORTANT industry is, most manufacturing and processing plants create noise pollution and emit harmful substances that threaten the environment and human health. The damage takes many different forms: emissions of particles, chemicals and gases into the air; liquid effluents leaked into land or water; the dumping of solid waste; and the unseen effects of high levels of radiation. With time and increasing output, the effects spread from industrial sites to agricultural and residential areas carried by the wind and via rivers, underground waterways, lakes and seas, harming humans and animals, damaging trees and crops, and ultimately destroying the Earth's atmosphere.

The down side of energy

When fossil fuels are burnt to generate electricity, harmful gases are formed. They include sulfur dioxide, nitrogen dioxide and carbon monoxide that pollute the air, and carbon dioxide and nitrogen oxides that contribute to the greenhouse effect – trapping the heat from the sun in the Earth's lower atmosphere.

The situation is worsened by the fact that since the 1970s' oil crisis, China and the countries of eastern Europe in particular have greatly increased their use of local, lower-grade coal for energy production. As a result, they have the highest levels in the world of sulfur dioxide emissions per unit area. By contrast, increasingly influential environmental lobbies in the United States and western Europe have had some impact in tightening state regulation on pollution.

Industrial pollution (*left*) One of the byproducts of the manufacturing process is pollution, usually atmospheric or waterborne. From the 1970s onward, many countries began to impose strict emission guidelines to minimize pollution. However, this cement works in Qatar shows no sign of changing its ways. Many of the world's developing countries have very limited controls over their industries' pollution output, putting economic development ahead of environmental concerns.

Poland's inferno *(above)* This pig-iron works in Nowa Huta, Poland, illustrates the dangers that many of the world's industrial workers face every day. Eastern European countries tend to emphasize output and production figures at the expense of modernization and workers' safety.

HIDDEN DANGERS OF THE ELECTRONICS INDUSTRY

In 1982 it was discovered that the booming electronics industry in the Santa Clara (Silicon) Valley of California is a source of major land and water pollution. The main cause is the use of two solvents used to remove greases from silicon microchips after their manufacture – 1,1,1–trichloromethane (TCA) and 1,1–dichloroethylene (DCE). A few parts per billion in domestic water supplies causes nervous and cardiovascular disorders, miscarriages and stillbirths.

Besides these degreasing agents, electronics companies also use hydrofluoric and hydrochloric acid to etch microchips, and poisonous gases such as arsine and phosphine to endow the chips with electrical properties. Arsine rapidly destroys the red blood cells, making blood transfusions necessary for human survival. In addition, some 3,500 other chemicals are used by these industries. In 1988 State of California tests found dangerous chemicals under the sites of factories belonging to 65 out of the 79 large firms surveyed – a high price to pay for their products.

Dangerous accidents

Exploiting and transporting natural resources is often a potentially hazardous process – the "Exxon Valdez" oil-spillage off the coast of Alaska in 1989 is only one illustration of how much harm such an incident can cause. Nuclear power, which could provide a way of cutting down the harmful emissions associated with coal-fired electricity generation, brings a threat of a different, more immediately devastating kind through radiation. The immense longterm damage caused by the explosion of the Soviet Chernobyl nuclear reactor in 1986 illustrated how appalling the consequences can be.

The production and use of chemicals give rise to great environmental concern, particularly in the West. Manufacturing processes and products are usually complex, highly concentrated, and involve harmful and hazardous gases, liquids and solids. Transportation of raw materials and finished products is often dangerous; and the problem of their safe disposal or recycling is yet to be tackled realistically.

Another concern arises from the use of highly toxic or corrosive substances such as chlorine, phosphorous compounds and alkylamines in making pesticides and herbicides. An accident at a plant using materials such as these can be devastating. An explosion at an insecticide plant in Bhopal, India, in 1984 released derivatives of phosgene (nerve gas) into the air, causing thousands of deaths and disabilities across a wide area. It was the worst factory accident in history.

Many other industries use chemicals in their manufacturing processes, and untreated waste products can add significantly to air and water pollution. Leather tanneries have been polluting rivers for over a century with chromium salts, and more recently with aluminum and zirconium salts. Textile mills release bleaching and dyeing materials into rivers while pulp, paper and cellulose industries are major sources of sulfide effluents in rivers, especially in the former Soviet Union, the Baltic area, the United States, Canada, Europe and Japan.

Moving Toward a Post-Industrial Society?

In the last quarter of the 20th century a clear pattern of deindustrialization has emerged in the world's more developed regions. The extraction industries, heavy engineering and manufacturing that were the cornerstones of industrial development in the 19th and early 20th centuries, are in a steady decline in Western nations. The United States, Europe and Japan in particular are transforming from goods-producing economies to service-based economies – the business of making things for sale is being replaced by selling various kinds of information or specialist knowledge.

Part of this pattern is that many key manufacturing industries have moved to bases in the developing countries where labor costs are lower, and where natural resources may still be abundant. The industries that remain in the West tend to be heavily automated, employing a minimal workforce in the actual manufacturing process, but requiring a relatively large number of service staff – managers, accountants, marketing personnel and technicians. These changes and developments are often regarded by economists as marking the rise of a post-industrial society in the West.

The period of transition

Many different factors combined to bring about this change of direction. The energy crisis caused by the soaring price of oil in the 1970s dealt a serious blow to many western companies, making them turn to relocation as a way of cutting overheads. Manufacturing processes that used to be integrated, especially in metal processing and in mechanical and electrical engineering, were segmented, with separate stages being dispersed to cheaper locations in rural areas. Often, whole sectors such as textiles, fashion garments or shoe manufacturing move to other parts of the world, particularly Asia and the Pacific region.

Better and faster travel facilities, and huge improvements in telecommunications throughout the 1970s boosted the pace of manufacturing dispersal in the United States and western Europe. Companies could stay in close contact even if they were located some distance apart, and finished goods could be transported

New industrial plants (*below*) operating at lower costs often rely heavily on women in the workforce to perform routine tasks. These female computer workers at a high-tech factory in Singapore supervise an automated assembly line.

Decline and fall *(left)* While service and high-tech industries are expanding, many traditional manufacturers are closing down. This iron and steel works in Longwy, France, is no longer able to compete effectively in the world market.

cheaply to distribution centers all over the world. Even within western nations, improved highways and high urban rents made rural rather than inner-city locations more attractive for new industrial plants. In many areas the trend away from traditional industrialization was paralleled by another significant trend to move new ventures out of the cities, commonly referred to as "counter-urbanization". Populations steadily declined in the older, larger cities and began to grow in smaller and medium-sized cities.

In many parts of the developed world, deindustrialization is an integral part of the transition to a different kind of industrial society where innovation will continue, but where industry will organize itself in a new way. As manufacturing moved out, the 1970s and 1980s witnessed a growing concentration of expanding advanced services in and around the major cities. Corporate headquarters and government administrative bodies attracted newly fashionable management consultancies and specialist business services to the city centers. At the same time research and development laboratories, universities, specialist technical, scientific and support services grew in suburban zones, alongside high-technology industries. This kind of service has been a major growth area in recent years.

Expansion in the service sector

One emerging feature of the post-industrial society is that distinctions between manufacturing, and a very wide range of services including quality testing, equipment maintenance, and training are becoming much more blurred within and between firms. Industry itself is entering an era where the focus is not just on making things, but on integrating research, design, production, marketing, maintenance, environmental controls and recycling waste. In most industrialized regions service industries are developing more advanced and environmentally friendly techniques for manufacturing. This kind of expertise is likely to play a future role in the reindustrialization of major innovating nations and the continuing development of some newly industrialized countries.

REGIONS OF THE WORLD

CANADA AND THE ARCTIC

Canada, Greenland

THE UNITED STATES

United States of America

CENTRAL AMERICA AND THE CARIBBEAN

Antigua and Barbuda, Bahamas, Barbados, Belize, Costa Rica, Cuba, Dominica, Dominican Republic, El Salvador, Grenada, Guatemala, Haiti, Honduras, Jamaica, Mexico, Nicaragua, Panama, St Kitts-Nevis, St Lucia, St Vincent and the Grenadines, Trinidad and Tobago

SOUTH AMERICA

Argentina, Bolivia, Brazil, Chile, Colombia, Ecuador, Guyana, Paraguay, Peru, Uruguay, Surinam, Venezuela

THE NORDIC COUNTRIES

Denmark, Finland, Iceland, Norway, Sweden

THE BRITISH ISLES

Ireland, United Kingdom

FRANCE AND ITS NEIGHBORS

Andorra, France, Monaco

THE LOW COUNTRIES

Belgium, Luxembourg, Netherlands

SPAIN AND PORTUGAL

Portugal, Spain

ITALY AND GREECE

Cyprus, Greece, Italy, Malta, San Marino, Vatican City

CENTRAL EUROPE

Austria, Germany, Liechtenstein, Switzerland

EASTERN EUROPE

Albania, Bosnia and Hercegovina, Bulgaria, Croatia, Czechoslovakia, Hungary, Macedonia, Poland, Romania, Slovenia, Yugoslavia

NORTHERN EURASIA

Armenia, Azerbaijan, Belorussia, Estonia, Georgia, Kazakhstan, Kirghiz, Latvia, Lithuania, Moldavia, Mongolia, Russia, Tadzhikistan, Turkmenistan, Ukraine, Uzbekistan

THE MIDDLE EAST

Afghanistan, Bahrain, Iran, Iraq, Israel, Jordan, Kuwait, Lebanon, Oman, Qatar, Saudi Arabia, Syria, Turkey, United Arab Emirates, Yemen

NORTHERN AFRICA

Algeria, Chad, Djibouti, Egypt, Ethiopia, Libya, Mali, Mauritania, Morocco, Niger, Somalia, Sudan, Tunisia

CENTRAL AFRICA

Benin, Burkina, Burundi, Cameroon, Cape Verde, Central African Republic, Congo, Equatorial Guinea, Gabon, Gambia, Ghana, Guinea, Guinea-Bissau, Ivory Coast, Kenya, Liberia, Nigeria, Rwanda, São Tomé and Príncipe, Senegal, Seychelles, Sierra Leone, Tanzania, Togo, Uganda, Zaire

SOUTHERN AFRICA

Angola, Botswana, Comoros, Lesotho, Madagascar, Malawi, Mauritius, Mozambique, Namibia, South Africa, Swaziland, Zambia, Zimbabwe

THE INDIAN SUBCONTINENT

Bangladesh, Bhutan, India, Maldives, Nepal, Pakistan, Sri Lanka

CHINA AND ITS NEIGHBORS

China, Taiwan

SOUTHEAST ASIA

Brunei, Burma, Cambodia, Indonesia, Laos, Malaysia, Philippines, Singapore, Thailand, Vietnam

JAPAN AND KOREA

Japan, North Korea, South Korea

AUSTRALASIA, OCEANIA AND ANTARCTICA

Antarctica, Australia, Fiji, Kiribati, Nauru, New Zealand, Papua New Guinea, Solomon Islands, Tonga, Tuvalu, Vanuatu, Western Samoa

North America

Central and South America

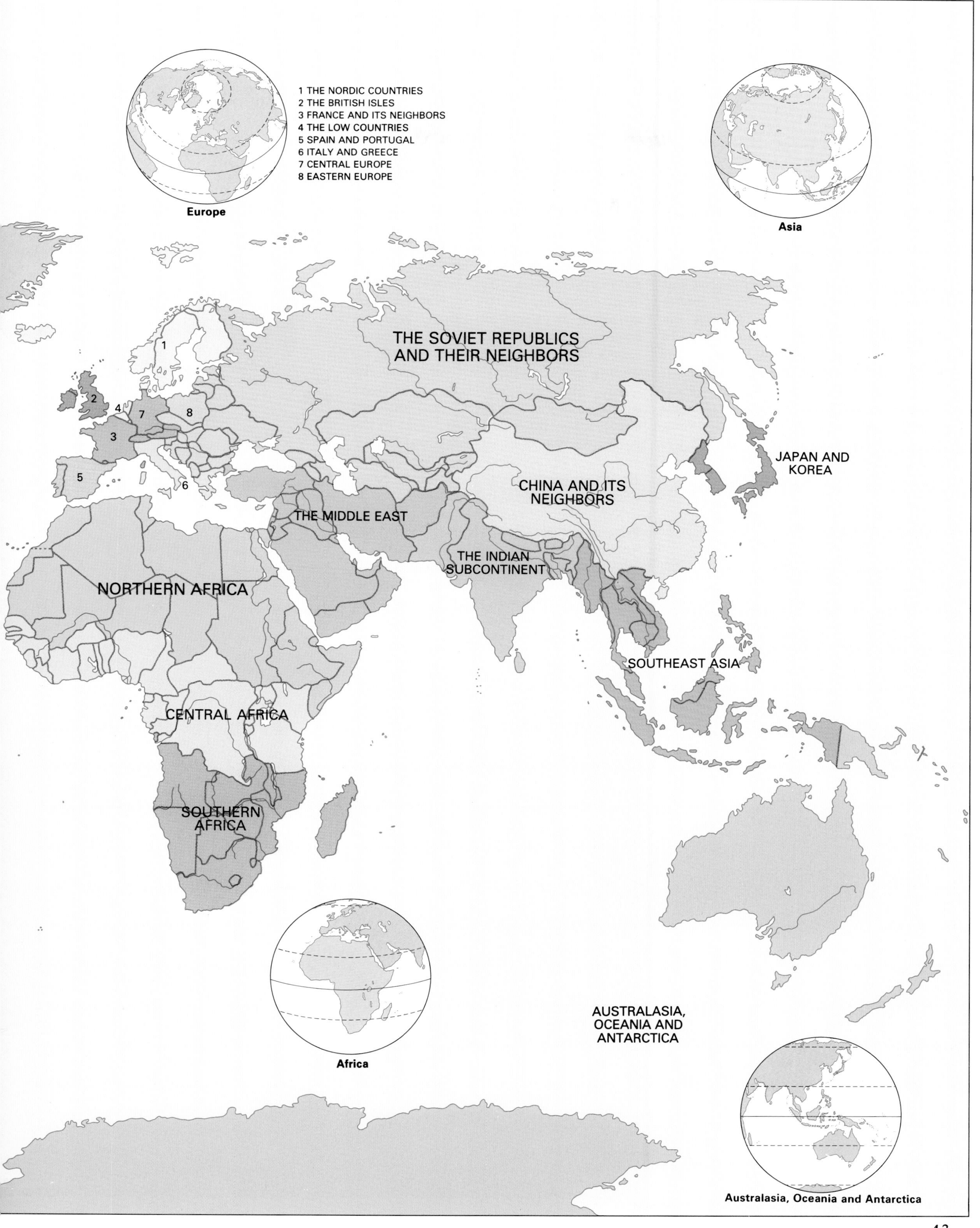

Europe
1 THE NORDIC COUNTRIES
2 THE BRITISH ISLES
3 FRANCE AND ITS NEIGHBORS
4 THE LOW COUNTRIES
5 SPAIN AND PORTUGAL
6 ITALY AND GREECE
7 CENTRAL EUROPE
8 EASTERN EUROPE
Asia
THE SOVIET REPUBLICS AND THEIR NEIGHBORS
JAPAN AND KOREA
CHINA AND ITS NEIGHBORS
THE MIDDLE EAST
THE INDIAN SUBCONTINENT
NORTHERN AFRICA
SOUTHEAST ASIA
CENTRAL AFRICA
SOUTHERN AFRICA
Africa
AUSTRALASIA, OCEANIA AND ANTARCTICA
Australasia, Oceania and Antarctica

GIANT INDUSTRIES IN A VAST LAND

A WORLD SUPPLIER OF RESOURCES · A CENTURY OF GROWTH AND DECLINE · SUPPORT STRATEGIES

Canada's huge expanses of territory offer a rich diversity of natural resources far beyond the needs of its small population. It has therefore had to look outside its boundaries for markets, and its leading role as supplier of raw materials to the rest of the world has shaped the pattern of its industrial development. The difficulties and heavy financial costs of extracting and shipping huge volumes of ore or timber across remote terrain have favored the involvement of giant corporations. Only in the most heavily populated areas of Quebec and Ontario is there significant manufacturing diversity. However, recent decline in the world demand for raw materials, and intensified competition from other suppliers, means that Canada today is having to redefine its industrial role and seek new challenges.

COUNTRIES IN THE REGION

Canada

INDUSTRIAL OUTPUT (US $ billion)

Total	Mining	Manufacturing	Average annual change since 1960
171.3	19.7	94.3	+3.5%

INDUSTRIAL WORKERS (millions)
(figures in brackets are percentages of total labor force)

Total	Mining	Manufacturing	Construction
3.3	0.2 (1.5%)	2.3 (17.1%)	0.8 (6.3%)

MAJOR PRODUCTS (figures in brackets are percentages of world production)

Energy and minerals	Output	Change since 1960
Coal (mill tonnes)	70.6 (1.5%)	+713%
Oil (mill barrels)	615.7 (2.8%)	+327%
Natural gas (billion cu. meters)	90.8 (4.9%)	+625%
Iron Ore (mill tonnes)	40.8 (7.2%)	+13.5%
Copper (mill tonnes)	0.8 (8.8%)	-8.5%
Lead (mill tonnes)	0.4 (11.5%)	+29%
Zinc (mill tonnes)	1.5 (20.9%)	+14%
Nickel (mill tonnes)	0.2 (24.6%)	-18%
Uranium (1,000 tonnes: U content)	12.4 (33.7%)	No data
Manufactures		
Aluminum (mill tonnes)	1.6 (7.2%)	+72%
Steel (mill tonnes)	15.1 (2.1%)	+286%
Woodpulp (mill tonnes)	21.0 (16.5%)	+16%
Newsprint (mill tonnes)	10.0 (31.5%)	+15%
Sulfuric acid (mill tonnes)	3.8 (1.4%)	+45%
Automobiles (mill)	2.0 (4.3%)	+509%

A WORLD SUPPLIER OF RESOURCES

To the first European settlers, Canada's vast spaces offered seemingly inexhaustible biological resources for exploitation – fur-bearing animals, plentiful fish in rivers and coastal waters, endless forests of fir and pine. At first fur was the most attractive of these resources, and fur traders, both French and British, penetrated the country's interior, setting up trading posts along the rivers from the Great Lakes to Hudson Bay. Extensive exploitation of Canada's forest reserves began when Britain's involvement in the American War of Independence (1775–83) and the Napoleonic Wars (1803–15) created an urgent demand for large quantities of timber for shipbuilding.

Throughout the 19th century the development of Canada's resources continued apace. From the 1880s the prairies were opened up for agriculture, and the production of lumber in British Columbia enjoyed rapid growth. Settlers also began to discover Canada's rich mineral resources. Deposits of nickel, silver, zinc

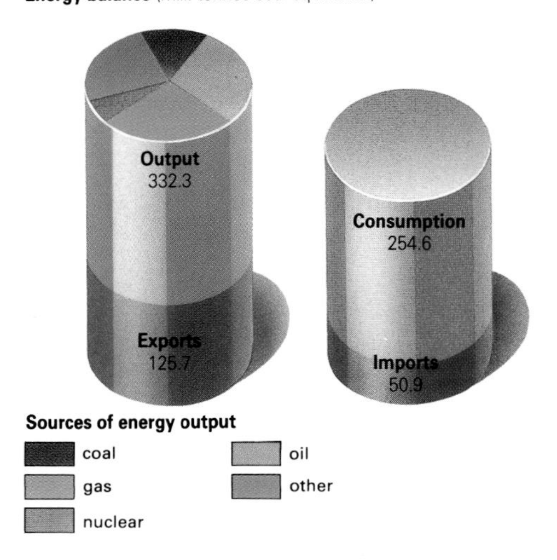

Energy production and consumption (*above*) Domestic gas, oil and hydroelectric power are the mainstays, though nuclear power is becoming increasingly important.

Map of principal resources and industrial zones (*right*) Canada's scattered natural resources include metals, coal, gas and oil. Manufacturing is concentrated in the Montreal–Toronto corridor.

A paper mill (*below*) on British Columbia's Gold river is ideally located. The river provides hydroelectric power and free transportation of logs.

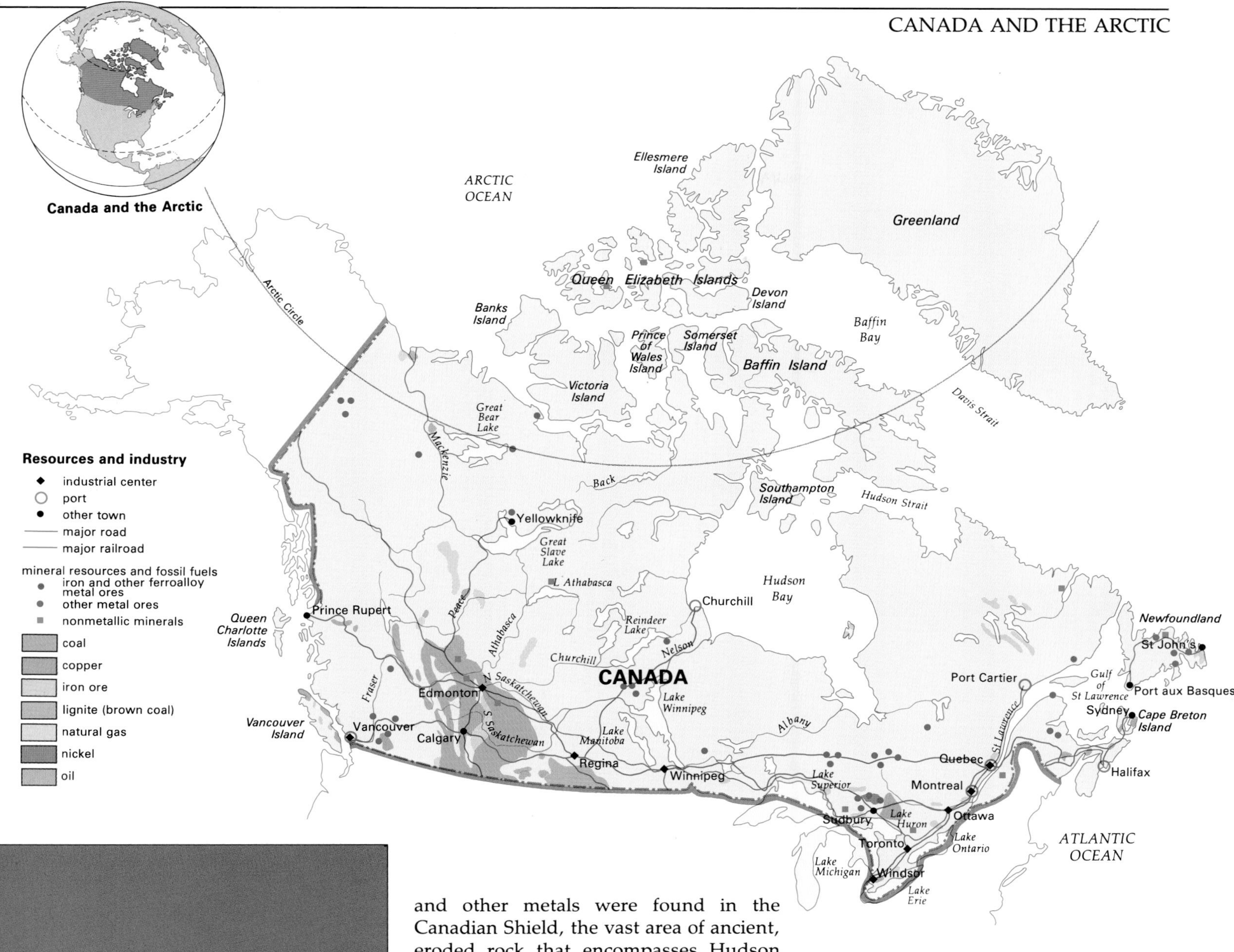

and other metals were found in the Canadian Shield, the vast area of ancient, eroded rock that encompasses Hudson Bay. In the late 1890s the discovery of gold nuggets in the Klondike, a tributary of the Yukon river in north British Columbia, sparked off the greatest outbreak of prospector fever ever recorded.

Many mineral deposits lie in the far north under permafrost, soil that is frozen for much of the year. This makes them difficult to extract, and before 1945 exploitation was sporadic. Since then resource development has continued to grow, especially in the west and north. Today Canada is a leading world producer of nickel, uranium, asbestos, zinc, iron ore and molybdenum.

It is richest of all in energy reserves. Coal (which it increasingly exports to Japan), petroleum and natural gas lie beneath the prairies, and there are petroleum fields in the Arctic Ocean and off the Atlantic coast. Rivers are used to generate hydroelectricity, though this is not so significant a source of energy as in the past. By contrast, nuclear power is on the increase. During the fuel crisis of the early 1970s Canada's energy exports became increasingly important, exceeding nonfuel mineral exports for the first time in 1975, though they have declined somewhat since then.

Neighborly concerns

Proximity to the economically powerful United States, always a major market for Canadian commodities, has often had a direct influence on resource development. For example, the northern forests of the Canadian Shield were first exploited at the end of the 19th century to supply newspaper publishers in the United States with woodpulp. After 1945, Canadian exports of raw materials to the United States grew and diversified as large American corporations, concerned about shortages of indigenous resources, sought new sources of supply.

One such corporation was the United States-owned Iron Ore Company of Canada. It began to mine the massive iron-ore

deposits of northeast Quebec in the 1950s to supply its mills in the United States. The town of Schefferville was built, as well as power stations, crushing plants, a 573km (355mi) railroad and port facilities at Sept Isle on the St Lawrence Seaway. Subsequent industrial decline in the United States led to measures to protect American industries by restricting imports of raw materials, and the Schefferville mines closed in 1983.

A prime reason for Canada's signing the Free Trade Agreement (FTA) with the United States in 1989 was to free Canadian exports from these protectionist restrictions. One likely effect of the FTA was to spell out for Canada a future role as the leading supplier of resources to the North American continent, rather than the world.

A CENTURY OF GROWTH AND DECLINE

By the end of the 19th century Canada's growing tide of exports had created an economic boom. Specialized machinery was needed to process minerals, lumber and agricultural products at their source. Railroads, ports and shipping facilities were built to transport these commodities to their markets in the United States and Western Europe. Canada's rapidly growing population created a demand for housing, domestic goods and food products. All this acted as a spur to the urgent development of largescale manufacturing industry.

Before Confederation in 1867, a few small manufacturing industries, supplying the local population with farming implements and equipment, had existed in southern Ontario and Quebec. It was here that a much more diversified range of industries now began to develop. Several factors encouraged the area's industrial preeminence – its relatively large population, its central location within Canada and its proximity to the industrial belt of the United States. The tendency for its industries to attract related industries, and a tariff policy designed to protect manufacturing, were also important.

The Quebec–Ontario corridor

Today an industrialized, urban belt runs from Quebec in the northeast to Windsor, Ontario in the southwest. Although manufacturing is highly diversified in the two large centers of Montreal and

Toronto, it is much more specialized within the numerous smaller towns of the area. Those in Quebec have a greater emphasis on textiles and clothing, while Ontario concentrates on automobiles, iron and steel, and electronics. This more favorable mix of industries helps explain Ontario's faster growth. Food processing and brewing are also important industries throughout the area.

A distinctive and significant element in Canada's industrial structure was created in 1965 by the Canada–United States Autopact. Under this agreement, tariffs on automobile export–imports between the two countries were lifted, allowing Canadian assembly plants – at that time entirely American-owned – to specialize more and produce a greater volume of vehicles for the United States. A required level of parts had to be purchased from Canadian sources. As a result, Canada now exports significant numbers of automobiles to the United States. Sales outside North America are rare, however, and imports are high. Research and development has been concentrated in the United States. Recent investment by Japanese automobile companies in plants in southern Ontario and Quebec seem likely to repeat this pattern.

During the 1980s unprecedented manufacturing job losses occurred in southern Ontario and Quebec as a result of three factors: technological change, overseas competition from both Japan and the developing countries and increasing economic integration with the United States. Canadian sales taxes and the FTA encouraged people to shop for retail goods in the United States, at the expense of local manufacturing. Hopes for the future prosperity of Canada's traditional manufacturing heartland rested on the attraction of more service industries and research-intensive activities.

Oil extraction in Alberta *(left)* This huge paddle dredge dwarfs the people working at its base but is scaled to the vast open spaces of the prairies, which conceal rich deposits of natural gas and petroleum. United States' companies have played a considerable role in developing the petroleum industry, which requires high capital investment. The Canadian government sponsors research and regulates the privately controlled industries that exploit these resources.

Cutting coats in Winnipeg *(right)* The distinctive white coats with multicolored stripes, called Hudson's Bay coats, are products of Canada's crafts industry. The original Hudson's Bay Company was founded in 1670 by French fur traders and London merchants. Today, the fur trade is still flourishing, and Hudson's Bay coats are popular local souvenirs.

The rest of Canada

Outside of the Quebec–Ontario corridor, industry has traditionally been based on local resources. Despite prolonged attempts to diversify the manufacturing base of eastern Canada, industry here is mainly restricted to fish processing and pulp and paper manufacture. Offshore oil reserves, however, encourage optimism for future development.

Western Canada has even richer resources than the east. In addition, both the petroleum processing industry in Alberta and the forestry industry in British Columbia have spawned a number of manufacturing industries making heavy equipment and machinery.

In the early 1980s, however, many of Canada's traditional resource-based industries began to experience decline. There were various reasons for this. The increased strength of the Canadian dollar lessened the competitiveness of exports to the United States. Technological changes and the development of new products and alternative sources of supply reduced the demand for certain minerals. A number of resources were becoming exhausted, and the exploitation of others was inhibited by widespread concern about environmental damage. As a result, attention turned from the bulk export of raw materials to finding ways of processing them at source that would add value to exports, and to establishing new markets for these products.

Attempts have also been made to diversify into new areas of enterprise and industry by expanding tourism, as well as developing sporting, educational, and arts and crafts activities. The construction industry concentrated on developing the new market potential of retirement housing. In addition, to meet the demands of British Columbia's rapidly increasing, affluent population, a number of industries oriented to this consumer market sprang up along the Pacific seaboard, including an electronics sector.

THE PETROLEUM BOOM IN THE PRAIRIES

Alberta, the most westerly prairie province, contains most of Canada's petroleum and natural gas. As recently as 1947 oilfields were discovered at Leduc, Redwater and Pembina, all within a radius of 120 km (75 mi) of the provincial capital of Edmonton. An extensive network of pipelines was quickly laid down, and these now deliver crude oil and natural gas from Alberta to eastern Canada and California in the United States.

During the 1950s a relatively small petrochemical industry was built up in Edmonton, producing fertilizers and industrial chemicals, but Canada's largest petrochemical complexes grew up closer to the major markets in central Canada, fed through the pipeline.

As crude oil prices escalated during the energy crisis of the early 1970s, Alberta's abundance of relatively cheap natural gas attracted much attention. As a result, a number of world-class, gas-based petroleum complexes were established in the area. Two giant ethylene plants at Joffre, near Red Deer, began production in 1979 and 1984, providing the raw material for a wide range of plastics and chemicals manufacturing processes.

The decline in international oil prices in the early 1980s removed Alberta's advantage, making further development of the petrochemical industry unlikely, but not before the boom had brought new prosperity to Canada's prairie capital.

SUPPORT STRATEGIES

Canada's rich resources are scattered across vast, often inaccessible territory. Operations to extract and process minerals, carry petroleum and natural gas long distances through pipelines, and fell and transport trees to processing plants on rivers and the coast are expensive. In the past, Canada lacked the manpower, the communications network and the financial capital to exploit its reserves of minerals and timber.

Foreign-owned companies were consequently pursuaded to speed up the rate of resource development and industrial growth by being offered cheap supplies of raw materials in return for capital, expertise and access to export markets. This open-door policy provided a means of generating export income, creating employment and encouraging regional development. Inevitably, however, it favored largescale industrialization in the hands of giant corporations.

Cement for construction *(left)*
Lime, sand and clay are found in several Canadian provinces, and cement is one of the main industrial products of Manitoba. Although materials tend to be produced close to sites, Manitoba is well positioned to deliver cement to a wide area, by rail or by the Trans–Canadian highway, or by water in summer and by sled and tractor in winter.

Maintaining the power lines *(right)*
Workers at the James Bay station, Quebec. Quebec province has more than 50 hydroelectric power plants, and Hydro-Quebec is Canada's largest producer of electricity. Federal control of resources is especially sensitive in the province, where separatist feeling frequently runs very high.

General Motors in Ontario *(below)*
Ontario has Canada's highest manufacturing output and employs half of the national workforce. With United States' firms in control, Canadian factories have been vulnerable to fluctuations in the American economy, and have also suffered from the decline of the steel and automotive industries from the mid 1970s onward.

Building the railroad

The interrelation between government policy, private investment and industrial development is well illustrated in the story of the Canadian Pacific Railway (CPR). The decision to build a transcontinental railroad was taken in 1872, just after the acquisition of British Columbia on the Pacific coast. This would link Canada's widely scattered territories, allow manufactured goods from Montreal and Toronto to reach new markets in the recently settled prairies, and carry wheat and lumber back to ports on the Great Lakes for shipment overseas.

However, economic depression called a halt to further construction soon after work had started. It was not until 1878, when the government granted land, money, the sections of the line already built and other concessions to the Canadian Pacific Railway Company of Montreal, that work pushed ahead rapidly. At the same time a new National Policy of tariff protection revived the manufacturing industry in Montreal and Toronto and encouraged investment in the enterprise.

The CPR quickly became the means of opening up mining areas in the Canadian Shield and western Canada, giving the company considerable influence over settlement patterns and resource development. It has benefited from this advantage to become, over the years, one of Canada's largest transportation and resource conglomerates. The building of the railroad also reinforced Montreal and Toronto's leading role in Canada's expanding industrial economy, placing them at the center of the country's banking system and other support industries.

The labor force that built the railroad was made up to a very large extent of Chinese immigrants. Until comparatively recently, Canadian governments consistently encouraged immigration to build up a work force to develop the country's natural riches. Particular ethnic groups tended to specialize in particular industries. For example, Eastern Europeans (Ukrainians, Czechs, Slovaks, Poles and Hungarians) were instrumental in opening up the agriculture of the prairies, the Germans established the shipbuilding industry in Nova Scotia, and the Japanese worked in the expanding fishing industry of British Columbia.

The growing role of the provinces

In the early years of industrialization, policy-making for resource development was regarded as the preserve of the federal government, located in Ottawa in southeastern Ontario. This emphasized

the dominant position held by Ontario and Quebec (representing the English and French-speaking cultures of Canada respectively) within the country's administrative and commercial structure. Until the early 20th century the federal government retained direct responsibility for developing resources in the Canadian Shield around Hudson Bay, and in the western parts of the country. The Shield areas were incorporated within Quebec and Ontario only in 1920, and it was not

A FORESTRY GIANT

The forestry industry of British Columbia produces two-thirds of Canada's lumber, half of its woodpulp, and important quantities of other forest products such as newsprint, paperboard and particleboard. Large, highly-efficient, integrated processing plants operate along the coast. In these complexes, logs supply either the wood processing or pulping operation. The residue of chips, sawdust or bark are used as fuel, as inputs for pulping, or in the manufacture of composite woods.

The biggest and most innovative of British Columbia's forestry companies is MacMillan Bloedel. In 1919 H.R. MacMillan (1885–1976), who had previously worked in a variety of jobs in the forestry and timber industries, founded a lumber trading company. It quickly became the province's largest lumber exporter, acquired plywood and sawmills, and in the 1940s built the province's first kraft (paperboard) pulp mill. It then merged with two other companies to create MacMillan Bloedel, Canada's largest forestry corporation.

In 1957 MacMillan Bloedel located its headquarters in Vancouver, building what was then the tallest skyscraper in the city – a decision that reflected Vancouver's growing importance as a regional center and also marked its rise as a city of metropolitan status to match Toronto and Montreal. In the recession of the 1980s, a controlling interest in the company was acquired by conglomerates based in Toronto, leading to considerable restructuring, but MacMillan Bloedel remains the giant of British Columbia's forestry industry.

until 1930 that the federal government transferred resource management rights to the prairie provinces.

After 1945 the governments of Canada's ten constituent provinces assumed increasing responsibility for the resources within their jurisdictions. In the main, priority was given to funding the building of pipelines, communications networks and roads to support large, export-oriented resource projects. Most provincial governments also tried to attract secondary manufacturing activities and to encourage new and rapidly expanding industries such as tourism. These provincial initiatives, usually supported by the federal government, were of great importance in promoting regional development throughout Canada.

Yet by the 1970s and 1980s many commentators had begun to argue that the decentralized and independent policy-making of the individual provinces with regard to the use of their resources had helped to create a national economy that served North American rather than Canadian interests. Barriers had been raised against trade between the different provinces, and too great a reliance placed on the United States as a market.

All this had occurred in a situation of growing tension between the provincial and federal governments. Control over resources played a crucial role in the debates over Canada's federal structure and the decentralization of power that divided opinion in the country during these years. The eventual outcome of the discussion will doubtless have profound implications for the development of Canadian industry, and for Canada's future as a nation.

Speaking to the world

Industrial competitiveness increasingly depends on the ability to diversify into new activities and the willingness to invest in new technologies. By comparison with other countries, industrial research and development in general in Canada has been underfunded. Typically, foreign-owned plants manufacture goods for the Canadian market that have been researched and developed by parent companies elsewhere. However, within the electronics and communications industry, Northern Telecom's longterm commitment to research and development is a prominent exception, showing that such commercial strategies for survival and growth are possible in Canada.

Canada's involvement in telecommunications is of long standing. Alexander Graham Bell (1847–1922) migrated to Ontario from Scotland before moving to the United States. In 1876 he made the first one-way, long-distance telephone call over a 13km (8 mi) line that he had set up in southern Ontario.

Thereafter, progress was rapid. Telephones were installed in Victoria on the west coast in 1878, the first telephone exchange was built in Vancouver in 1888 and the first transcontinental telephone exchange took place between Vancouver and Montreal in 1916.

However, from the very beginning the telephone companies in Canada (both those supplying services and also those manufacturing equipment) were subsidiaries of American-owned companies. It was not until the 1950s that Bell Canada became a wholly owned Canadian corporation. In 1956 it bought out the equipment-manufacturing company of Northern Telecom (then called Northern Electric) and channeled considerable investment capital to its research and development laboratories in Ottawa, in southeastern Ontario.

Today Northern Telecom is easily Canada's largest telecommunications firm with an impressive record of innovation. Its chief activity is in the manufacture of sophisticated telephone and telephone-switching technology, and in 1989 it had sales of over US$6 billion and a payroll in excess of 47,000 employees. It spends over 11 percent of its revenue on research and development. As a result, Northern Telecom has widely expanded its original market by becoming a leading global supplier of fully digital telecommunication systems. Northern Telecom products are used in over 90 countries and it has established many manufacturing operations in Canada, the United States, Europe and Asia.

Satellite Earth station, Alberta (*above*) spread out against the sky like a giant fan. The Telesat system in Canada links isolated areas such as the Northwest Territories to the rest of the country; innovations have included broadcasts in the Inuit languages. In 1958 the first ever satellite operated for 13 days on batteries; now they last for years and are retired due to obsolescence, not failure.

Computerized telephone chips (*left*) The Novatel factory in Calgary is an example of Canada's new high-tech industry. Its location in Canada's western corridor is a counterbalance to the domination of the older southeastern industrial sector. Computer chips play an important role in satellite communications. For example, largescale integrated chips help to cancel echoes on long-distance telephone calls.

Communication over longer distances
This success is not altogether typical. Canada's electronics industry is small by international standards and of the 100 largest firms, almost 75 percent are foreign owned. However, in recent years Canada has developed a wide range of internationally recognized competitive strengths, especially in the field of long-distance communications. Most significant have been the contributions it has made to the development and manufacture of communications satellites. These make possible the exchange of live television programs, and of telephone and radio communications, between countries and across continents and oceans. Signals from an Earth station are sent to an orbiting satellite, which amplifies them and then transmits them to a station in another region of the Earth.

The federal government has played a significant role in developing expertise in satellite technology by initiating a series of research projects over the years. The Alouette satellites launched in 1962, the Anik satellites launched a decade later (developed in conjunction with private industry and built in the United States) and the Hermes satellite launched in 1976 all significantly extended long-distance broadcasting capability.

In the private sector, Spar Aerospace of Toronto is Canada's leading developer and manufacturer of satellite communication systems. It was responsible for developing Canadarm, the remote manipulator system used on the United States' space shuttle.

Treasure Island

The spectacular and intricate scenery of Canada's Maritime Provinces was created by glacial action during the last ice age. These same glaciers invested a wealth of mineral and metal resources in the islands and mainland of the area. One of the most abundant natural metal resources in Canada is copper.

Found in quantity mainly in Nova Scotia around the Cap d'Or, but also scattered throughout the country, Canada's reserves of copper make it one of the world's leading producers. Because copper is so important in manufacturing, the mining industry in Canada is very large and profitable.

Copper may have been the first metal used by humans. Today its use in manufacturing ranks second, exceeded only by iron. Copper has several special properties that contribute to its extensive use throughout the world. Apart from silver, which is much more expensive, copper is the best conductor of thermal and electrical energy. This makes it an ideal lining for furnaces, pans and other utensils that need to be heated and cooled rapidly, as well as being essential in the electronics industries. Copper is also the base for several alloy metals, including bronze (copper and tin) and brass (copper and zinc).

Although it is an essential biological element, concentrations of the metal can be toxic, one of the hazards of working with copper. Waste from mining and refining damages the mine's surrounding environment. This is particularly true of the fragile ecosystems off some of Canada's island mines.

A copper mine off Canada's northeast coast. Most of the copper mined in Canada is used by the country's growing high-tech and electronics industries.

LEADING THE WORLD IN SERVICES

MASSIVE ENERGY CONSUMPTION · MINERAL RESOURCES · THE GROWTH OF MANUFACTURING · INDUSTRY'S CHANGING STRUCTURE · NEW PATTERNS OF EMPLOYMENT · THE ROLE OF GOVERNMENT

The United States (US) has been the world's leading industrial nation since the late 19th century. Extensive mineral and energy resources were discovered across the country's huge land area, and from these heavy industries developed, turning iron and steel into automobiles and machinery. From the mid 20th century a major change began, with increasing emphasis on light and high-tech manufacturing and service industries, now the most sophisticated in the world. Lately American manufacturing has been affected by increasing competition from Japan and other Pacific countries, particularly in high-tech products and automobiles. Nevertheless the sheer wealth of mineral and energy resources in the United States, its huge land area, skilled workforce and developed infrastructure assures it a prominent world position.

COUNTRIES IN THE REGION

United States of America

INDUSTRIAL OUTPUT (US $ billion)

Total	Mining	Manufacturing	Average annual change since 1960
1,249.5	86.4	1,032.9	+2.1%

INDUSTRIAL WORKERS (millions)
(figures in brackets are percentages of total labor force)

Total	Mining	Manufacturing	Construction
32.96	0.85 (0.7%)	23.1 (19%)	8.0 (6.6%)

MAJOR PRODUCTS (figures in brackets are percentages of total world production)

Energy and minerals	Output	Change since 1960
Coal (mill tonnes)	897.6 (19.2%)	+127%
Oil (mill barrels)	3173.9 (14%)	+24%
Natural gas (billion cu. meters)	488.5 (25.4%)	+36%
Copper (mill tonnes)	1.86 (13.7%)	+28%
Lead (mill tonnes)	1.05 (10.3%)	+72%
Manufactures		
Commercial jet aircraft	2251 (84.5%)*	No data
Domestic/catering ovens, incl. microwaves (mill)	20.5 (21.7%)	No data
Cement (mill tonnes)	67.4% (7%)	+20%
Steel (mill tonnes)	90.1 (12%)	-1%
Automobiles (mill)	8.0 (25%)	+20%
Televisions (mill)	13.6 (16%)	+133%
Synthetic rubber (mill tonnes)	2.05 (23%)	-19%
Nitrogenous fertilizers (mill tonnes)	9.5 (13%)	+4%
Semiconductors (US $ billion)	22.3 (40%)	N/A

N/A means production had not begun in 1960

** Applies only to noncommunist world in 1989*

MASSIVE ENERGY CONSUMPTION

The United States is richly endowed with energy resources, the driving power behind the nation's industrial growth. Coal, petroleum and natural gas account for two-thirds of the country's total mineral production, making the United States one of the world's largest producers of energy. Production is far surpassed, however, by the country's huge energy consumption, which is vastly in excess of any other nation. Consequently the country is dependent on imported energy supplies, especially petroleum.

The earliest source of energy in the region, both during the colonial era and before, was wood. In the 19th century wood was displaced by coal, particularly the bituminous coal found in the area around Pennsylvania, West Virginia and Ohio. Mining in this area was soon conducted on a huge scale. However, this kind of coal contains high amounts of sulfur, which pollutes the air, and its use was restricted by the Clean Air Act of 1970. After that the mining industry shifted to the vast low-sulfur coal reserves of the Great Plains and Rocky Mountain states. Anthracite, a hard low-sulfur coal, was an important source of energy for more than a century, mainly as a home-heating fuel. Since peak production in

Economic lifeline (*below*) Alaskan oil production was not substantial until 1961, when the Kenai field was discovered. Distribution links, such as the pipeline below, helped Alaska's oil industry to develop and it is now ranked second behind Texas.

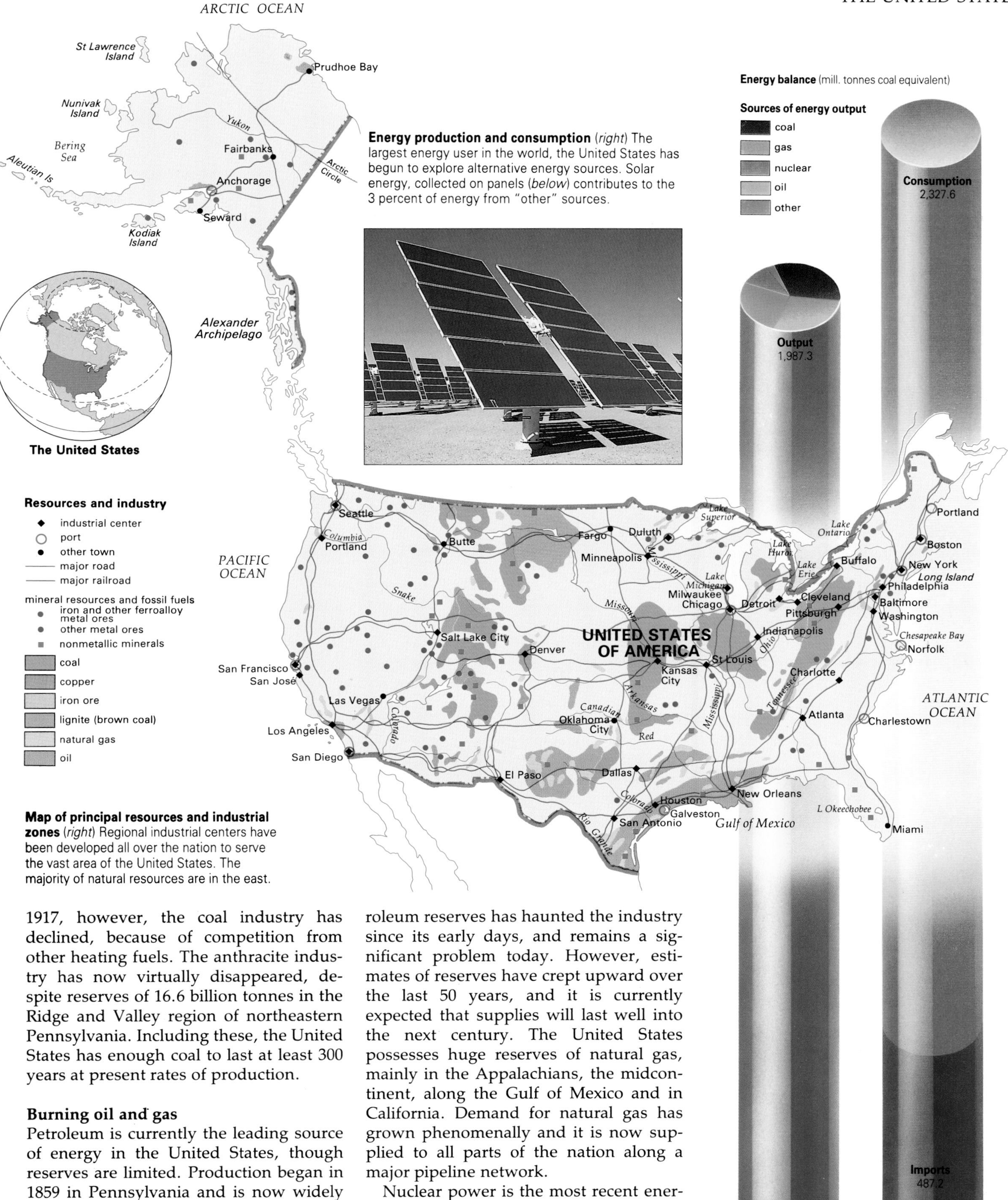

Energy production and consumption (*right*) The largest energy user in the world, the United States has begun to explore alternative energy sources. Solar energy, collected on panels (*below*) contributes to the 3 percent of energy from "other" sources.

Map of principal resources and industrial zones (*right*) Regional industrial centers have been developed all over the nation to serve the vast area of the United States. The majority of natural resources are in the east.

1917, however, the coal industry has declined, because of competition from other heating fuels. The anthracite industry has now virtually disappeared, despite reserves of 16.6 billion tonnes in the Ridge and Valley region of northeastern Pennsylvania. Including these, the United States has enough coal to last at least 300 years at present rates of production.

Burning oil and gas

Petroleum is currently the leading source of energy in the United States, though reserves are limited. Production began in 1859 in Pennsylvania and is now widely distributed. Although exploratory drilling continues, the last major oil discoveries were in 1933 (the East Texas Field) and 1968 (the Prudhomme Field of Alaska). The threat of exhaustion of petroleum reserves has haunted the industry since its early days, and remains a significant problem today. However, estimates of reserves have crept upward over the last 50 years, and it is currently expected that supplies will last well into the next century. The United States possesses huge reserves of natural gas, mainly in the Appalachians, the midcontinent, along the Gulf of Mexico and in California. Demand for natural gas has grown phenomenally and it is now supplied to all parts of the nation along a major pipeline network.

Nuclear power is the most recent energy source in the United States, with the first power plant built in 1957, at Shippingport in Pennsylvania. Unlike other kinds of energy that were developed by private capital, nuclear power has always been controlled by the federal government. Uranium is the chief fuel, mined from large deposits in western Colorado, eastern Utah and northern Arizona. By 1988 the country had 108 operable plants, producing a significant percentage of the country's electricity, and one-third of the world's nuclear energy. However, the future of nuclear power is now uncertain, as public confidence was jeopardized by the reactor accident at Three Mile Island, Pennsylvania, in 1979.

In search of renewable energy
Hydroelectric power is currently the only significant source of renewable energy in the United States. As long ago as the 17th century streams were used as a source of water power in New England. The invention of the turbine allowed rapid development of electricity generation, and it is estimated that about one-third of the nation's potential hydroelectric resource has now been developed. The Pacific coast states of Washington, Oregon and California are major sources of hydroelectric power, as are the mountain states of the west and the southeast. The Tennessee Valley Authority operates no fewer than 32 dams. Wind power has been developed in California, though on a much smaller scale. To put it in perspective, California generates more wind power than the rest of the world combined, but even on windy days it meets only 1 percent of the state's needs.

Coal by rail Although most natural resources are found west of the Mississippi river, the majority of American coal is in the east. Existing rail networks were set up early in the 20th century to transport coal to the western and central states.

MINERAL RESOURCES

The United States is rich in a great variety of metals and other minerals, and these have played an important role in the country's industrial achievement. However some reserves, notably iron, are now greatly depleted. This has created increasing reliance on new technology – so minerals can be extracted more efficiently – as well as on imports.

Iron and steel
Iron ore is widely distributed in the country, but most deposits are small. Only in the upper Great Lakes are the iron ore reserves large enough to supply the American iron and steel industry. Traditionally the best source is hematite, an iron oxide with an average metal content of 51 percent. However, reserves are now nearing exhaustion and taconite, a lower-grade ore with 15 to 20 percent metal content, is also being exploited.

Several other metals are mined in conjunction with iron to make steel. Normally they are mixed with iron to make ferroalloys, and used by the steel industry in that form. Chromium, tungsten and molybdenum are the most important ferroalloy metals, used to give steel toughness and resistance against corrosion. Tungsten is mined in California, Washington, Oregon and Montana; chromium in Colorado, Nevada and California, and molybdenum is widespread, with enough deposits to make the country self-sufficient. Vanadium is also mined in the United States, and is used in the production of vanadium steel, which is outstandingly resistant to fatigue and shock. The United States also has scattered deposits of manganese – vital to the refining process – though the ores are too low grade for production to be viable.

Metals not used in steelmaking
US production figures for copper – a metal important both in the electronics industry and in the production of military weapons – were surpassed only by Chile in 1988. Three-quarters of United States copper output derives from largescale, low-cost, open-pit deposits. Arizona accounts for about 60 percent of total output and New Mexico, together with Utah, 38 percent. In addition, the country is the world's third largest producer of lead: output in 1988 was exceeded only by Australia and the Soviet Union. Lead is particularly important in the making of storage batteries, and is not destroyed when used. Consequently some 50 to 60 percent of annual production is accounted for by recycling. The main areas of production are located in the flat plains of the Mississippi valley between northern Oklahoma and southern Wisconsin, and in the Coeur d'Alene district of Idaho.

Zinc, which is used as a coating to prevent steel erosion, used to be pro-

Cooper canyon This is the Bingham mine in Utah, the world's largest open-pit copper mine. The United States is second only to Chile in the production of copper. Major deposits are spread throughout the western part of the country.

duced mainly where the borders of Missouri, Kansas and Oklahoma meet. Now, however, the metal is mined mostly in eastern Tennessee and the Rocky Mountain states, where it is generally associated with lead, silver and gold ores.

Silver and gold have been mined in the United States for about 150 years, but surface supplies of the metals in their pure forms have been exhausted and neither is now of major economic importance. In the past bauxite – the ore from which aluminum is derived – was mined in significant quantities, notably in Arkansas, although this too has declined greatly in recent decades. There is, however, still a significant refining industry that uses imported bauxite, mostly from Jamaica. Because the refining process is such an energy-intensive operation, refineries are generally built close to cheap electricity supplies, notably in Washington, Kentucky and Texas.

The nonmetallic minerals

The United States is well provided with sulfur, a hugely important mineral that is widely used in the making of explosives, military materials, fertilizers and insecticides, pulp and paper, dyes and coal-tar products, rubber, paint and food products. Deposits are distributed all across the country, with the greatest concentrations along the Gulf of Mexico. Reserves are large enough for mining to continue for several hundred years.

Building materials such as stone, sand and gravel, clays and limestone are relatively abundant and widely distributed all across the country. Transporting such bulky resources is costly, so they are normally used locally. The most important types of stone are granite, marble, sandstone and limestone, with limestone making up two-thirds of stone output. Leading clay-producing states are Pennsylvania, Kentucky and Missouri.

The United States possesses about a half of known world reserves of phosphate, mostly used in fertilizers. Mining began in South Carolina in 1867, and is now conducted in Florida, Tennessee, Idaho, Montana and Wyoming. Before World War I, potash was imported from Germany, but since then the United States has grown into the world's leading supplier, with production concentrated in Kansas, Oklahoma, Texas, New Mexico, Utah and Michigan.

THE LURE OF GOLD

Gold has played a major role in shaping the destiny of the United States. The discovery of the metal in California in 1848 led to the gold rushes to the west in 1849 and the 1850s. These influxes of people were central to the early development of the city of San Francisco. Without the discovery of gold the great westward migration might have been delayed by decades.

After the California goldrush, gold was discovered in the Black Hills of South Dakota (1874) and at Cripple Creek, Colorado (1891). Since then surface deposits of the metal have been exhausted, and gold mining has declined in importance. Most US gold produced now is a by-product of ores of other metals such as copper, lead and zinc. A major exception to this is the remaining mine at Lead in South Dakota, the largest of its kind in the Western hemisphere. The deposit has been worked since 1876, and mining now occurs at depths below 1,500 m (5,000 ft). The seam, which appears to be limitless, contains slightly less than 14 g (0.5 oz) of finely divided gold per tonne of ore. The United States is the world's fourth largest producer of gold, exceeded only by South Africa, the former Soviet Union and Canada.

THE GROWTH OF MANUFACTURING

By the early 20th century the United States had become the world's industrial giant, leading the other industrial nations in output of iron and steel, motor vehicles, machinery and a range of other goods. Manufacturing began on the Atlantic coast; the sea routes to Europe offered access to an important market and brought a constant supply of cheap immigrant labor. As the home market grew, industry began to spread inland to the Mississippi river between Minneapolis-St Paul and St Louis. For some time this northeastern quadrant of the country remained the main manufacturing center, and as late as 1940 it still housed two-thirds of the nation's factories.

Smoke stack industries

Iron and steelmaking first developed on a large scale around Pittsburgh in Pennsylvania and Youngstown in Ohio. Production was close to the coal reserves of western Pennsylvania and eastern Ohio, while there was easy access along the Monongahela, Ohio and Allegheny rivers to the iron ore deposits by the Great Lakes. The area soon became the largest iron and steelmaking center in the world.

From 1875 another form of manufacturing began to develop in eastern Pennsylvania when the Lehigh district became a center for the production of Portland cement. The area possessed large quantities of limestone – a natural cement rock – as well as having convenient access to the eastern market. Consequently it achieved a near monopoly of production for the next 25 years. Pennsylvania was also the site of an early oil-refining industry. Oil was discovered at Titusville in the west of the state in 1859, and refineries developed locally, as well as at market centers such as Pittsburgh and Cleveland.

The age of the automobile

By the turn of the century a wholly new form of industry began to develop in southern Michigan. Growth of motor production in the state was prompted not by access to raw materials but by a local technological breakthrough. In 1899 the Olds Motor Works of East Lansing produced what was probably the first profitable automobile: the Oldsmobile. Success attracted other inventors to the state, in-

CHICAGO – FROM STEEL TO SERVICES

Manufacturing first developed in Chicago to supply the vast midwestern market with food and to export it to Europe via the St Lawrence Seaway. The city's earliest industries – meat packing and grain milling – specialized in processing the area's mainly agricultural produce. By the late 19th century the city had become a center of iron and steel production as well, and developed heavy industry along the same lines as Detroit and Pittsburgh. Since the 1970s, however, manufacturing has declined. The meat-packing companies have now moved away from the city and the stockyards closed in 1971. Employment in the iron and steel industries plummeted as foreign imports increased.

However, Chicago had other potential resources to develop. As the third most populous city of the United States, it has a thriving workforce. Its position at the southern tip of Lake Michigan is a pivotal one that has made it a vital communication center – the city boasts the world's largest rail terminal, ocean-going liners dock at its harbor and O'Hare airport is among the nation's busiest. The city is also the hub of the federal interstate highway system in the Midwest. These advantages have enabled Chicago to develop expanding transportation and service industries, particularly in the areas of health, finance, law, accountancy, business consultancies, marketing, catering and retailing. Most new jobs are skilled and command salaries comparable with those in manufacturing industry. The greatest single problem for the city has been retraining and redeploying former factory workers.

cluding David Buick (1854–1929), Henry Leland (1843–1932) and Elwood Haynes (1857–1925). The greatest of them all, Henry Ford (1863–1947), was born there.

Detroit, the leading industrial city of southern Michigan, rapidly became the motor capital of the world. In 1909 Henry Ford built the world's largest manufacturing plant in Highland Park, Michigan. His innovation of mass production along a simple assembly line gave the Ford company undisputed leadership in the industry for nearly two decades.

The spread of manufacturing

The first major manufacturing area to develop outside the country's northeast quadrant was in the Piedmont region in the southeast. From the 1880s a cotton textile industry began to establish itself there, helped by lower southern wages and the efficiency of new factories. In addition, the southern railroads offered special shipping rates on finished cotton products to offset the disadvantage of being so far from the northern markets.

This trend became increasingly pronounced in the 20th century, as the factors that had favored the northeast quadrant began to change. From the 1960s Europe ceased to be the main source of cheap labor, displaced by Latin America and east Asia. Immigrants – many of them illegal – tended to arrive in California and the southern states. The mature industrial areas found themselves at a disadvantage, as their skilled and organized labor force commanded higher wages than labor in developing areas.

Another force for change was the dramatic rise in fuel costs. Industrial plants had to become more energy efficient, making local availability of coal or iron less significant. Low-cost transportation boosted national distribution and export abroad, favoring a shift from heavy to light industry, and encouraging the production of high-value items such as electronic equipment. As major heavy industry declined in the region, access to iron supplies near the Great Lakes ceased to be such an advantage.

The consequences of these changes were profound. From the 1940s onward manufacturing has spread throughout the United States. The most rapidly growing region has been the Pacific coast, which now employs about a sixth of the nation's industrial workforce. California has been one of the world's great manufacturing growth areas in recent decades, particularly in computer technology. Developments such as Silicon Valley have attracted some of the country's most skilled personnel. Several of the southern states, too, have seen a major expansion in light industry, helped by low local costs and the area's growing role as a center of the nation's communications.

Smoke-stack skyline Blast furnaces (smelting iron ore from the Great Lakes) and steel mills dominate Cleveland, Ohio. Sited on the banks of Lake Erie and the Cuyahoga river, the city is well placed to make use of cheap water transport.

Taking to the air

The aerospace industry has been one of the great success stories of manufacturing in the United States since World War II. It has grown into one of the most profitable sectors of production while also playing a major role in national power and prestige.

The aircraft industry began with the first flight of the Wright Brothers at Kitty Hawk, North Carolina, in 1903. Production of aircraft fluctuated with military and economic demands, but by World War II the industry was well advanced, playing a major military role in the conflict, with machines being produced on a huge scale. The succeeding years of the Cold War between the United States and the Soviet bloc ensured continued demand, while emphasis shifted from bulk production to technological sophistication, with research playing an ever more important role. Commercial aircraft production also grew rapidly and by the late 1960s the United States was the world leader, supplying the great bulk of aircraft to the commercial airfleets of noncommunist nations.

Missiles and space launches

It was after World War II that the United States began to develop missile tech-

The final frontier (*right*) The Space Shuttle was developed to be a reusable transportation vehicle servicing scientific and commercial operations in space. In particular it was intended to carry parts to build a space station in orbit. This 18 m (60 ft) cargo bay can carry payloads of up to 30,000 kg (65,000 lbs).

Poised in the assembly line (*below*) The Boeing 747, here in its final assembly stages in Everett, Washington, was the first large-capacity passenger aircraft. Launched in 1966, the only 747 still in production is the largest model, 747–400.

nology. From the 1950s long-range and short-range missiles began to play an increasingly important role in the country's nuclear defense systems, and the federal government made large investments in the field. Production mushroomed during the following decades, with growing emphasis in the 1980s on perfecting missiles capable of intercepting incoming missiles, as well as on developing the precision weaponry used during the 1991 Gulf War. By the following year, however, the demise of the Soviet Union and government spending cuts at home made the future of the industry less certain.

Development of space technology on a large scale began in the United States in the 1950s. In 1957, the launch of the first space satellite, Sputnik, demonstrated that research in the Soviet Union was clearly ahead of the rest of the world. In response, American funding for research and development grew from practically nothing in 1958 to $6 billion in the mid 1960s. During the 1970s emphasis in the space program shifted from planning launches to the moon to developing the Space Shuttle, a vehicle designed to transport people and cargo between the Earth and orbiting spacecraft. Aircraft now account for about 55 percent of total aerospace output in the United States, missiles for 20 percent and the Space Shuttle for 25 percent.

A hierarchy among suppliers

The US aerospace industry is supplied with parts and components by some 4,000 firms, organized in a clear hierarchy. At the top are about 60 prime contractors, many of which have developed some specialist knowledge, such as propulsion, avionics or nuclear power. In the second tier are subcontractors who deal with items such as landing gear, life-support systems, airborne computers and auxiliary power systems. The third and fourth levels are subcontractors who provide subassemblies for the prime companies. This complexity has meant that system management has become a corporate specialty, with a systems manager coordinating production flow from several companies. Nowhere else in industry is there a program of such technical complexity and competitive urgency.

About a third of the aerospace industry is in California, with two leading companies – Lockheed and McDonnell Douglas – based in Los Angeles. Washington state is the second major center, with Boeing in Seattle. The US aerospace industry, which is the largest in the world, competes and cooperates with aerospace producers in western Europe and Japan. Until its demise in 1991 the Soviet Union was the second largest manufacturer followed by the United Kingdom and France, though Japan is also emerging as a major world producer.

INDUSTRY'S CHANGING STRUCTURE

During the 1980s manufacturing in the United States came under increasing pressure from foreign competitors using cheaper labor. High wages at home encouraged some US companies to transfer production abroad, which contributed to unemployment. At the same time, management methods were challenged by new and effective foreign techniques, especially those employed in Japan, emphasizing cooperation between management and the workforce. Competitiveness was also affected by a longterm lack of investment in new plant.

The overall effect of these changes has been a fall in the number of workers employed in US manufacturing. From an all-time peak of approaching 21 million in 1979, the workforce declined to 19.8 million by 1988. The industries most affected have been those that were dominant in the 19th and early 20th centuries: textiles, clothing, tobacco, leather, petroleum products, mining and processing metals and transportation equipment. More recent industries such as electronics and machinery have survived better on the world market. Different sectors of industry have chosen different methods to adapt to this challenge from abroad.

The decline of iron and steel

The US iron and steel industries dominated world production until the 1960s. After that, in spite of competition from overseas, American mills did not modernize, and continued to award large wage increases. Consequently productivity dropped against investment, and the United States became a high-cost producer. By the 1980s the US steel industry was facing crisis and, in order to survive, underwent dramatic changes.

Increasingly since the mid 1970s leading American steel producers had diversified and even ceased steel production altogether. US Steel – the symbol of steelmaking strength during much of the 20th century – had removed half of its assets from the steel business by 1980. In 1982 it acquired Marathon Oil, and then in 1985 Texas Oil and Gas. Parts of the corporation became heavily involved in the chemicals and plastics industries normally associated with the petroleum business. By 1986 it had dropped the

Big facilities for a big business *(above)* The chemical industry in the United States is one of the most important sectors of the national economy. The growth of US petroleum-based chemicals, including plastics and man-made textiles, reflects the global domination of American oil companies.

Creating synthetic fibers *(right)* Many synthetic fibers, including polythene and nylon are formed from a basic building block, a monomer, linked together in long chains to form polymers. The basic building block for polythene is the ethylene monomer. The nylon polymer is similar to polythene, but formed from different monomers.

Plastic molding methods *(below)* The structure of polymers means that they can be shaped or molded during manufacturing. There are four basic industrial methods used to do this: injection, extrusion, vacuum and blow molding. Injection molding is used for solid objects, finished on both sides; extrusion is used for tubes; and blow and vacuum molding are used for bottles, bowls and cylindrical products.

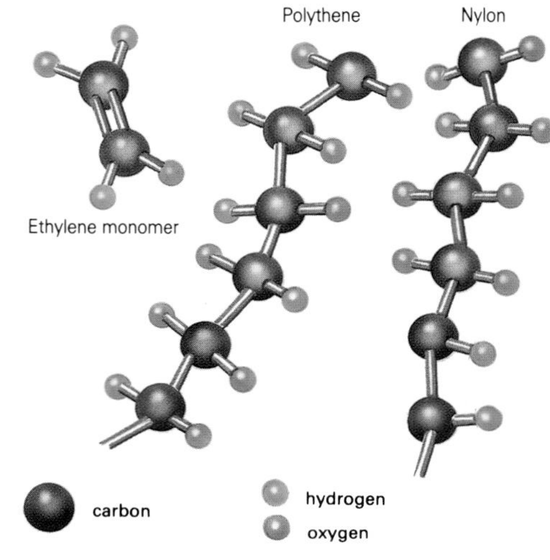

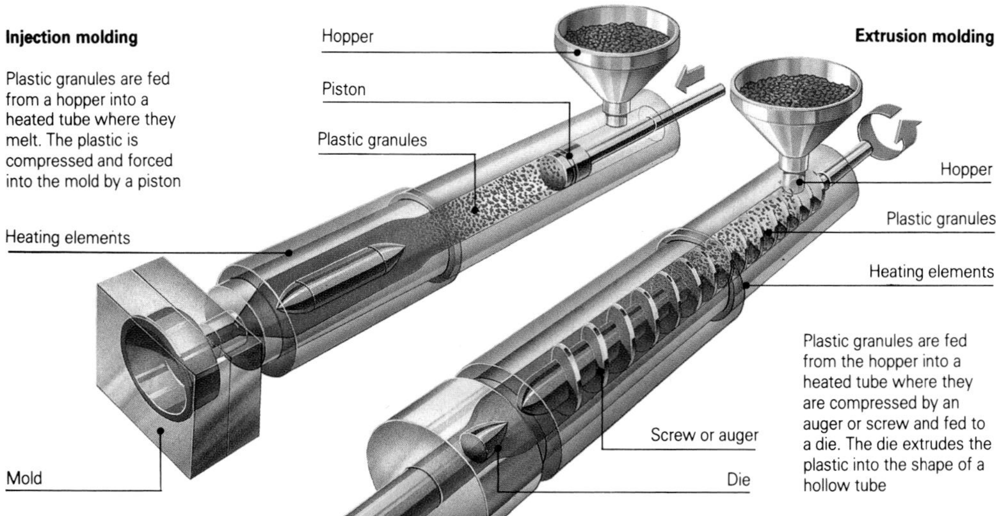

word "steel" from its name, becoming USX. By then, energy accounted for two-thirds of its revenues and all profits.

By 1987 the American steel industry, though reduced in size, had stabilized and the surviving older mills had modernized. More important was the growth of the smaller, more efficient mini-mills, which produced quality steel at competitive world prices.

Responding to Japanese influence

For most of the 20th century the United States led the world in motor vehicle manufacture. In the 1950s and 1960s production was dominated by "gasguzzlers", large automobiles with high fuel consumption. After the energy crises in the early 1970s smaller, foreign automobiles – particularly Japanese ones – began to compete effectively in the US market. The three largest American motor companies – General Motors, Ford and Chrysler – were slow to change traditional attitudes. Though change did eventually come, the Japanese managed to capture a share of the biggest automobile market in the world. In response, General Motors built a new factory to produce a small automobile named the Saturn.

Another remarkable development since the 1970s is the number of foreign (especially Japanese) automobile factories operating within the United States. In 1990 foreign companies produced 11 percent of US output, up from 5 percent in 1986. The US automotive industry is now part of a world integrated industry.

SATURN'S RADICAL LABOR AGREEMENT

By the 1980s General Motor's share of the US market had been devastated by foreign competition. In order to recover its earlier advantage, the corporation opened a new factory at Spring Hill, Tennessee, in 1990 to produce a new automobile: the Saturn.

While the plant has many technical innovations, the real revolution at Spring Hill is its labor-management agreement, which is the most radical ever developed in the United States. In 1984 a team of Saturn workers toured the world to discover the most effective factory practices. They concluded it was vital to gain consensus between the management and the workforce, while employees should have a sense of ownership within the company. Consequently the coordinator of the United Auto Workers (UAW) and the company president cooperate closely, even working from the same office. The workforce of 165 teams is more powerful than any other assembly workforce in the United States, or even Japan. They interview and approve new personnel and run their own areas and budgets. All decisions are based on consensus.

Production of the Saturn began in October 1990. After some initial problems the car seems to have been accepted by the American public and shows every sign of proving a success.

Nissan in the US (*left*) Since the 1970s Japanese automobile manufacturers have been building a strong base in the United States as well as challenging the domestic market with smaller and cheaper imported models that are more economic to run.

Blow molding

1 A length of hot plastic tubing is placed in the open mold. The mold is then closed, sealing the bottom of the tube

2 Air is blown into the tube, "inflating" it to the shape of the mold

3 The plastic cools and solidifies, the mold is opened and the finished article is removed. Excess plastic is trimmed off

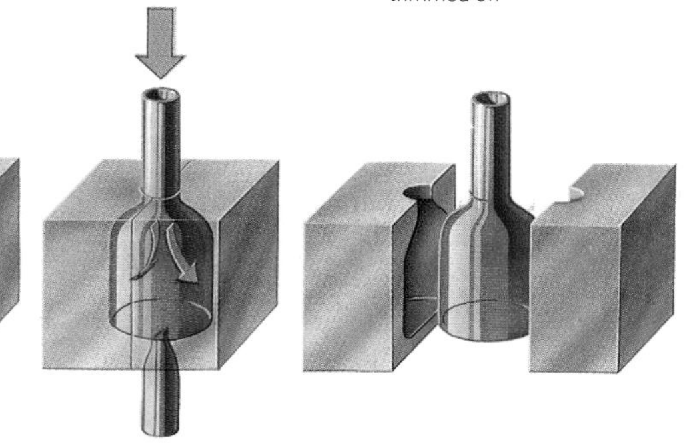

Vacuum molding

1 A sheet of plastic is placed over the mold

2 A heating element seals the top of the mold and heats the plastic. Air is drawn out of the mold, creating a vacuum, sucking the plastic into the shape of the mold

3 The plastic cools and the finished article is removed from the mold. Excess plastic is trimmed off

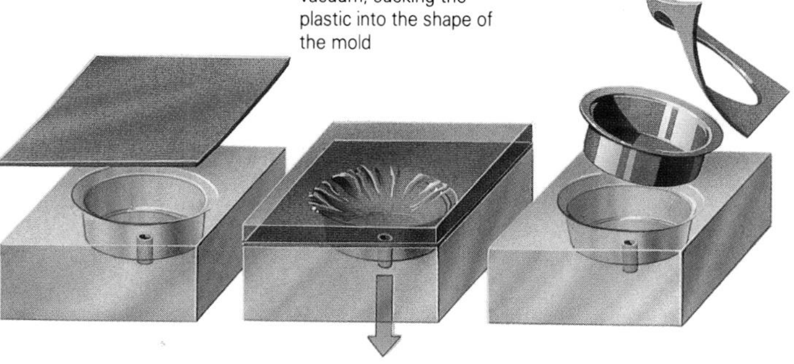

Long-haul freight (*above*) Railroads are still an important part of the transportation system in the United States. The large distances between manufacturer and consumer make transportation and distribution one of the largest service industries.

NEW PATTERNS OF EMPLOYMENT

Industry in the United States has undergone a transformation in recent decades. At the beginning of this century, it was synonymous with manufacturing; now it is dominated by service industries, with three-quarters of the working population employed in this sector. Service industries include transportation, health and medicine, finance, management, education, law, advertising, wholesale and retail trade, public relations, library and information services, architecture and engineering. Unlike manufacturing, services do not produce any tangible goods, and their value may be lost if not used quickly. A ticket for an airplane seat left vacant yesterday is of no value today. However, service industries do have a vital role in creating wealth.

An expanding job market

The service industries have been the catalyst for growth in the job market in the United States. Since the mid 1960s the country has created more employment than anywhere else in the world. Between 1975 and 1985 21 million new jobs were generated for the expanding population, also providing employment for the unprecedented number of women in the labor force. During this same period of time western Europe, in contrast, experienced a net loss of jobs.

Between 1979 and 1987 service industries created no less than 14 million jobs while the manufacturing sector declined by 1.9 million. Wages have also been affected. Low-wage employment is now present in all areas of the region, but is especially common in the Midwest, as well as among displaced workers who lack the skills that would enable them to move into another type of work. By contrast, service jobs often command high and increasing salaries.

Unemployment and restructuring

During this period of transition to a service economy, national unemployment rose by nearly half a million. The increase was the result of four main factors. Some of it was caused by the workers voluntarily quitting their jobs to look for new ones. Unemployment caused by the ups and downs in the business cycle also played its part. Structural unemployment caused by a mismatch between jobs and workers has been a problem, as has the effect of corporate restructuring, which has led to drastic cuts in the labor force.

Corporate restructuring has become

Designer products *(above)* Computer assisted design (CAD) is rapidly becoming commonplace in product development and manufacture in the United States. With widespread use of CAD, a whole range of computer support industries have developed to train new users and increase the role of information technology within business.

Making the news *(below)* Information has become a major commodity in the United States, and television interview shows, such as NBC's *Meet the Press*, supply consumer demand. The US has the highest number of independent television stations and networks and the highest ratio of newspapers per head of population in the world.

THE RISE AND FALL OF LABOR UNIONS

The first labor union was established by leather workers in Philadelphia in 1792, but the idea was slow to take root. Not until 1869 was an organization formed that hoped to represent all workers: the Knights of Labor. They achieved a number of spectacular strike victories and by 1886 – when they were renamed the American Federation of Labor (AF of L) – membership had reached 700,000. Gradually a rival organization formed – the Congress of Industrial Organization (CIO) – which grew to control unions in steel, automobiles, textiles, rubber, and packing-houses. By 1940, the CIO and the AF of L each claimed 5 million workers.

Labor cooperated fully with the government during World War II. At its conclusion, however, a series of strikes by the labor unions brought reprisals that culminated in the Taft–Hartly Act of 1947, which sought to restrict the unions. This triggered a revival of labor's political activities, prompting the CIO and AF of L to combine in 1955, adding to their influence. Recently, however, American unions have suffered a sharp decline in membership. They relied heavily on the manufacturing workforce, and the potential number of members fell during the deindustrialization of the 1980s, weakening the unions' bargaining position. Significantly, most workers in the service sector are not union members.

common in US industry. In order to withstand fierce global competition, many American companies have been forced to make fundamental changes to the way they conduct their business. Executives are relentlessly examining the efficiency of their companies' procedures. Unsatisfactory product lines are being sold and plants closed at a rate never seen before. In 1988 the telephone company AT&T cut its workforce by a tenth in an effort to save an annual $1 billion, with a third of the jobs lost being managerial.

The avalanche of corporate mergers and acquisitions has been another force for change in the employment market. In 1988 alone there were more than 4,000 mergers, worth a record $199 billion. These have led to job cuts, with merged corporations eliminating unprofitable divisions and staff whose jobs were duplicated. During the 1980s, for example, General Electric spent $11.1 billion on buying 338 businesses, at the same time shedding 232 businesses worth $5.9 billion, and closing 73 plants and offices. Middle managers and executives have been hit particularly hard by this new trend. Between 1983 and 1987 some 60,000 of these – out of a national total of 1.2 million – lost their jobs.

Retraining the workforce

Employee training has an important place in US business plans for the 1990s. It is seen as playing a vital role in allowing companies to overcome skill shortages in the workforce and to keep abreast of technological advances. Training is usually concentrated among the more educated and skilled employees, such as managers, engineers and sales people. The least likely to be trained are unskilled and semiskilled employees. Introductory training for new employees is the most popular type of course. Other areas include management, communication and computer literacy.

Retraining is needed where current skills are rapidly becoming obsolete. The larger employers, such as those engaged in hospitals, communications, finance and manufacturing, are especially involved in such programs, which are expanding rapidly within the nation. Remedial training is required for those who do not possess even the basic skills of reading, writing and arithmetic. Experts believe that much more remedial training will be needed in the future.

THE ROLE OF GOVERNMENT

Three tiers of government – local, state and federal – together form the country's largest employer, responsible for 13 percent of the workforce. Through its sheer size and the extent of its political control, government has played a central role in the development of industry. In some cases it has had a beneficial effect, offering incentives to promote growth; in other instances high-taxing administrations have proved an intolerable burden to struggling businesses.

Exerting direct influence

The US government first set up its own factories to supply US military forces. As early as 1794 the federal government built the Springfield Arsenal in Massachusetts. During World War I and in its aftermath, administrations began to intervene outside the sphere of defense, playing an important part in the location and development of industry. Factories were set up to produce materials seen as essential for national security, such as synthetic rubber and aluminum. In some instances the federal government intervened to promote growth in depressed areas – this was the aim of the 1961 Area Redevelopment Act and the 1965 Appalachian Development Act. Federal research and development funding were also crucial to the growth of high-technology industries during the 1980s, notably in the field of missiles, aircraft and space research.

Indirect government influence

The US government has played an important role in indirectly shaping manufacturing growth through its import policies. It has encouraged free trade across the world in the hope that this will allow greater access to foreign markets for American goods. Recently a free trade policy has been pursued with Canada and Mexico. At the same time some industries – such as steel and motor vehicles – have been partially protected with import restrictions, since buoyancy of trade in these areas is considered to be of vital national importance.

Indirect intervention has also been used to promote growth in areas of the country unusually struck by hardship. The most notable example of this is the creation of the Tennessee Valley Authority (TVA). This was set up by the federal government in 1933 to encourage development in an area that – though rich in resources – was suffering from depleted soil, periodic flooding and a depressed economy. Rather than build factories with its own funds, the TVA tried to create the right conditions to attract private corporations. The authority had considerable success, providing cheap and plentiful electricity as well as controlling floods and expanding river navigation. A significant amount of industrial growth in the Tennessee valley owes its existence to the TVA's efforts.

Another indirect way for the government to promote industry is through transportation. Since 1887 the federal government has exerted considerable control over national transportation, with responsibility for developing inland and coastal waterways and for setting the rates for moving freight between states. As a result, many industries have been attracted to flood-free locations beside navigable waterways. In 1956 the federal government instituted the Interstate Highway program, building 66,000 km (41,000 mi) of limited-access highways. These have been important in influencing the location of manufacturing, with many new businesses and production lines growing up beside key interchanges of the federal highway system.

Government is also in a position to influence the development of industry through legislation in any one of several areas from taxation to unionization, to

Men at work *(above)* During the Great Depression of the 1930s, President Franklin Roosevelt (1882–1945) introduced measures to get the population and economy working again. The Works Progress Administration gave jobs to thousands.

A crushing blow to waste *(right)* Environmental groups, particularly in land-scarce states, have successfully lobbied for compulsory local recycling. Issues such as this often get cross-party support.

The Mexican connection *(below)* The US and Mexican governments have jointly developed tax incentives for companies on both sides of the border. This El Paso company finishes textiles begun in Mexico.

LOBBYING FOR POWER AND INFLUENCE

The basic aim of a lobbyist is to influence legislation that will affect the business he or she represents. Every major industry – and some minor ones – have a staff of lobbyists in Washington DC to present their positions on proposed legislation in Congress. Position papers are prepared and sent to the legislators, though effective lobbying is often very personal. Members of congress are taken to dinner or on trips, or contributions are made to their campaign funds in order to secure their support for a particular measure. At times a lobbyist may support a proposal that is beneficial for the entire nation, such as the 1986 Freedom of Information Bill. At other times his or her interests may be more specialized, defending cigarette advertising or gambling regulations, for example.

A second tier of lobbying takes place between congressmen and the president. Negotiation and bargaining are important elements in the American political process, frequently cutting across party lines. A representative may offer support for presidential legislation only if the president, in turn, gives federal financing for some project the representative is pursuing in his or her state. At the same time the representative is also a vehicle for the industrial lobbies they represent, passing on their concerns to the president.

health and safety regulations. In recent decades increasingly stringent rules have come into effect over health and safety in workplaces, as well as control of labor. While beneficial to the workforce, these measures sometimes have a restraining effect on the growth of manufacturing. Government aid in one area can also have a detrimental influence elsewhere, exerting greater taxation pressure, and increasing competition.

State and local inducements

Since the 1930s, state and local governments have become increasingly involved in promoting manufacturing. During the 1960s and 1970s this mainly took the form of competition. State administrations fought hard with one another to lure industry to their parts of the country ("Smokestack Chasing") using combinations of grants, loans and tax abatements. More recently, state governments have adopted a rather different policy, trying to establish the right local conditions for high-technology innovation. North Carolina and Pennsylvania led the way when they set up research projects, seed and venture capital funds, equipment subsidies and endowed university chairs. Other states were quick to follow this example and in 1988 alone, 44 states invested a total of more than $800 million in new science and technology partnerships among government, industry and in local schools and colleges. Every state now has some form of science and technology initiative.

A further innovation has been state action to encourage the export of locally manufactured products. Most states have established foreign trade offices, mainly in Europe and the countries of the Pacific rim. Seed money and loans have also been made available, along with seminars on foreign economies and cultures, language-teaching sessions, and information on import–export licensing.

The silicon revolution

The silicon chip has been at the core of a world technological revolution that will touch every institution in society. It evolved as a result of more than a century of efforts to produce a "computing machine" that is small, fast and cheap. The cost of performing a given operation on a computer has declined more than a millionfold since the 1960s. The microchip has become a vital element in a whole series of US products, from digital watches to televisions. It has transformed the design of many goods, with mechanical or moving parts being replaced by electronic functions, so reducing size, assembly problems and transportation costs. At the same time wholly new products have been created for the market, newly affordable to a large part of the population. The desktop computer is probably the most well-known example.

Microelectronics has shown a seemingly limitless capacity to merge with other technologies. Computerized automatic switchboards have combined with computer-controlled transmissions. Satellite technology has been interwoven with automated billing, and the development of fiber optics has revolutionized the world of communications. While a copper circuit can carry up to 22 telephone conversations, the equivalent fiber-optic "circuit" can carry 8,000. Problems of distance have been greatly reduced, extending the global potential of manufacturing and other industries.

The high-tech revolution has also transformed manufacturing methods. Robots are playing an increasingly important role in US production, especially of motor vehicles. In many respects robots are better suited to the modern factory environment than human beings. They specialize in repetition and never suffer boredom, absentmindedness or fatigue. Also, they can be built to withstand hostile environments and can be operated without break for 365 days each year. Estimates suggest that first generation robots could be used to perform a maximum of 3 percent of all industrial work. Robots of the next generation, however, will be considerably more sophisticated, and are likely to carry out a much larger share of production.

The growth of industrial parks

High-tech industries usually congregate in specialist complexes, of which the United States boasts more than 45. One of

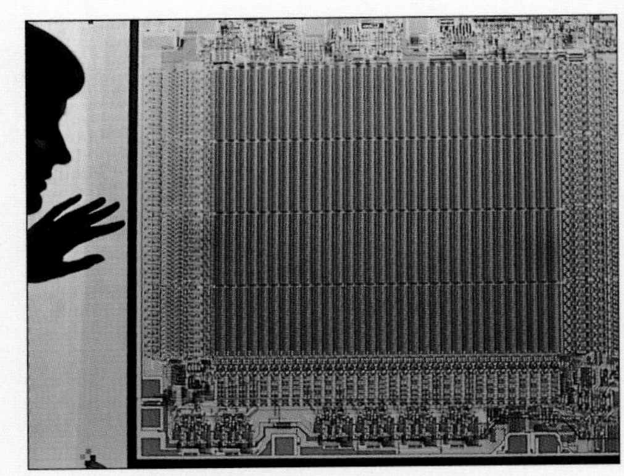

Searching the circuitry (*above*) A technician scrutinizes an enlarged photograph of a circuit plan. Circuit boards are made up from hundreds of silicon chips, each performing its own separate function. Circuit plans act as a board's road map.

Wafer maker (*right*) The highly skilled workforce, based mostly in California, wear protective clothing in the sterile environment of a silicon chip factory. Cleanliness is very important as a single speck of dust can ruin a chip.

Chip processing (*below*) US scientists and researchers developed the integrated circuit chip in 1959, and changed the course of electronics. Manufacturing chips begins with the main base, crystaline silicon. Wafers, only 0.1 mm (0.004 in) thick, are cut with diamond saws from a silicon ingot, then inplanted with impurities to give them a positive charge, before being processed in the way described below. The forecasted annual revenue from chip making in the year 2000 is in excess of $60 billion.

Stages of chip manufacture

1 A positive-type wafer of silicon is baked in an oxidization oven to 1,000°C, which results in a thin insulating layer of silicon dioxide forming on the surface

2 A light-sensitive coating is applied to the silicon dioxide layer and a photo mask of the circuitry is laid on top. The wafer is exposed to ultraviolet light, which only reaches the unmasked areas

3 The photo mask is removed and the wafer is passed through a developer which dissolves away the areas of coating that were exposed to the ultraviolet light, leaving the pattern of the mask on the wafer

4 The wafer is etched in hydrofluoric acid. This eats away the silicon dioxide where the light-sensitive coating has been removed, leaving two channels. The rest of the light-sensitive coating is removed and the wafer is placed in an oven where phosphorus vapor is absorbed into the exposed silicon, forming a negative layer in the channels

5 Back in the oxidization oven, a second layer of silicon dioxide is added

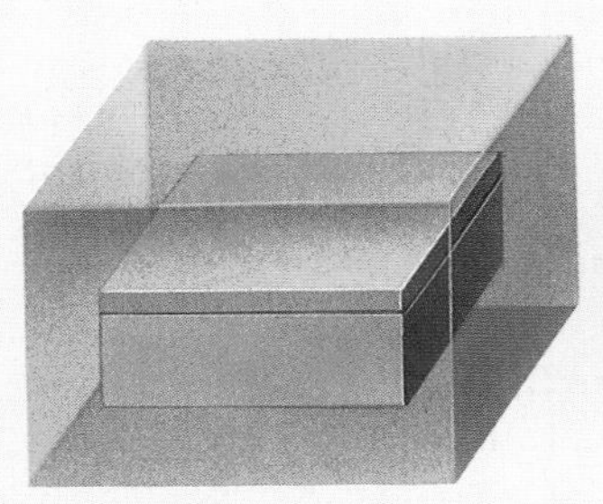

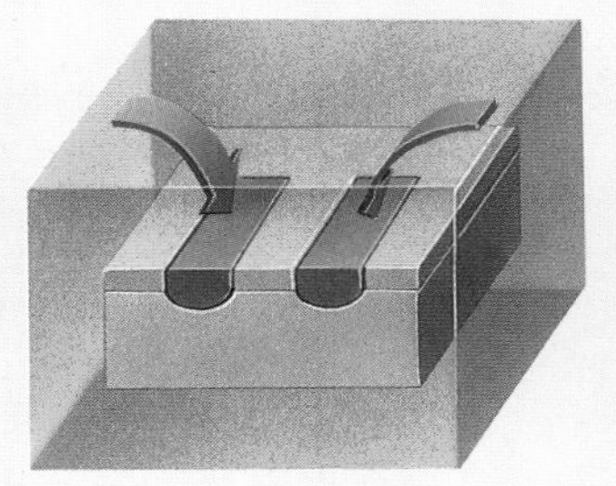

6 The wafer is given another light-sensitive coating and goes through further masking and etching stages

7 It is returned to the oxidization oven once more for another layer of silicon dioxide to be added

8 The wafer goes through further masking and etching stages, which deepen and narrow the channels

9 A layer of aluminum is added by placing the wafer in a vacuum chamber and passing vaporized aluminum over it

10 The wafer goes through a final masking and etching process which forms the electrical connections for the complete circuit

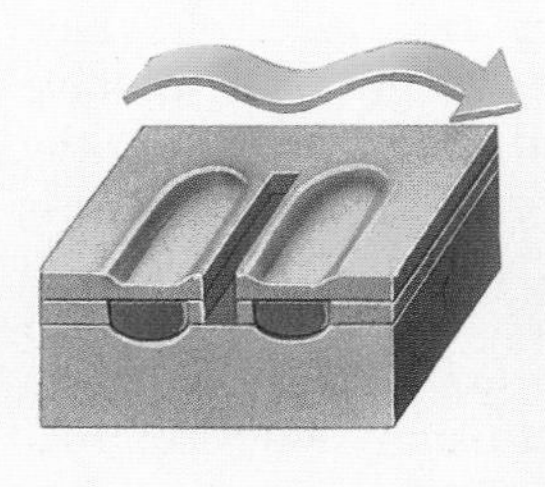

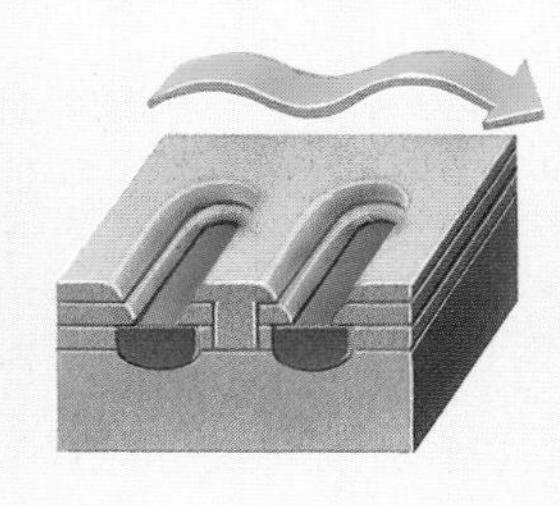

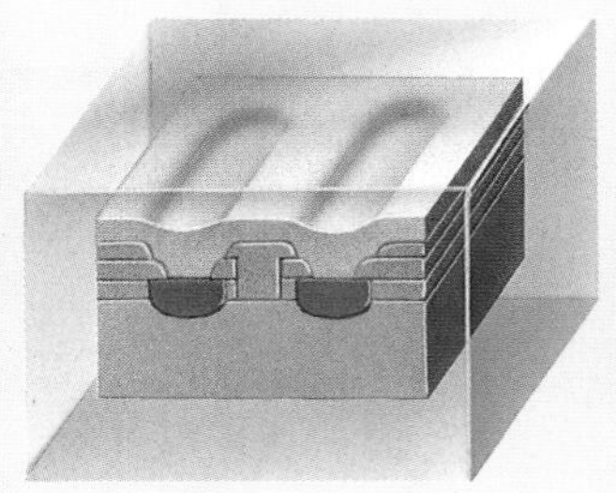

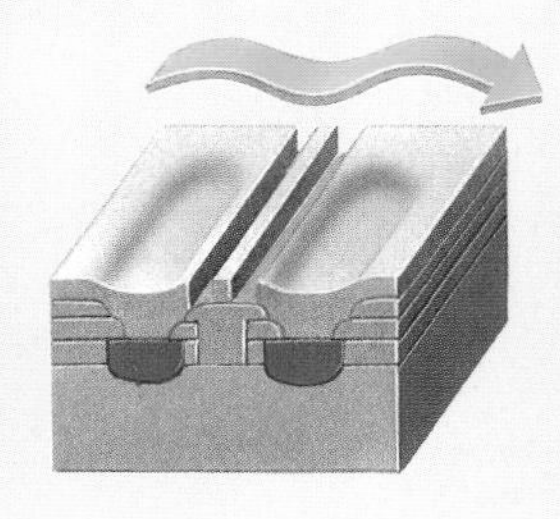

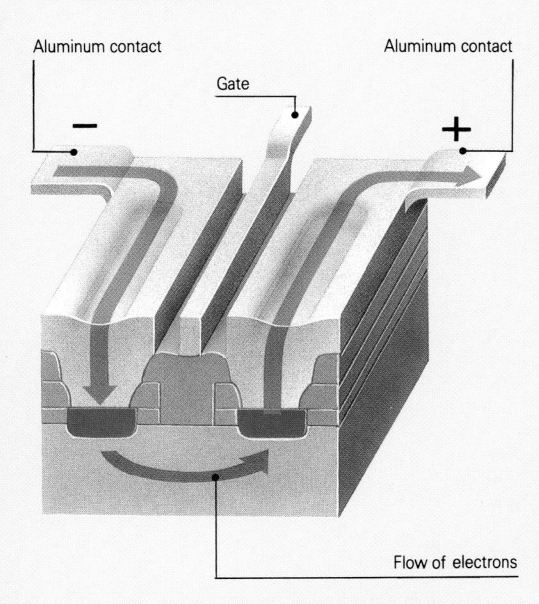

The transistor works as an on-off switch by allowing current to flow or not. In the on position, a small positive current is fed to the gate, which attracts free electrons from the positive silicon base and allows the current to flow between the terminals

the best known is the Santa Clara ("Silicon") Valley in Palo Alto, California, which is famous for its semiconductor industry and mini-electronic products. Other high-profile complexes are in Boston in the northeast – near highways 128 and I-495 – specializing in computers and electronic devices, and in the Research Triangle of North Carolina, producing biological products.

High-tech industries, unlike traditional manufacturing, are not attracted to a location by cheap availability of power or easy access to communications. Instead they tend to develop in areas where government authority is supportive of research, and where significant funding is available. Nearby universities are particularly important, providing a body of skilled scientists and engineers, as well as scientific information and research programs. There are six major universities near the Boston high-tech complex, with three of these – Massachusetts Institute of Technology, Harvard University and Boston University – playing key roles in its growth. Likewise, Stanford University and the Universities of California at Berkeley and San Francisco have been hugely important for the development of the Silicon Valley complex.

A major characteristic of high-tech items is their rapid obsolescence. Massive funds have to be continually injected into research and development projects to update old products and develop new ones. Most of this capital comes from federal sources, though private and state money has also played a part in many technological centers. The state of North Carolina actively sought to involve federal funds and companies in the Research Triangle. Likewise, local and state governments supported the Boston complex by creating a suitable environment for high-tech growth, through improvements in transportation and the creation of industrial parks.

A drop in the ocean

Upset stomachs and heavy aircraft frames have more in common than motion sickness; magnesium can be used both as a liquid digestive aid and as a lightweight metal.

A metallic element and the eighth most abundant element in the Earth, magnesium never occurs in a pure form in nature, only as a compound with other elements. The process for extracting magnesium is unique among metals. Instead of being mined it is pumped from the sea. Seawater contains about 0.13 percent of magnesium and nature gives a constant supply of the resource. One of the largest magnesium processing plants is in the United States, on the Gulf of Mexico in Texas. Huge pumps on sea platforms suck water out of the Gulf and mix it with calcium hydroxide squeezed from the plentiful oyster shells around Galveston Bay. This produces magnesium hydroxide – the base from which milk of magnesia, magnesium metal and other compounds are made.

Magnesium metal is extremely light and has a very low density, making it ideal for use in airplane construction. A part made of steel weighing 32 kg (70 lbs) would only weigh 7 kg (15 lbs) made with magnesium metal. However, because magnesium has such a low density, it has little strength, and so it must be alloyed with other metals. Magnesium metal is usually mixed with aluminum to form a very lightweight but strong and flexible metal alloy.

Metal from the sea These pumps in the Gulf of Mexico have an inexhaustible source of seawater from which to extract magnesium.

BORDERLINE BOOM

OIL-RICH MEXICO · TRADITION MEETS THE MODERN WORLD · VAST WORKFORCE, LITTLE WORK

Central America has sufficient resources to support an independent industrial sector. However, most governments lack capital to invest in training and technology, relying heavily on the United States for funding and as a market for exports. The region exports oil (particularly from Mexico), natural gas, precious metals and other minerals. It also processes sugar, fruit, crude oil and metallic ores. Its cheapest resource is the labor force. Employees earn a fraction of equivalent wages in Western Europe or the United States. Low wages and large populations have encouraged labor-intensive manufacturing, but industrial expansion is constrained because the market for manufactured goods is fragmented by political boundaries. A thriving tourist industry has also developed, notably on the Caribbean islands.

COUNTRIES IN THE REGION

Antigua and Barbuda, Bahamas, Barbados, Belize, Costa Rica, Cuba, Dominica, Dominican Republic, El Salvador, Grenada, Guatemala, Haiti, Honduras, Jamaica, Mexico, Nicaragua, Panama, St Kitts-Nevis, St Lucia, St Vincent and the Grenadines, Trinidad and Tobago

INDUSTRIAL OUTPUT (US $ billion)

Total	Mining	Manufacturing	Average annual change since 1960
101.3	9.9	82.2	+5.3%

INDUSTRIAL WORKERS (millions)
(figures in brackets are percentages of total labor force)

Total	Mining	Manufacturing	Construction
9.4	0.7 (1.4%)	6.06 (12.0%)	2.62 (5.2%)

MAJOR PRODUCTS (figures in brackets are percentages of world production)

Energy and minerals	Output	Change since 1960
Oil (mill barrels)	1097.2 (4.8%)	+444%
Bauxite (mill tonnes)	7.5 (7.7%)	-56.3%
Antimony (1,000 tonnes)	3.7 (5.2%)	-18.9%
Silver (1,000 tonnes)	2.4 (12.4%)	+200%
Sulfur (mill tonnes)	2.1 (14.3%)	+33%
Fluorspar (1,000 tonnes)	756 (16.7%)	No data

Manufactures		
Residual fuel oil (mill tonnes)	39.7 (5.4%)	No data
Cement (mill tonnes)	33.7 (3.1%)	+992%
Steel (mill tonnes)	7.6 (1.1%)	+406%
Fertilizer distributors (1,000)	176 (45.8%)	N/A
Rubber footwear (mill pairs)	38.8 (4.3%)	N/A
Soft drinks (mill hectoliters)	56.1 (9.8%)	N/A

N/A means production had not begun in 1960

OIL-RICH MEXICO

Petroleum is the region's most important natural resource and its major source of energy. Reserves are concentrated in Mexico, where oil was first discovered in the early 20th century. By 1920 the country was second only to the United States as a world producer. In 1938, however, output was slowed when the government controversially nationalized the oilfields, provoking an international embargo. In the early 1970s huge new reserves of oil were discovered in the Bay of Campeche and in the southeast of the country. Exploitation of these oilfields has transformed the Mexican economy and placed the country among the world's leading five oil producers. In the early 1990s, it was estimated that Mexico had more than five percent of the world's oil reserves, as well as substantial supplies of natural gas. Trinidad is the only other country in the region to have significant oil and gas resources. Mexico also has a modest coal-mining industry in the Sabinas basin north of Monterrey.

Hydroelectric power is becoming increasingly important in Central America. The mountains and heavy rainfall in parts of the region have been harnessed to generate significant amounts of hydroelectricity, which in some countries now meets up to a third of national needs. In addition, geothermal heat is being tapped in volcanic areas of Mexico, Nicaragua and El Salvador. These power stations are efficient even when they are very small and are well suited to serving rural areas. Mexico has experimented with nuclear power but found it too expensive to produce. In Cuba and other sugar-growing parts of the region bagasse – the dry, pulpy residue left after the extraction of juice from the sugar cane – is used to make fuel, providing a further local source of power. Another plant, sisal, has also been harnessed for industrial use, especially in the Yucatán in Mexico. Its fibrous parts are used for making cord and rope.

Precious metals and minerals

Gold, zinc, silver and nickel are mined and processed in many areas of Central America. Long before the European invasion the local population was using gold for ornamental purposes, and the search for the precious metal was one of the main forces behind subsequent Spanish exploration. Since then many veins have been exhausted, and as long ago as the 16th century silver overtook gold in importance. The Dominican Republic is now the only country in the region with significant gold production. Mexico remains the world's leading source of silver, with fresh sources continuing to be discovered. The Reál de Angeles silver mine in the central province of Zacatecas, opened in 1983, is the world's largest.

During the 20th century Mexico has become an important producer of other metals too. Copper succeeded silver in importance in the 1900s, lead replaced copper in the 1930s and was itself over-

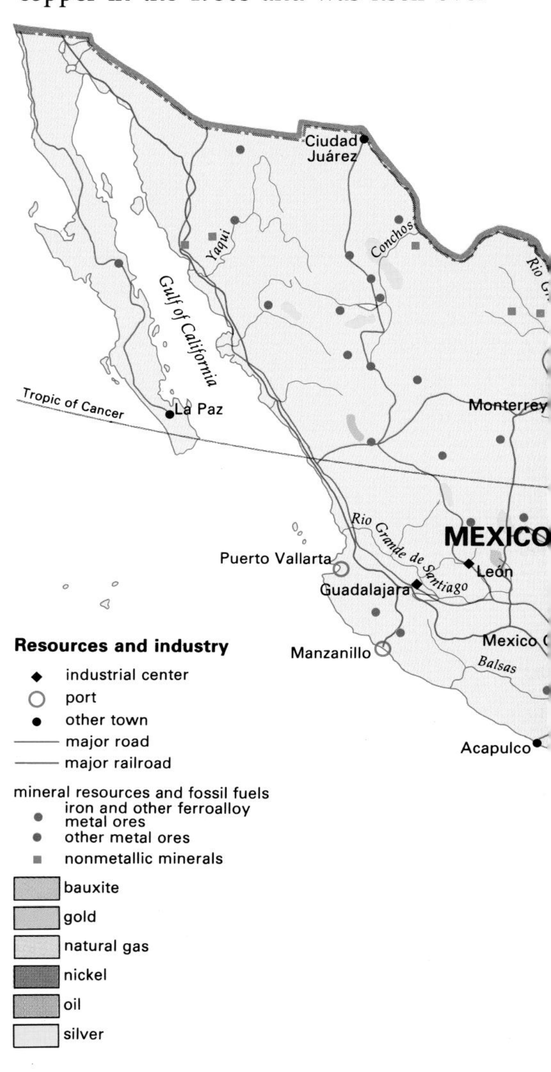

Map of principal resources and industrial zones (*above*) Most of the resources in the area are in Mexico, the world's leading producer of silver and a member of OPEC (Organization of the Petroleum Exporting Countries). Jamaica, Haiti and Dominican Republic have large reserves of bauxite and Cuba is rich in both gold and nickel.

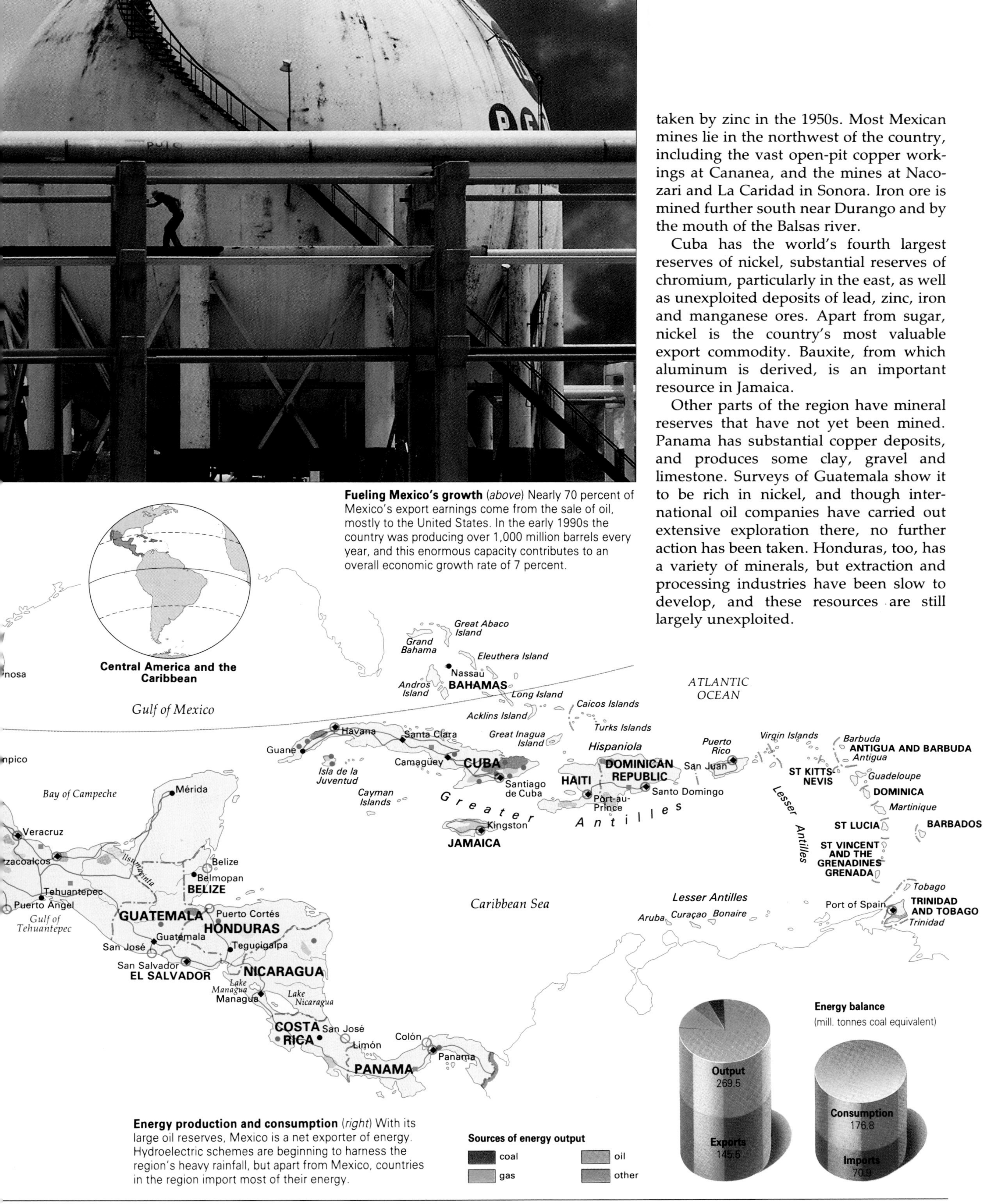

Fueling Mexico's growth (*above*) Nearly 70 percent of Mexico's export earnings come from the sale of oil, mostly to the United States. In the early 1990s the country was producing over 1,000 million barrels every year, and this enormous capacity contributes to an overall economic growth rate of 7 percent.

Energy production and consumption (*right*) With its large oil reserves, Mexico is a net exporter of energy. Hydroelectric schemes are beginning to harness the region's heavy rainfall, but apart from Mexico, countries in the region import most of their energy.

taken by zinc in the 1950s. Most Mexican mines lie in the northwest of the country, including the vast open-pit copper workings at Cananea, and the mines at Nacozari and La Caridad in Sonora. Iron ore is mined further south near Durango and by the mouth of the Balsas river.

Cuba has the world's fourth largest reserves of nickel, substantial reserves of chromium, particularly in the east, as well as unexploited deposits of lead, zinc, iron and manganese ores. Apart from sugar, nickel is the country's most valuable export commodity. Bauxite, from which aluminum is derived, is an important resource in Jamaica.

Other parts of the region have mineral reserves that have not yet been mined. Panama has substantial copper deposits, and produces some clay, gravel and limestone. Surveys of Guatemala show it to be rich in nickel, and though international oil companies have carried out extensive exploration there, no further action has been taken. Honduras, too, has a variety of minerals, but extraction and processing industries have been slow to develop, and these resources are still largely unexploited.

TRADITION MEETS THE MODERN WORLD

Almost everywhere throughout Central America traditional handicrafts exist side by side with modern, capital-intensive manufacturing industries. These traditional craft workshops produce goods mostly for local consumption, drawing on the region's resources and requiring no imported technology. Many of the building materials for ordinary homes are made locally by small family concerns, which can respond to the ebb and flow of demand more readily than sophisticated, largescale manufacturers.

In some places, production has grown to supply areas outside the immediate locality. In highland Guatemala, for example, Amerindian villages specialize in household products made to a local design. One might make a particular type of pottery, another specialize in a style of furniture or a particular textile weave, and each village trades its speciality in the regional markets. Other smallscale manufacturing such as tilemaking, tanning, candlemaking and carpentry are usually carried out by farmers in their homes. Sometimes regional specializations become national ones. León, in central Mexico, has become the country's footwear capital through combining local handicraft skills with the marketing and organizational methods of modern industry. Much of the leatherwork in León is subcontracted to smaller operators, reducing overheads and risk, as well as allowing rapid responses to changes in style. This system is effective, too, in garment manufacturing. In the region as a whole, about half the manufacturers employ no more than five workers.

Processing and progress

The region also supports largescale industries processing raw materials, whether for export or as the first stage in a longer manufacturing chain. Particularly important is the refining of oil, metallic ores and agricultural produce, as well as the manufacture of iron and steel, petrochemicals and sulfuric acid. These industries, which are well developed in Belize, Costa Rica, Panama and Mexico, are capital-intensive, rely heavily on imported expertise, make only modest demands on unskilled labor, and generate relatively little employment.

WEALTHY NEIGHBORS

According to the United States' trade policy, goods made from American components but assembled in Mexico can be exported to the United States without paying duty (except on the value added to the article by the assembly process). This gives a huge boost to Mexican industry. Wages are so much lower in Mexico – in the early 1990s they were only one-tenth of those in the United States – that many American companies have found it worthwhile to set up subsidiary production just across the border. The Mexican government has encouraged this trend by doing all it can to make the newcomers welcome: it has removed restrictions on employment and property ownership by foreigners by liberalizing planning laws.

By the early 1990s no fewer than half a million Mexicans – four times as many as ten years earlier – were working at "in-bond" factories (*maquiladoras*), assembling imported parts for the United States market. The most important products are electrical and electronic accessories, closely followed by automobile components and vehicles for export. The towns in the border zone – Tijuana, Ciudad Juárez, Neuvo Laredo, Reynosa, Matamoros – now employ one in eight of all Mexicans working in manufacturing and generate export earnings of US$2 billion.

Assembling in new industrial areas (*above*) To lure industry away from its traditional base around Mexico City, the government began the *maquiladora* program with the United States, offering companies incentives to export parts and import finished goods.

Economy up in smoke (*below*) Hand-rolled Cuban cigars, highly prized around the world, have been a mainstay of the island's economy. Tobacco is grown in the fertile northwest and central areas, and cigars are made in small factories and homes.

The alchemy of oil (*right*) Oil is a versatile resource yielding byproducts used to make asphalt, soap, gasoline and many drugs and chemicals. To separate these products from the crude raw material the oil must be heated and distilled in a fractional distillation unit. The crude oil is pumped into a heating tower. Here the different chemicals in the oil begin to boil and rise as vapor. As the vapor rises it cools and begins to condense again. Because each chemical has a different boiling point, they condense at different levels. Natural gas is quite light and condenses at the top level, while asphalt lies at the bottom.

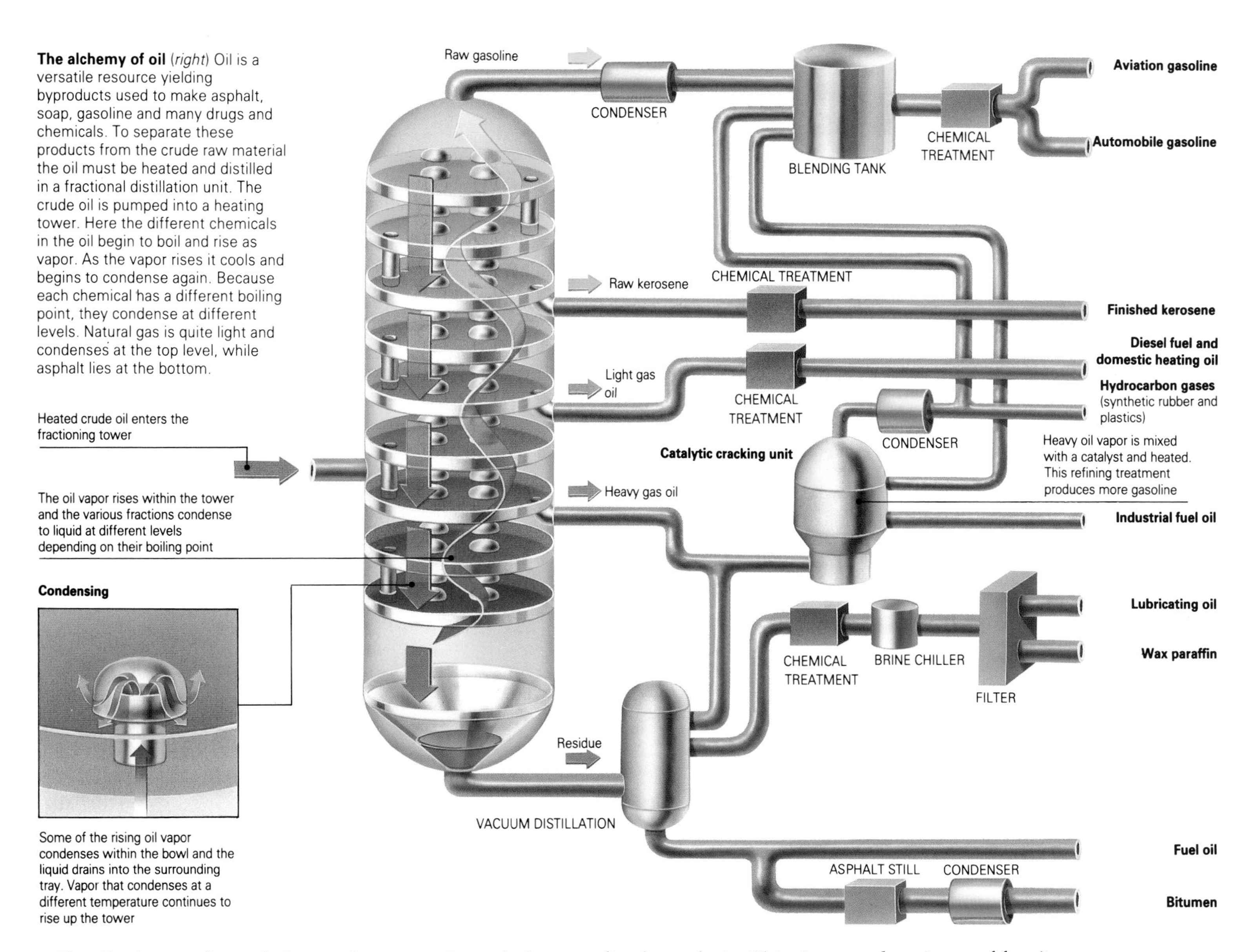

Usually they are located close to the raw materials they use. Refineries are often built near the mines. In Mexico an iron and steel complex has been established at the Pacific port of Lazaro Cordenas, near the large iron-ore deposits at Las Truchas and Pena Colorado and the cheap hydroelectric power produced at the nearby Infiernillo and La Villita dams. Also in Mexico, vast petrochemical complexes have been built at Cactus and La Cangrejera in the southeast, close to the country's newly discovered oilfields. These manufacture a whole range of organic chemicals, including detergents, acrylic resins, polyester fibers and emulsifying agents, as well as the raw materials for fertilizers. Oil refining, by contrast, is often located close to the industrial plant where it is needed, as tankers and crude-oil pipelines offer cheap bulk transportation of the unrefined product. This is particularly true of Mexico, where refining capacity has increased dramatically in recent years, and by 1990 two-thirds of the country's oil production was being processed near industrial sites.

The flourishing food industry

Food processing is a significant activity in many Central American countries. In the tropical parts of the region, preparing and roasting coffee beans is a valuable export industry. Cigarettes and alcohol are produced in most countries, as are textiles, canned food and soft drinks. Belize and Guatemala possess a large number of sawmills for their lumber trade, while Costa Rica is an exporter of paper products to neighboring countries. Tires are made in Guatemala, and both Costa Rica and Panama exploit their dense forests to produce items of furniture.

Cotton gins, beef slaughterhouses, fruit canneries and other factories are spawned by the region's growing range of agricultural produce. Sugar mills have been industrial features of the rural landscape in much of the region since the 17th century. Today their numbers are reduced but individual mills are larger and serve wider plantations. Cuba, in particular, has an extensive processing capacity. It is still the world's leading sugar exporter though in the years following the socialist revolution in 1959, the industry was heavily dependent on export to the former Soviet bloc, which paid artificially high prices. The major political upheavals that took place in the former Soviet Union in the early 1990s left the future of Cuba's sugar processing industry in a very uncertain condition.

VAST WORKFORCE, LITTLE WORK

European exploitation of the region's natural resources had a devastating effect on the peoples of Central America. The Spanish conquerors were anxious to lay claim to the region's rich deposits of gold and silver, and to gain these they drove many Amerindian people from their traditional way of life to work in the mines. Output increased rapidly, and by the late 18th century Mexico alone was producing more than half of the world's silver. However, the wealth created by the mining industry was shared by only a small number of Mexicans, most of them descendants of the Spanish. By far the larger portion was exported, providing two-thirds of Spain's revenue. The mining workforce of Amerindians and Mestizos (part Spanish, part Amerindian) remained very poor.

Poverty is still a serious problem among the industrial workforce today. Even during the 1960s and 1970s, when Mexican industrial output increased fourfold, workers' living standards were very low. Recession in the early 1980s made poverty among the workforce very much worse. At the root of the difficulty was the region's rapid population increase. In the period between 1950 and 1990 the inhabitants almost tripled in number, from 53 million to about 150 million. This pushed down industrial wages, and also encouraged inefficient overemployment. The demographic trends also led to the emergence of a young workforce whose expectations cannot be fulfilled in their own countries. Significant numbers of Mexicans, Guatemalans, Salvadoreans and others cross illegally into the United States every year in search of better paid jobs, often in the service industries. The exploitation of people as a resource is probably the most remarkable industrial feature of the region. The border between Mexico and the United States draws a line between an abundance of cheap labor and a vast market in which to sell it.

Government intervention

The ruling parties in the region have had a significant influence during the 20th century over how and where their country's industry has developed. In Mexico thoughout the 1960s and 1970s manufacturing growth was disproportionately concentrated in the capital. As a result, Mexico City, which in 1990 had a population of 19 million, may soon be the world's largest urban center, housing half the country's industry. Partly this is due to the capital's location at the hub of Mexico's transportation system, but government policy has also been important. Administrations exempted factories in Mexico City from the country's otherwise strict antipollution laws. This was done in an attempt to find employment for the city's swelling population, but it only encouraged more people to join them.

In Cuba the pattern has been very different. There the socialist government has deliberately promoted smaller cities, using capital investment to create jobs away from Havana, to prevent urban growth being concentrated in the capital. The policy has had some success. The governments of other Caribbean islands, too, have exerted strict control over where industries establish themselves – an important consideration in countries where land is scarce.

Recently even Mexico has begun to adopt this approach. Alarming pollution and congestion in the capital have finally led to a complete change of policy, and further industrialization is now discouraged, with growth being promoted instead in other cities, particularly the coastal ports.

The trade policies of the region's governments have also had important effects on industry. During the 1960s and 1970s protectionist policies were adopted in an attempt to encourage increased manufacturing in fields such as cigarette

Modern pirates of the Caribbean (*above*) Tourism is the leading industry in most islands of the Caribbean. Much of the attraction stems from the area's maritime history and its scenic beauty.

Made in the USSR (*below*) Cuba and the former Soviet Union had close trade and economic links dating from the 1960s. Over 75 percent of Cuba's exports, mostly sugar, were to Eastern Bloc countries. In return, the Soviet Union supplied equipment, money and cheap oil.

ANCIENT CITIES, SILVER SEAS

The Caribbean Islands are in a prime position to benefit from tourism. As well as a hot, sunny climate and fine beaches, they are conveniently close to the huge market of the United States. On the whole, they have made the most of this potential. In the Bahamas tourism now accounts for 60 percent of national income and 70 percent of the population are employed in the industry. In Barbados and the Dominican Republic tourism has now replaced the sugar industry as the most important source of foreign exchange.

The mainland countries have often had greater difficulty in developing tourism; many of them are struggling to overcome reputations for political instability. Mexico, however, has a large tourist industry, helped by easy access to the United States. Remains of the Mayan civilization and other pre-Spanish relics have proved an important attraction for visitors, notably sites such as Teotihuacán near Mexico City and the jungle city of Tikal in Guatemala. Belize has exploited its remarkable coral reefs, which are among the most spectacular in the world. In addition the region's exotic plants and animals have proved popular attractions, and both Guatemala and Costa Rica have established national parks to safeguard their natural heritage.

making and rice processing. The policies proved effective in some ways, but many countries have since taken the opposite approach, encouraging agriculture rather than industrialization. Another area of experiment was in extending the market for industrial goods. Although the population of the region is large, the lack of wealth in many countries has meant that the buying power per capita is very little. In the early 1960s the West Indian Federation (later the Caribbean Community: CARICOM) and the Central American Common Market (CACOM) were formed in an effort to widen domestic markets, but these were by no means wholly successful. Commercial ties have always been stronger with third parties – especially the United States – than within the region.

Further attempts at overcoming this problem were made more recently by the United States itself. President Ronald Reagan's Caribbean Basin Initiative (1983) and President George Bush's Enterprise for the Americas Initiative (1990) were both designed to reduce trade barriers throughout the region and to encourage countries to develop those activities, such as tourism, in which they have a competitive edge. Though these policies have not been without effect, the problem remains a serious one.

Budding trade unions

In those countries where the trade unions have managed to gain a position of power, they have had an important influence in shaping the region's industry. This was true of Honduras, where in 1954 the trade union movement won agreement to a labor code that is regarded as one of the most all-encompassing in Central America. The code regulates the relationship between management and the labor force, attempting to protect both workers and businessmen. In other countries unions have played a less positive role, and during the 1960s and 1970s many countries suffered disruptive and economically damaging strikes.

It is in Mexico that unions have become most important. The mighty oil union, for example, holds great power over its members' jobs, and such control makes the strike weapon particularly effective. It also exerts important influence directly over the government, through the ruling Institutional Revolutionary Party (PRI), in which chiefs of the oil union hold office. Some wealthy unions are also powerful bodies. From the 1970s the oil union gained substantial revenues from its members' wages, from contracts awarded to outside companies, and – until prevented by legislation in 1984 – also from its ability to subcontract half of all drilling arrangements. Naturally this generated a significant reserve of wealth, and in some cases this was used to the clear benefit of its members. Workers employed at the Pajaritos petrochemical complex had an entire town – Nanchital – built for them by their union, complete with paved streets, hospital and church. It is generally recognized, however, that significant sums of money also disappear through simple corruption.

Aluminum – mining and refining

Mining bauxite, the ore from which aluminum is obtained, plays an important role in the region's economy, most notably in Jamaica, which is one of the world's leading producers. Aluminum, a light, strong and malleable metal, became industrially important in the 20th century. Jamaican production dates from 1942 when bauxite was first discovered on the island. The international company Alcan was the first to set up operations, and mining began in 1952. Alcan went on to pioneer the refining of alumina (a purified form of the ore) in Jamaica, and by 1960 bauxite and alumina had become the country's leading export commodities. A second company, Alcoa, started operations in 1968.

Despite this promising start, Jamaican bauxite production soon ran into difficulties. From the early days of the industry, the island's government showed a tendency to intervene in operations, particularly when it encouraged domestic refining by placing a levy on bauxite exports. In 1974 state intervention was extended, with disastrous consequences. The government under Michael Manley (1972–80), which is committed to "democratic socialism", decided to take a controlling interest in the bauxite industry, an action that caused aluminum companies to divert investment to other suppliers, notably in Australia. Matters worsened when the recession of the 1980s reduced world demand for aluminum. The Jamaican industry virtually collapsed as output fell and refineries closed.

Matters began to improve, however, when a program backed by the International Monetary Fund was implemented to counter the country's economic crisis. Under this program, state involvement in mining was reduced and private investment revived. The effort proved effective and by the beginning of the 1990s state-owned companies were responsible for only a quarter of bauxite production on Jamaica. Mines and refineries were reopened and new joint ventures were negotiated between the government and various companies.

Refinery works (*right*) In Jamaica, bauxite is only refined into alumina. Turning it into aluminum requires huge amounts of electricity, so refineries have to be located near sources of cheap power. Jamaica has poor energy resources and exports raw alumina.

Ore to alumina (*below*) Aluminum-bearing bauxite ore is moved by conveyor belt through refining processes. First the ore is ground, mixed and filtered in a chemical solution that separates out impurities. When the solution dries out, it leaves powdered alumina.

The refining process in Jamaica

The first stage in the conversion of bauxite to aluminum is the creation of a pure aluminum oxide, known as alumina. This reduces the material's bulk by half – tripling its value in the process – and is done as close to the mine as is practical. Control of transportation costs is a major

factor in profitable bauxite extraction. By contrast, the next stage of the process – the reduction of alumina to aluminum metal – is highly energy intensive and, refining ideally takes place close to cheap sources of electricity.

Jamaica enjoys easy bauxite extraction and transportation. The deposits lie close to the surface, while mines are located near coastal ports, allowing easy access to the North American market. The only disadvantage is that some of Jamaica's tourist resorts have suffered from their proximity to the industry. In terms of processing, however, the country is at a disadvantage, especially compared to South American competitors. Brazil and Venezuela have recently begun to exploit bauxite reserves in the Amazon Basin, making use of the huge hydroelectric power supplies that are located nearby. In Jamaica, by contrast, energy has to be imported at high cost, making the final stage of processing – from alumina to aluminum – uneconomic.

As a result, the country exports no aluminum. This is a disadvantage, as refined aluminum is worth four times as much as alumina. Nevertheless the bauxite industry remains of great importance to Jamaica's economy, and is set to continue that way. In 1990 bauxite reserves were estimated at approximately 230 million tonnes. Although production brings some wealth to the island, it creates few jobs. Bauxite mining and processing are both labor intensive, and together they employ only 4,000 people in a country rife with unemployment.

PRECIOUS METALS AND WATER POWER

EXTRACTION ON A GRAND SCALE · SPORADIC MANUFACTURING · INDUSTRY AND THE STATE

In a region as diverse as South America it is hardly surprising that industry presents a widely varied picture. Brazil – which contains close to half the continent's land area and population – boasts one of the world's largest industrial economies. Mining, food processing and the manufacturing sector account for most of the activity. Exports include arms and passenger jet aircraft. Argentina industrialized much earlier than the rest of the region. Although its manufacturing sector has declined, agricultural products and mining remain important. Elsewhere, industry is dominated by extraction, often of only one or two minerals (copper in Chile, oil in Venezuela and Ecuador, tin in Bolivia, bauxite in Surinam, and oil and copper in Peru). Such dependency, however, causes over-reliance on the fluctuating world market.

COUNTRIES IN THE REGION

Argentina, Bolivia, Brazil, Chile, Colombia, Ecuador, Guyana, Paraguay, Peru, Surinam, Uruguay, Venezuela

INDUSTRIAL OUTPUT (US $ billion)

Total	Mining	Manufacturing	Average annual change since 1960
231.5	22.9	181.7	+6.4%

INDUSTRIAL WORKERS (millions)
(figures in brackets are percentages of total labor force)

Total	Mining	Manufacturing	Construction
23.8	0.56 (0.54%)	16.7 (16.3%)	6.6 (6.4%)

MAJOR PRODUCTS (figures in brackets are percentages of world production)

Energy and minerals	Output	Change since 1960
Oil (mill barrels)	2674.0 (11.9%)	+109.6%
Iron Ore (mill tonnes)	119.1 (21.1%)	+91.8%
Bauxite (mill tonnes)	12.9 (13.2%)	+15%
Copper (mill tonnes)	1.8 (21.2%)	+89%
Tin (1,000 tonnes)	59.1 (29.4%)	+72%
Silver (1,000 tonnes)	2.5 (12.9%)	+54%
Manufactures		
Tanning extracts (1,000 tonnes)	57.1 (55.3%)	No data
Coffee extracts (1,000 tonnes)	64.5 (11.6%)	No data
Rubber footwear (mill pairs)	124.7 (13.9%)	No data
Ladies' blouses and underwear (mill)	891.9 (27.9%)	No data
Cement (mill tonnes)	51.6 (4.7%)	+406%
Steel (mill tonnes)	33.9 (4.6%)	+930%
Locks and Keys (mill)	271.7 (69.1%)	No data
Electrical fuses (mill)	88.8 (12.7%)	No data

Energy balance (mill. tonnes coal equivalent)

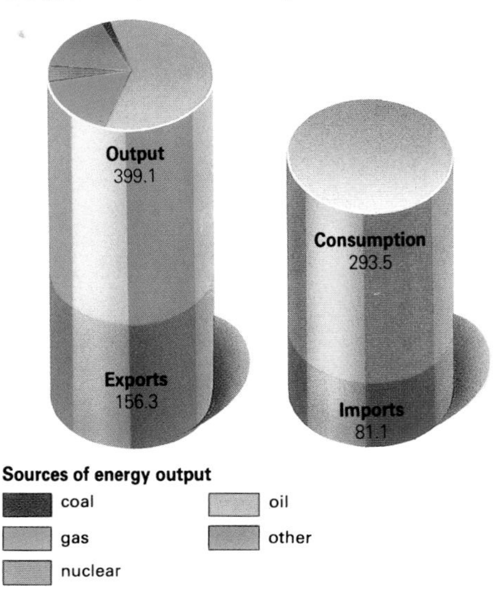

Sources of energy output
coal
gas
nuclear
oil
other

Energy production and consumption Oil is by far the most important of South America's energy resources, and it is a major oil-exporting region. The potential for hydroelectricity is great, but has so far been developed only in Brazil.

EXTRACTION ON A GRAND SCALE

The South American region possesses a wealth of natural resources. The continent has some of the world's largest deposits of metals and significant reserves of oil and coal. In addition it has tracts of excellent farming land that makes it a significant producer of food for export all round the world. However, the development of resource-based and manufacturing industry to process these resources varies widely from country to country.

South America has been famous as a source of precious metals from the earliest days of the Spanish empire, when the wealth of its Andean gold and silver mines was the envy of Europe. Although these metals are now mined less prolifically (many Andean seams have long been exhausted, and Brazil is now the only major exporter of gold) others are now

Raw development (*left*) Refined metals fetch a higher price on the world market than ores, so many mines have processing plants on site. Smelters at a mine in Peru release untreated effluent into a river.

Map of principal resources and industrial zones (*right*) South America has a wealth of natural resources, including a fourth of the world's copper reserves.

being exploited. The continent has one-quarter of the world's copper reserves, mostly in Chile, but also in Peru.

Surinam has important reserves of bauxite, while Bolivia is an exporter of tin and zinc. A number of countries have reserves of iron ore, particularly Brazil, whose "Iron Quadrangle" in Minas Gerais in the east and huge reserves in the Carajas mountains in the Amazonian basin have raised it to the rank of the world's leading exporter. More than any other South American country, Brazil has developed support industries to exploit its resources. Increasing amounts of Brazilian iron ore are processed prior to export, while a sizable steelmaking sector has emerged. This, in turn, feeds the manufacturing industries producing vehicles and components. Venezuela has also developed significant steel and aluminum production, while there is some smelting of zinc, copper, lead and silver in Peru and Bolivia.

Energy for export

The region is well provided with energy resources, with more than half South American countries able to supply their own energy needs. The rich oilfields around Lake Maracaibo, the southern inlet of the Gulf of Venezuela, have allowed Venezuela to become one of the world's leading oil producers and a founder member of the Organization of Petroleum Exporting Countries (OPEC). Oil reserves are also present in Argentina, Colombia, Peru, Ecuador, Bolivia and Brazil. Many of these countries also have natural gas, and both Peru and Ecuador have petroleum-processing industries.

The continent is less well supplied with coal. Seams are mined in a number of countries, but it is only lately that substantial reserves have been discovered in Colombia. The coal complex in Guajira peninsula in the extreme northeast, which was first developed in the mid 1980s, promises to be one of the world's leading sources of coal. Uranium is found in Argentina and Brazil, both of which use it to generate nuclear power.

Lastly, South America has great potential for hydroelectricity, though the degree to which this has been realized varies greatly from country to country. The huge Itaipu and Tucuru complexes, both opened in 1984, have added so greatly to Brazil's electricity production that power generation far outstrips national needs. In the Andes, however, the resource remains far from fully exploited.

One of the region's most significant resources, which has led to important – if uneven – industrial development, is food production. The meat-canning industry of Argentina, Uruguay and Brazil dates from the early part of this century. Chile has long been a producer of quality wine, while it has benefited from its long coastline to develop one of the five largest fishing industries in the world.

An orange juice coup

Brazil boasts one of the region's leading success stories of the 1980s: the mass production of orange juice. A succession of frosts in Florida severely reduced the United States' production of concentrated orange juice. Brazil stepped in to make good some of the shortfall in world demand and since then has made a major investment of time, money and resources into developing this new product.

SPORADIC MANUFACTURING

Manufacturing is spread very unevenly throughout the South American region. The level of concentration ranges from Brazil, which has extensive and sophisticated industries, to Paraguay, where there are only a few factories, most of them concerned with agricultural processing. In the poorer countries of the region – those of the central and northern Andes (Colombia, Ecuador, Peru, Bolivia) and also Paraguay – manufacturing has mostly developed since 1945.

The bulk of manufacturing tends to be in textiles and food processing, though some other industries are represented. Multinational companies such as Ford and Renault have set up vehicle-assembly plants in Ecuador. The industrial sector in Peru is more varied, thanks partly to its larger market, and partly to government intervention in the early 1970s. In addition to its steel industry, the country began to produce chemicals, petroleum products, smelted metals, and engineering and electrical goods. However, many of these industries had difficulties when tariffs on imports were reduced in the late 1970s and early 1980s.

Changing systems of manufacturing

Throughout South America, small and medium size enterprises – many of which are "informal" and operate outside the regulations – predominate in numbers. However, large concerns (many of them owned by multinational companies), though far fewer, account for most of production value. Unlike the Andean countries, Brazil, Argentina, Uruguay, Chile and Venezuela now have little artisan-based industry.

All of these countries, except Venezuela, have a history of manufacturing dating back to the last century. Between 1880 and 1914 coastal cities such as Buenos Aires in Argentina and Montevideo in Uruguay expanded rapidly, boosted by the development of their hinterlands and largescale immigration from Europe, until they were among the world's largest urban centers.

Such concentrated markets provided the basis for many industries such as food and textiles. In Argentina, for example, they accounted for over one-fifth of the country's total earnings by 1900. Many facilities for industrial growth, however, such as gas and water supplies, sewerage, electricity, tramways, railroads and ports, were developed by foreign-owned companies, and for some years local manufacturing did not extend beyond goods for the consumer market.

It was the Great Depression (1929–39) that radically changed this state of affairs. The sharp drop in demand for traditional exports (coffee in Brazil, copper in Chile, meat and grain in Argentina and Uruguay), meant that South American countries could no longer afford to import manufactured goods from Europe and North America. Throughout the region new industries sprang up making the kind of product that had previously been imported. At the same time, governments began to subsidize industrial growth and encourage manufacturing diversity. The 1940s and 1950s saw the foundations laid for future development, and industry grew apace during the 1960s and 1970s. Brazil and Argentina, in particular, saw great setbacks in the 1980s, but much of the new development is still in place.

Liquid wealth (*left*) Barrels of petroleum are stacked high for export in the Terpel state petroleum factory, Colombia. Until quite recently, the emphasis of the region's industry was on exporting raw materials. As a result, manufacturing was slow to develop, and remains patchy and sporadic, though every country is now making efforts to diversify its output.

Bobbing and weaving (*right*) Alpaca wool is a source of income for many small communities in Chile and Peru. The animals are sheared every two years, and each produces about 3 kg (6.5 lb) of very high quality wool, which is spun and woven by local workers. The suri breed gives a light, fluffy wool, often used in sleeping bags and parkas. The huacaya's wool is coarser, ideal for making the ponchos that are worn throughout South America.

Manufacturing in the 1990s

Both Brazil and Argentina today have a wide range of industries that include vehicle production, pharmaceuticals, cosmetics, cement, synthetic fibers, and the production of consumer goods including washing machines, refrigerators and televisions. In Venezuela a number of consumer and metal-processing industries have been established, including aluminum production and steelmaking. Venezuela is now a significant producer of processed metal. In marked contrast with most governments of the region the regime of General Augusto Pinochet in Chile (between 1973 and 1989) reduced the role of the state in industry. The "shock treatment" of cutting tariffs and state subsidies produced a sharp decline in manufacturing, though there was regrowth later and by the 1990s the country had rebuilt its production of textiles, vehicles, chemicles and consumer goods.

Brazil still boasts the most diverse and sophisticated range of manufacturing in the region. As well as a consumer-goods sector large enough to supply the country's ever-growing population, Brazil has developed a number of impressive high-technology industries. Turbines, generators, transformers and reactors for electricity production are made there, while Petrobras – the state petroleum company – has been able to use its expertise in deep-sea exploration to win contracts in the Middle East, providing drilling platforms and other equipment.

There is also a large weapons industry; the two state production companies ENGESA (military vehicles) and IMBEL (ordnance) allow Brazil to be a leading arms exporter to the developing world. The country has also developed a successful aerospace industry (EMBRAER), manufacturing a range of aircraft including the Tucano trainer and the Bandeirante small passenger jet. Since the early 1970s an extensive vehicle industry has developed. However, production is almost entirely in the hands of multinational companies, though domestic firms are prominent in the manufacture of components.

LOCAL INNOVATION – FUEL ALCOHOL IN BRAZIL

Brazil has been a major producer of sugar since the early colonial era. It was during the Great Depression of the 1930s, when the country's coffee earnings slumped, that interest first arose in the use of sugar alcohol as a fuel, thus reducing reliance on costly imports of petroleum. The Institute for Sugar and Alcohol (IAA) was set up to develop methods of processing. Production and distribution plants were established with government subsidies, but despite the success of these efforts alcohol was used only on a limited scale, usually in the form of ethanol, which was blended with petroleum.

The oil price crisis of the 1970s reactivated interest. Fuel costs tripled at a time when Brazil had yet to develop any significant oil production and the number of vehicles in the country was rising rapidly. Alarmed by the cost of oil imports, the government set up the National Alcohol Program (PNA or "Proalcool") in 1975, and sugar-cane production was once again subsidized. For the first time automobiles were manufactured to run solely on alcohol whereas before engines had had to be specially adapted. These measures were to prove effective, and by 1984 as much as 45 percent of Brazil's fuel consumption was of ethanol alcohol. Although the program could reasonably claim considerable success, by the end of the decade it was beginning to come under pressure, with Brazil's serious debt problems forcing severe reductions in state subsidies.

INDUSTRY AND THE STATE

The role played by the state has always been a distinctive feature of industry in South America. Governments have intervened in the whole evolution of their countries' manufacturing industry, both by the creation of import tariffs and directly through setting up large state corporations. Usually these corporations are established with foreign involvement, and they may dominate a whole industrial sector.

The origins of state intervention in the industries of the region lie in the 1930s. Before 1929 governments had made little effort to promote industry, but from the 1930s they began to adopt policies of greater intervention.

These were increased after 1945, especially in Argentina, where in the 1940s and 1950s the regime of Juan Perón (1895–1974) attempted to extend the country's industrial base, mainly at the expense of agriculture. From the 1960s even the government of Venezuela – which, uniquely in the region, had seen rapid growth in the Depression years as a result of its rising oil exports – was working to build up a manufacturing sector. State control of industry became most marked in the postwar period up to 1980, particularly in many countries experiencing military rule – specifically Brazil, Peru, Uruguay and Argentina. Chile, in struggling to reduce the state role in industry, was very much the exception.

The Triple Alliance

The cooperation between governments and foreign companies is often referred to as the "Triple Alliance" (the third and usually smallest part being local private companies). The arrangement has reaped important benefits for the region. By bringing in foreign expertise governments have been able to make use of technologies that were lacking in their own countries. Foreign investment also allowed the creation of an improved infrastructure of roads, airports, and electricity, all vital for further development.

The price of intervention

State intervention has had undoubted drawbacks. The strength of the Triple Alliance made it hard for local companies to compete and grow, and foreign ownership of industry has become extensive. In Brazil, where the Triple Alliance has been particularly important, close to half the country's industrial assets were in foreign hands by 1980, as well as almost all of the automobile industry. Nor has high technology spread outside the Triple Alliance sector. Most of the population continues to work in the numerous small enterprises that, added together, make up only a small part of national production.

In addition governments have not always shown themselves best suited to the task of directing industry. The Peronist regime in Argentina after 1946 has been accused of building up manufacturing at the expense of agriculture, to the detriment of the country. While some of Brazil's great state corporations have

The blue waters of Chile (*right*) Hydrometallurgy, or leaching, extracts metals from ore by using liquid solutions, often simply water. Sulfuric acid is here being used to separate copper from its oxide ores. The blue color is characteristic of copper leaching.

A modern assembly line (*below*) Brazil has the largest air force in South America, and the country's thriving aircraft industry supplies most of the force's planes. Exports of civilian and military aircraft contribute significantly to Brazil's national income.

While-you-wait industry The street tailors and seamstresses in Cartagena, Colombia are typical of the country's reliance on cottage industries. Colombia's high taxes and an unreliable infrastructure make large manufacturing industries virtually nonexistent.

proved successful, others have provoked charges of corruption and incompetence. The nuclear program ground to a halt in 1985, having already cost $2.5 billion, and with only two of the eight planned power stations ever likely to be completed. It was military governments that took on much of this high level of debt, with the encouragement of Western banks, eager to lend the sums accrued after rises in the price of oil.

The Triple Alliance now seems to be in decline. Following the debt crisis and ensuing recession, the military regimes that had dominated the region had by the mid 1980s (1990 in the case of Chile) largely given way to elected regimes. At the same time there has been a marked turning away from the notion of state enterprise, with growing pressure in many countries for bureaucracies to be reduced and state assets to be privatized. Both in terms of industrial policies and politics, the region seems to be moving toward a European–North American model. The long experiment in state-run industry has certainly had its costs, both in terms of debt and unviable industries. At the same time it has led to the creation of an extensive and varied industrial base, which may prove no small asset when the economic climate improves.

CHILE'S COPPER BONANZA

Chile is estimated to have at least 25 percent of the world's copper reserves, and copper alone accounts for almost half the country's exports. Mining began in the early 19th century, and soon supplied some 46,000 tonnes a year, almost half of world production at that time. By the 1890s, however, technical developments in the United States had reduced Chile's market share to as little as 6 percent.

The copper mining industry in Chile was dominated by a few United States-owned companies until the 1950s, when the Chilean government began to take a more active role to ensure that the country reaped more of the benefits by reinvesting revenues in local refining plants. The nationalization of the industry in the early 1970s by the leftwing government under Salvador Allende (1908–73) was accompanied by expansion; a process that was extended still further and more rapidly by the politically opposed regime of General Pinochet, using both state and foreign investment. In 1991 the world's third largest copper mine was opened at La Escondida, with output planned to reach 760,000 tonnes a year by 1995 – equal to 4 percent of the world's copper production alone.

From water to electricity

Interest in developing hydroelectric power from the water resources of South America dates back some decades – President Kubitschek of Brazil (1902–76) ordered a feasibility study for a project at Itaipu as early as 1956. Despite great potential in the region, it was some time before significant development began. One reason for this was the relatively low price of oil during the 1960s, and the fact that several South American countries were able to develop their own oil resources. At the same time foreign oil corporations discouraged the creation of alternative energy resources, and proposals for hydroelectric projects were often faced with stringent conditions by foreign banks.

All this changed during the 1970s, when oil prices rose sharply in reaction to conflict in the Middle East, action by the OPEC oil cartel and – in some cases – oil nationalization. South American countries without substantial domestic oil resources had incentives to seek cheaper alternative energy, and many decided to invest in the plant and construction necessary to tap into hydroelectric power. One other factor was also important: the price rises in oil had caused the world's leading banks to build up large reserves, and they were suddenly willing to lend money to develop other methods of generating electricity. Consequently the 1970s and 1980s saw the development of numerous hydroelectric projects, with large schemes in Chile, Brazil and Argentina. Many of these were joint ventures, such as Yacyreta (Paraguay and Argentina) and Itaipu (Brazil and Paraguay).

Building the Itaipu dam

The Itaipu scheme, on the Parana River between Paraguay and Brazil, is the world's largest hydroelectric installation, producing more than 12,600 megawatts of power. Following lengthy negotiations between the two governments, the project was finally established in 1974. Construction started in 1976, the first turbines began turning in 1984, and the scheme was finally completed in 1991, three years behind schedule.

The project should supply 40 percent of Brazil's energy needs as well as most of those of Paraguay, which is set to become one of the world's leading electricity exporters. Thanks to Itaipu and other projects such as Tucurui in Amazonia, and also the development of offshore oil, Brazil is already self-sufficient in energy and exports the surplus.

The project is an example of the Triple Alliance of co-operation between government, foreign capital and expertise, and domestic concerns. It was set up by the Brazilian and Paraguayan governments through the government-owned Brazilian national electricity company, Centrais Eletricas Brasileiras (Eletrobras)

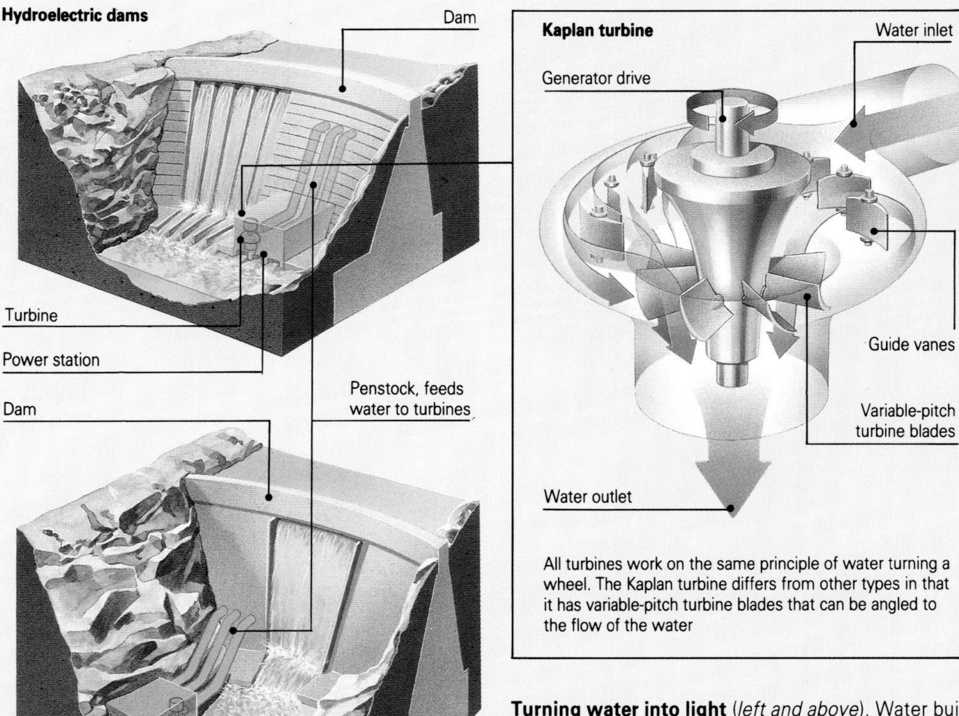

All turbines work on the same principle of water turning a wheel. The Kaplan turbine differs from other types in that it has variable-pitch turbine blades that can be angled to the flow of the water

Turning water into light (*left and above*). Water builds up in a reservoir behind a dam, raising it to the height of the damworks, and is then piped down to a series of turbines. The water catches and pushes the turbines' blades, which spin to create energy. A generator driven by the turbines converts this energy into electricity. This is then passed to a transformer and changed into a high voltage direct current that is suitable for transmission over long-distance cables.

The flow of electricity (*above*) Brazil has plentiful rainfall and one of the most extensive river systems in the world. This makes hydroelectric powerhouses, such as Itaipu on the Parana river, a significant source of Brazil's electricity needs.

and the Administracion Nacional de Electricidad (ANDE), the Paraguayan state electricity board. Eletrobras raised most of the necessary finance from banks in Europe, the United States and Japan, together with a smaller proportion from Brazilian banks. At least 80 percent of the generators and other heavy equipment were to be Brazilian made. This requirement brought important contracts to Brazilian heavy industry, including Bardella and Mecanica Pesada, and also the Brazilian subsidiaries of leading European multinationals, such as Siemens, Asea Brown Boveri (ABB), Alsthom and Voith. Most of the construction materials, such as steel and cement, were also from Brazilian sources, whether state-owned or private.

Although a considerable achievement, the Itaipu project has not been without problems. Costs soared during its construction, from $2 billion in 1972 to an estimated $12.2 billion by the time of its opening in 1991. One-third of this cost was merely servicing the accumulated debt. At the same time oil prices – the original impetus for the scheme – have fallen, and the recession (itself prompted by debts accumulated by projects such as Itaipu) has reduced the demand for electricity and cut the profits of the industry. Nevertheless, Eletrobras remains confident that the economic climate will improve at some time in the future, proving Itaipu and other similar projects to have been more than justified.

Eldorado's slum

To the first European settlers, the newly discovered lands of South America were an Eldorado where fabulous reserves of gold and other precious metals were believed to exist. The cry of "gold" was first heard in Brazil in 1695, and soon the gold rush was underway. Settlers and slaves moved in mass to the gold-rich interior from the agricultural coastal areas. By 1760 gold mining had reached maximum output, with Brazil supplying 80 percent of the world's gold – over 900,000 kg (2 million lb) was produced in the 18th century.

The period that followed saw a decline in production and interest in the mines, with only individual prospectors left working the gold. In the past 50 years, spiraling inflation and poverty in Brazil forced millions of poor peasants to look anywhere for work and money. Many turned to the mines and a new migration to the interior began.

Conditions in the open-cast mines were and are very primitive. Without the aid of machinery or even tools, thousands of people fight for a small patch of ground to work. Others are beaten or robbed of their meager takings and some are pressed into a form of slavery. The Brazilian government has recently begun to take over, mechanize and regulate the free-for-all mines. But for the poor Brazilian the fight for a living continues.

Uphill struggle Prospectors at the Sierra Pelada gold mine head toward an uncertain future as the Brazilian government takes over the once free-for-all mine.

IRON, TIMBER AND NORTH SEA OIL

A WEALTH OF RESOURCES · AN EMPHASIS ON ENGINEERING · THE DRIVE TO EXPORT

Natural resources – including rich deposits of mineral ores, especially iron, forests, water to supply hydroelectric power, fisheries, fertile land (in Denmark and southern Sweden) for farming and, since the 1970s, oil from the North Sea – form the basis of the Nordic Countries' industry. Today the people of the Nordic Countries produce more than 4.5 percent of the world's exports. Yet their total population is a mere 23 million. The limited nature of domestic demand for manufactured goods led Nordic companies to establish themselves in the international market, and to maintain their position by raising industrial efficiency. Exports now account for 40 percent of the region's total national income, and manufacturers know that their survival is dependent on successful competition in the world market.

COUNTRIES IN THE REGION

Denmark, Finland, Iceland, Norway, Sweden

INDUSTRIAL OUTPUT (US $ billion)

Total	Mining	Manufacturing	Average annual change since 1960
172.1	23.3	95.7	+3.2%

INDUSTRIAL WORKERS (millions)
(figures in brackets are percentages of total labor force)

Total	Mining	Manufacturing	Construction
3.6	0.36 (2.9%)	2.4 (19.8%)	0.82 (6.7%)

MAJOR PRODUCTS (figures in brackets are percentages of world production)

Energy and minerals	Output	Change since 1960
Oil (mill barrels)	589 (2.6%)	N/A
Natural gas (billion cu. meters)	32 (1.7%)	N/A
Iron Ore (mill tonnes)	22.5 (3.9%)	-13.5%
Copper (mill tonnes)	0.3 (3.7%)	No data
Zinc (mill tonnes)	0.4 (6.2%)	No data
Manufactures		
Woodpulp (mill tonnes)	18.9 (15%)	+1%
Newsprint (mill tonnes)	4.4 (13.8%)	+36%
Steel (mill tonnes)	7.4 (1.1%)	+86.8%
Ships (mill gross tonnes)	0.5 (4.5%)	-59.4%
Automobiles (mill)	0.5 (1.1%)	+397%
Telecommunications equipment (US $ billion)	7.5 (8.1%)	No data

N/A means production had not begun in 1960

Red heat *(above)* Metalworking has a long tradition in the region. Sweden's ironworks alone, fueled by charcoal from the abundant forests, produced a third of the world's iron until coal-derived coke replaced charcoal in the mid 19th century.

A WEALTH OF RESOURCES

The exploitation of mineral wealth began a long time ago in both Sweden and Norway. Huge deposits of silver, copper and iron are to be found in central Sweden, and metal processing developed around the rich mines in Bergslagen in the 13th century. The first mines to be exploited for the international market were at Sala, where production of silver peaked in the 16th century, and at Falun, which was the largest producer of copper in the world for more than two hundred years, until 1830. Norway's deposits are smaller than Sweden's, but silver and copper were both being mined in Norway in the 17th century.

Sweden's forests provided a ready and cheap supply of charcoal to fuel the blast furnaces used for iron smelting. Only a few regions in Europe were able to match

Energy production and consumption *(below)* Oil accounts for the highest energy output, followed by hydroelectricity and natural gas. With high output and low consumption all the countries of the region can export energy and still maintain reserves.

Energy balance (mill. tonnes coal equivalent)

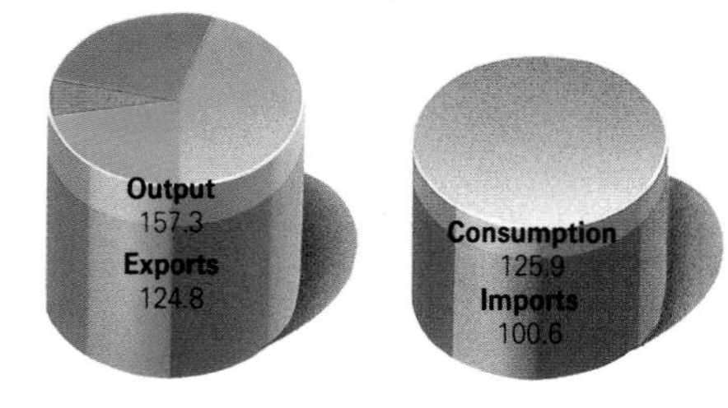

Sources of energy output

its natural resources, and during the 18th century the country became a world leader in the production of iron. However, lacking any substantial deposits of coal, it began to lose place once coke replaced charcoal as a more efficient fuel, and Swedish iron production consequently declined from about one-third to 3 percent of world production by the mid 19th century.

Low-cost methods of steelmaking on a mass scale were developed in Europe and the United States in the course of the 19th century. The improved version of the original Bessemer process (which used compressed air to convert pig iron to steel) developed in 1875 by Sidney Thomas (1850–85) made it possible to use pig iron with a high phosphorous content (typical of north European deposits) to make steel. Its introduction into Sweden enabled the huge iron-ore deposits in the north of the country to be exploited for the first time.

Once again Sweden's export of iron and steel products soared, especially after the opening in 1902 of a railroad linking the vast deposits at Kiruna, Gällivare and Malmberget. These lie within the Arctic Circle close to the Atlantic harbor of Narvik in Norway, free of ice all the year round. Norway's largest iron mine, at Kirkenes, in the northeast, was opened in 1906. After the closure of the mines in Norway and Bergslagen, Sweden, in the 1970s and 1980s, the Swedish Lappland mines remain the only producers of iron in the Nordic Countries.

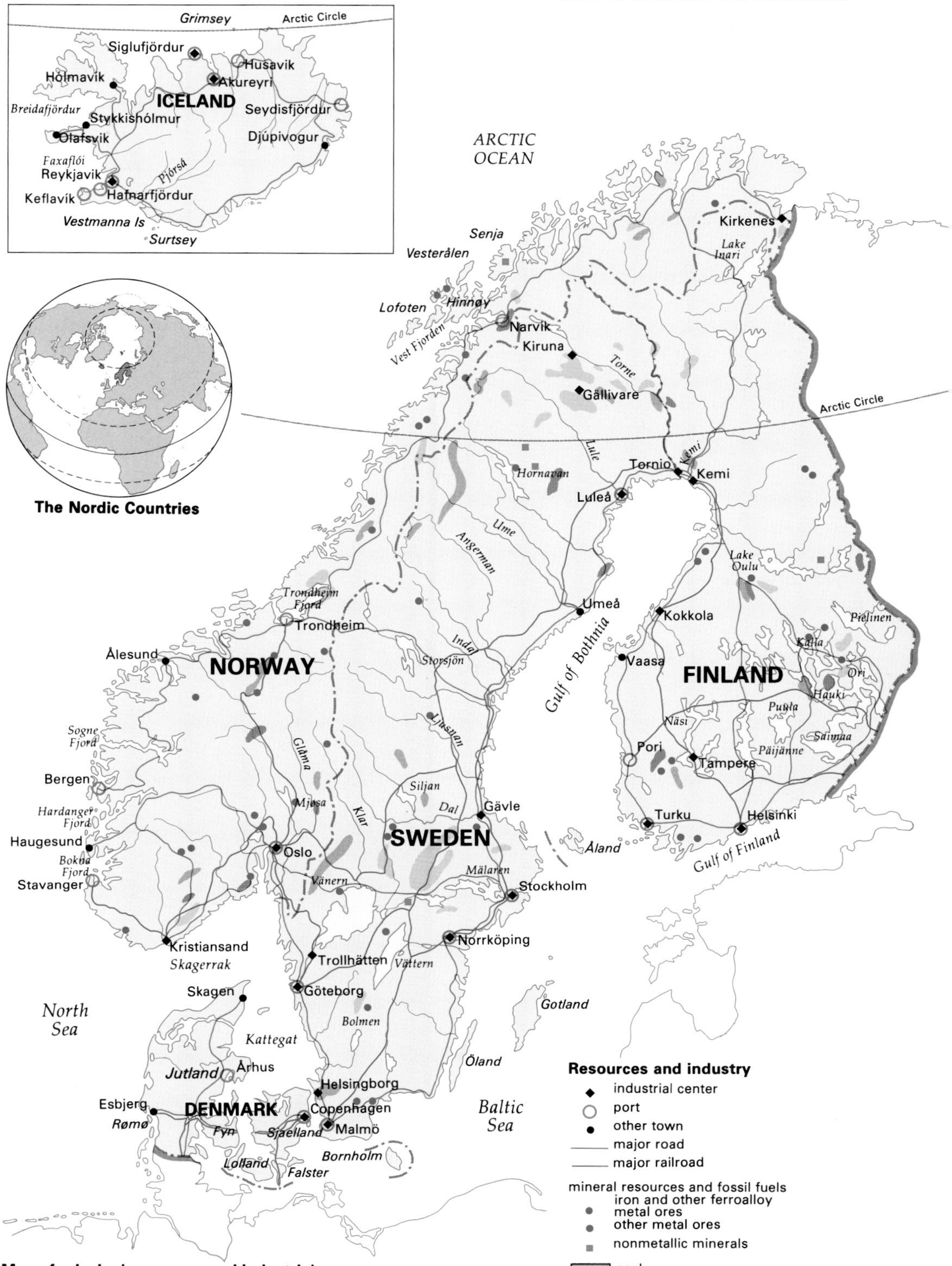

Forests and water

Another important ingredient in the industrial development of the Nordic Countries, with the exception of Denmark and Iceland, was the existence of vast forest resources. First exploited to provide charcoal for smelting, lumber was also much in demand in the mines to make pit props, and it was also the region's main building construction material and source of fuel. By 1870 Sweden had become the world's largest exporter of lumber and wood products, and Norway also had a substantial forestry industry.

Timber pulp replaced linen rag as the main raw material for papermaking at the end of the century. Only the Nordic Countries had sufficient forest reserves to meet the soaring demand in Europe. The abundance of fast-flowing rivers played an important role in developing the pulp industry. It provided a means of floating logs from the central forested areas to the processing plants located along the coast of the Gulf of Bothnia, from where pulp could be exported to the rest of Europe. The rivers also supplied hydroelectricity for the industry.

Denmark and Iceland do not share the mineral and timber reserves enjoyed by the other Nordic Countries. Agriculture and fishing lie at the center of their industrial activity. In Denmark, intensive cereal growing, dairy herds and pig rearing have long supported a flourishing food-processing industry. Denmark has also benefited from the discovery of oil in its sector of the North Sea, which now supplies an ever-increasing share of its energy needs. Fishing is Iceland's main natural resource, and is still the mainstay of its industry, though ways are being sought of diversifying into other areas.

The Nordic Countries

Map of principal resources and industrial zones
Sweden and Norway have significant deposits of metal ores, and both Denmark and Norway have offshore oil and gas fields. Most industrial zones are located on the coast, while mining is concentrated in the interior.

AN EMPHASIS ON ENGINEERING

From very early days industrial development in Sweden and Norway was led by the mining companies. They began the industrialization process, and opened it up in new directions. Commercial sawmill production was originally started by the mining companies near their metal and mining centers in the 17th century, and it quickly became the driving force behind the region's exporting industries. Unlike the iron industry, where competition from other countries was severe, the Scandinavian sawmills enjoyed a paramount position in the European market, owing to the size of their forest reserves and competitive prices.

The iron and steel companies – by virtue of their control over huge areas of forest – were also intimately involved in the move that took place in the industry from lumber to pulp production when the demand for the former stagnated in the 1890s. Early examples of companies involved in both iron production and pulp processing were STORA and Uddeholm. STORA today is one of two giant companies (the other being SCA) that dominate Swedish forestry, and is one of the largest forest companies in the world.

Change in the metal industries

In the 20th century the expertise gained in refining steel and other metals was successfully extended to the production of aluminum. This metal is refined from bauxite into alumina, a powder which is then smelted by electrolysis into aluminum using a very high input of electricity. The industry therefore developed in Norway after World War I, when cheap sources of hydroelectricity were readily available. Alumina was imported from Jamaica and most of the plant was foreign-owned. Norway is now the largest aluminum producer in Europe, with refineries in the west and southwest. In its drive to find new areas for industry in the late 1960s, Iceland also started an aluminum refinery. This is supplied by alumina from northern Australia.

Both Norway and Sweden now also specialize in the refining of ferro-alloys, metals such as manganese, chrome and silicon that are alloyed with iron and used in the manufacture of alloy steels. Sweden was an early pioneer in this sector of the steel industry, with companies such as Sandvik and Avesta leading the way. Sandvik continues this tradition by making products in cemented carbide, a combination of very hard materials used in mining and drilling.

Engineering gains primacy

In recent years industrial production in the region has been dominated by the manufacture of machinery, along with transportation vehicles and equipment: today this sector is by far the largest contributor to exports. The development of the engineering industry in the late 19th century was closely linked to the mining and metal industries. Traditionally there has always been a high degree of cooperation between companies involved in different stages of the extraction and manufacturing processes; for example, between those involved in mining iron

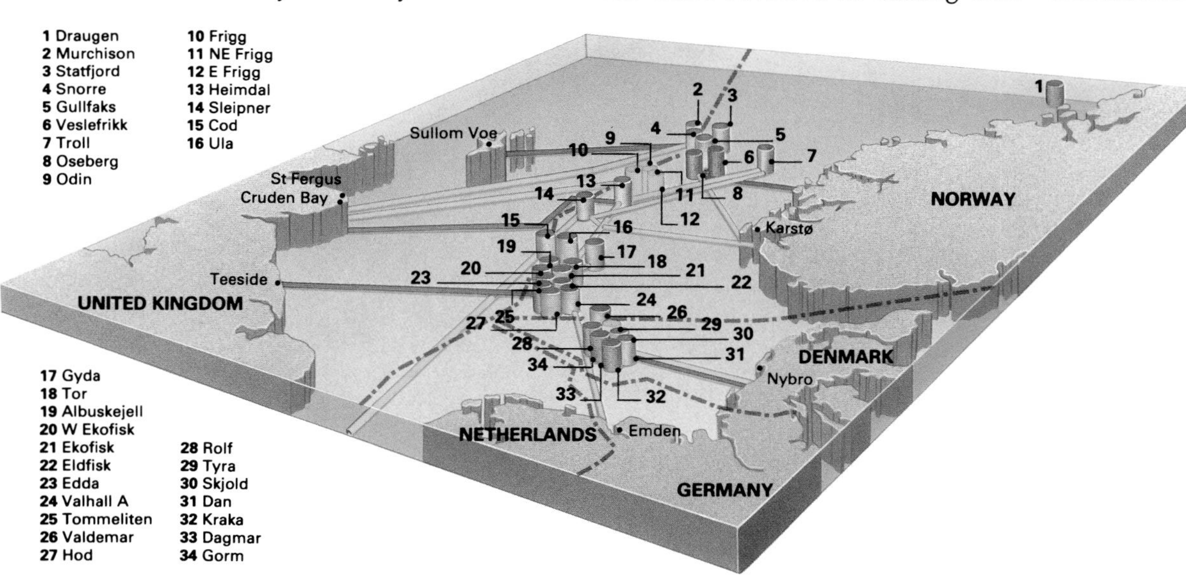

Staking their claims in the North Sea *(left)* The North Sea contains the ninth largest reserves of oil in the world. Although the British sector covers the widest area, the largest known reserves of both natural gas and oil are in the Norwegian sector. Norway and Denmark now supply most of the region's oil. The offshore fields are particularly valuable to Denmark since the mainland lacks other fossil fuels and minerals.

ENERGY FROM THE ROCKS

Iceland lacks mineral deposits, but it possesses vast natural resources of energy. Only about 10 percent of the hydroelectric potential of its rivers has been tapped, and it is able to harness energy from the many hot springs and geysers that occur naturally.

Geologically young – much of its rocks were only formed during the last 10,000 years – Iceland is a volcanically active area. This means that heat from natural radioactivity, which keeps rock in a molten state below the Earth's crust, usually at depths of about 30 km (18 mi), rises close to the surface to heat underground water. It then escapes in the form of hot springs, geysers or fumaroles (vents of gas or steam).

Iceland contains more of these features than any other country in the world, and the heat from their waters is used to provide domestic space heating for nearly 80 percent of the population. As steam it heats greenhouses for commercial fruit and vegetable farming, and provides the energy for Iceland's developing industries, including a plant that refines sedimentary deposits from lake Mývatn. These contain the shells of microscopic aquatic plants, or diatoms, which are treated to make pure diatomite, a material that has many uses – as an industrial filter, as an extender in paints, ceramics, bricks and other products, and to insulate boilers and blast furnaces.

ore, in steelmaking, in the production of carbides, and in the manufacture of mining and drilling equipment.

After World War I the presence of a steelmaking industry producing high-quality sheets promoted the rapid development of shipbuilding in Sweden and Norway, and of an automobile, truck and aircraft industry run by companies such as Volvo, Saab and Scania-Vabis. Sweden also specialized in the production of armaments, such as the Bofors antiaircraft guns. In Denmark, local demand gave rise to an engineering industry manufacturing specialist diesel engines for ships and other vessels. The Danish also specialize in producing machinery for the dairy, meat and cement industries.

Finland did not develop any significant heavy industry until after World War II when, faced with paying heavy war reparations to the Soviet Union, it rapidly expanded its industrial base. Close economic ties were forged with the Soviet Union at this time. For example, Finland specialized in building heavy icebreakers for use by Soviet fleets in Arctic waters. There are also links with the Soviet oil refining and petrochemical sectors. The Finnish state-owned company Neste was founded to process and distribute fuels produced by the Soviet oil industry. Today it is an international conglomerate.

In the absence of coal reserves, hydroelectricity was the region's main source of energy for industrial and domestic consumption until the second half of the 20th century. Since World War II there has been a significant nuclear power program, especially in Sweden and Finland, where nuclear power is now more important than hydroelectricity. In the mid 1970s Norwegian and Danish oil and natural gas reserves in the North Sea were opened up, and these fields now supply most of the oil consumed within the Nordic Countries.

The growth of the petroleum industry has had far-reaching effects in Norway. By 1980 the revenue from exports of crude oil and natural gas had come to equal the combined value of its traditional exports. The oil industry also helped to keep alive the heavy-engineering sector, which was then facing a decline in demand for shipbuilding – Norway's traditional engineering strength – by enabling a switch to be made into manufacturing equipment such as oil rigs and platforms.

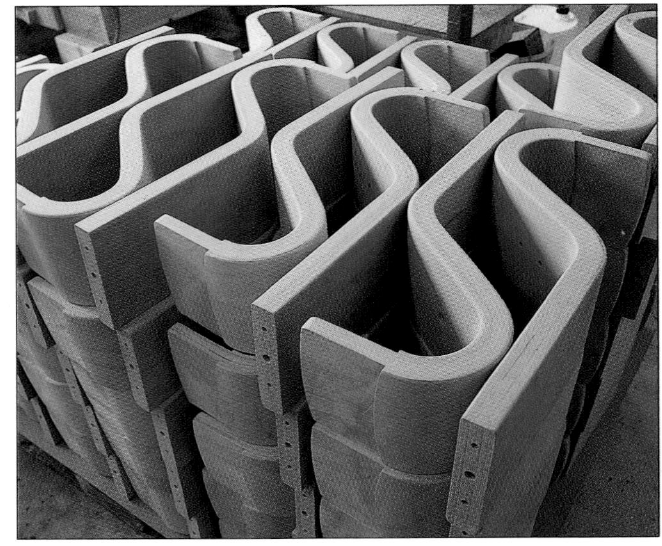

Oil rig construction at Stavanger (*above*) Norway's major port and fourth largest city was once an important shipbuilding center. In the early 1970s 10 percent of the world's tonnage was carried on Norwegian vessels. Now Stavanger is the heart of Norway's oil industry and a related manufacturing sector has sprung up, specializing in drilling equipment, oil rigs and platforms.

Wood, a versatile resource (*right*) Scandinavian furniture is prized by affluent consumers all over the world for its sleek contemporary lines. Timber is plentiful in Sweden, Norway and Finland, where fast-flowing rivers make its transport easy and cheap. However, it is Denmark that has developed the most prestigious furniture and design industry.

THE DRIVE TO EXPORT

With their comparatively small populations, the Nordic Countries have always had to rely on exports in order to run their industries at full capacity. They were not early centers for technical innovation within the mining and metal industries. Sweden's preeminent position in world steelmaking was due to the suitability of local ores, and other conditions, for mass production methods. Technical know-how, as well as the capital to develop the industry, were originally imported from elsewhere in Europe, particularly Britain, Germany, France and Belgium. The provision of capital generated by banks and trading houses involved in forestry and metal exports further enhanced industrial expansion, and as skill and expertise grew a talent for innovation developed.

Successful Swedish technical patents taken out about the 1890s laid the foundation for many of today's giant export companies. These include the milk separator and the steam turbine, which were developed by Alfa Laval and ASEA respectively, the safety match by Swedish Match, the self-aligning ball bearing by SKF, the gas accumulator for automatic lighthouses by AGA, the air compressor by Atlas Copco, instruments for precision measurements by C.E. Johansson and the welding technique by ESAB.

Famous Nordic industrial entrepreneurs include the Swedish chemist and engineer Alfred Nobel (1833–96), who invented dynamite and other powerful explosives and bequeathed his considerable fortune to establish the international Nobel awards. Norwegian entrepreneurs were particularly active in the founding of the Finnish forest industry, for example in the development of the largest company, Enso-Gutzeit. The Norwegian Hans Gutzeit (1836–1919) established his sawmill at Kotka, southern Finland, in 1872. In 1896, it became a Finnish company, and in 1918 the Finnish government assumed control.

An integrated industry *(above)* Pulp mills are often located on the coast, where hydroelectric power is easily harnessed to run them. Since wood pulp replaced linen in papermaking, much Nordic timber has been sent to papermills. Finland's mills are the largest and most modern in the region.

Packing frozen fish *(right)* Fish is a staple of the Nordic diet and a long-standing export product. Refrigeration and freezing now keep this perishable product fresh. Norway alone now exports more than 100,000 tonnes of fresh fish annually, including 15,000 tonnes to the United States by air.

THE PAPER INDUSTRY

Paper is made by separating the cellulose fibers present in plants; these are wetted to produce pulp, which is suspended in water and filtered on a wire screen to form a sheet of fiber. This is then pressed and dried. Since the end of the 19th century, wood pulp has been the chief constituent of paper, especially newsprint, and Sweden and Finland have been major exporters of wood pulp for the world paper industry. Sweden's pulp manufacture peaked in the 1950s, with more than one-third of the world's total pulp exports and 10 percent of world production. The combined Nordic share of world pulp production is now 14 percent.

Until the 1970s, tariff barriers were imposed by the rest of Europe to restrict exports of paper from the Nordic Countries in order to protect their own paper production. Papermills in Sweden, Norway and Finland supplied mainly the local market, but with the lifting of these barriers they began to manufacture paper, particularly newsprint and paperboard, for the international market. By 1990 the Nordic share of the world production of newsprint had risen to 14 percent, the same level as for pulp exports, and was 8 percent for paperboard.

Much of this production is centered in central Sweden and southeast Finland. Finland's forestry industry developed later than that of Norway or Sweden, but grew tremendously in the 1950s and 1960s.

Increasing confidence

As the engineering and manufacturing industries expanded in the first half of the 20th century, many foreign innovations were adapted and then developed by Swedish companies. Among these a three-phase system for transmitting electrical power more efficiently over long distances was developed by ASEA; Scania established an expertise in heavy truck manufacture and Ericsson specialized in telephone exchanges. Capital for such research development and expansion was provided by European investors.

Some companies, such as the Swedish automobile company Volvo when it was founded in 1926, were dependent on foreign materials as well as technology. Volvo gradually substituted imported goods with Swedish manufactured engines, transmissions and other components and by the 1950s it had launched an export drive. Because of heavy competition in European and, later, North American markets, Volvo was forced to maintain top quality together with high productivity. It also faced strong competition in the home market from Saab automobiles and Scania buses and trucks. This situation further strengthened the competitive edge for Volvo and Saab–Scania, who are now both world leaders in heavy truck manufacture.

The drive to increase the volume of foreign trade from industrial exports is strong in all the Nordic Countries, but since the 1960s Denmark has made particular efforts to decrease its reliance on its agriculture and food-processing industries by expanding its manufacturing output in areas such as machinery and transportation equipment, clothing, footwear and furniture.

A skilled, well-organized workforce

The industrial and exporting success of the Nordic Countries has grown side by side with the creation of a well-motivated workforce. Industrial welfare clauses, such as the provision of accident and health insurance for workers and fixed-hour working days, have long had legal enforcement in the region. Most countries have well-organized workers' and employers' associations. In Iceland, for example, 9 out of 10 workers belong to trade unions or employee organizations.

Most industries are privately owned. Sweden has many cooperative enterprises, such as SSAB (Swedish Steel AB) and LKAB (iron-ore mines). Its more important wholly state-owned industries are now organized into one giant concern, Procordia, co-owned with Volvo, giving the state considerable influence in industry. In late 1991 the new nonsocialist government launched an extensive privatization program.

Norway has only a few large state-owned concerns, notably Statoil and Norsk Hydro. These enjoy considerable independence in their management. Many small private businesses are family-owned; the rest are joint-stock companies. About 10 percent of production comes from foreign-owned companies. Fewer than 5 percent of companies have more than 100 employees. This 5 percent employs half the industrial labor force and generates more than half the country's industrial production.

The Volvo model

Volvo's automobile factory at Uddevalla, 80 km (50 mi) north of its main operations in Göteborg, southwestern Sweden, is very unusual. Its production techniques, which were radically new when the factory opened in 1988, are based on a novel form of work organization designed to motivate staff by giving them personal involvement in all stages of making a complete automobile. The more usual conveyor-belt method of automobile production, in which each worker repeats one or two specialized tasks all day long, has been rejected.

Within the Uddevalla factory are six separate "factories", known as product shops. These are more independent of each other than are traditional departments along a conveyor-belt production line. In each product shop, there is a product leader and eight different work teams, each producing a separate automobile. There are usually ten people in every work team, and each member performs a wide range of tasks, including administration. Each work team plans together how the work is to be organized, so every member takes responsibility for their own part of the operation. This procedure reduces the amount of stress and boredom that assembly-line workers typically experience.

If a completed automobile is found to be faulty, it is put right by the team that built it. This ensures that every employee is quality conscious, creating pride in his or her work. Every member takes it in turn to be spokesperson for their particular work team. In this way, all emloyees are involved in decision making,

A model product shop (*far right*) Workers at Uddevalla demonstrate the success of Volvo's social and industrial experiment. Their tasks promote teamwork, personal accountability and superior quality control. Assembling complete vehicles involves workers with the total product. The factory is clean and pleasant, and interaction prevents boredom and isolation.

Modern technology at work (*right*) The worker who uses the computer may also be the one who fits wheel bearings. Training in all aspects of production lessens the division between highly skilled and less skilled workers. Volvo has also introduced a sorting machine to make up component kits, believing that the workforce should not be wasted on tasks that can be handled by machines.

A revolutionary assembly plan (*right*) On an old-fashioned production line the chassis moves along to different fitting stations. At Uddevalla, however, the chassis is delivered to one of six separate teams and remains stationary. Parts are brought to it on an automated trolley. Finished cars are tested at one of the two test stations on the upper level. Final surface treatment includes paint finishing, rustproofing and undersealing.

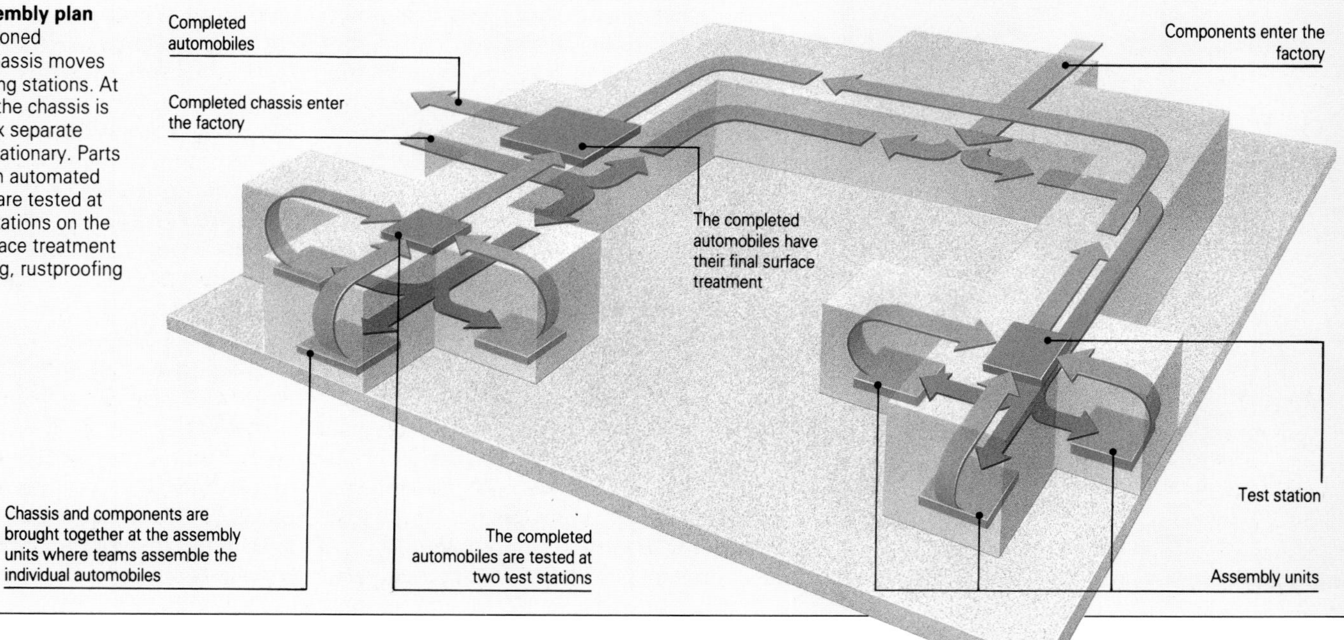

and there are relatively few levels of management between the plant manager and the automobile workers.

Care is taken to create work teams that mix men and women of different ages according to their compatibility and range of skills. The right balance is vital to give work teams a sense of identity and promote individual job satisfaction. Every employee is gradually trained to master all, or nearly all, the different tasks needed to produce an entire automobile.

Smaller is better

Increases in efficiency of the labor force have been one of the key successes to result from these innovatory production methods. Greater job satisfaction reduces the high rate of labor turnover that factories using conveyor-belt production lines normally face, and injuries at work are fewer once workers are released from boring, repetitive production-line tasks that do not allow individuals to realize their potential and think for themselves.

Sweden's school-leavers are now educated to such a level that they are reluctant to accept the kind of work tasks that the automobile industry traditionally offers. The success of the Volvo model at the Uddevalla factory in attracting and keeping workers shows that people at work want to be able to use their brains as well as their hands. In 1990 workers at the Uddevalla factory built only 20,000 units, though it has the potential capacity for 60,000 a year. However, the better utilization and motivation of its human resources has resulted in better quality products with a worldwide reputation, an advantage that needs to be balanced against the supposed economic advantages of largescale production.

Volvo's current approach is to provide customers with automobiles that give them maximum quality and the option of different finishes and equipment. This aim is incompatible with largescale production, where economy of scale is achieved by manufacturing a uniform product. Lower productivity results when different products have to share the same assembly line. Once customer satisfaction (and the willingness to pay for a superior product) is placed at the center of the manufacturer's operation, then smaller plants are more efficient than larger ones. The Uddevalla plant may become a model for the future organization of manufacturing production.

Putting the pieces together

Lost among the momentous political and social changes that were taking place in Europe in the wake of World War II, a small Danish company in 1949 introduced a toy that would revolutionize children's play. Ole Kirk Christiansen's toymaking company had produced its first plastic LEGO brick. Today the LEGO Group comprises 36 companies on 5 continents; since 1949 more than 300 million people will have played with the famous LEGO bricks.

The basic bricks are molded plastic blocks that have built-in studs and tubes. The clutching power of the studs allows children to build the LEGO bricks into shapes of their own design. Two 8-stud bricks can be put together in 24 different ways and 6 bricks have a remarkable 102,981,500 combinations.

At LEGO's five molding factories, the plastic granules are heated to about 225°C. Molding is done under pressure that can vary from 25 to 150 tonnes, depending on the components. It takes between 7 and 10 seconds to mold, cool and extrude one batch of LEGO components out of their molds, and then the process starts again.

After the extrusion process, the molded items are sent for decoration or assembly. Components are decorated in automatic printing machines, and the decorated components are joined together in the assembly departments by specially designed and constructed machines. Assembled components are packed and shipped to local distributors around the world. The largest LEGO factory can produce more than 60 million sets each year.

Head and shoulders above the rest One of LEGO's 1,300 different molded product models marches off the assembly line. A truly international company, most of the machines used are made in LEGO's German and Swiss factories.

AN EARLY START TO INDUSTRY

FUEL FOR GROWTH · CHANGING PATTERNS OF INDUSTRY · PEOPLE FOR THE JOB

The key to Britain's early industrialization was the technical ingenuity of men such as Richard Arkwright (1732–92), who pioneered new methods of spinning cotton, and James Watt (1736–1819), improver of the steam engine. Their innovations opened the way to largescale production, leading to the rapid rise of industries such as heavy engineering and textiles in the north of England, and in Scotland and Wales, where there was a ready supply of coal. Britain's agricultural society was transformed as people deserted farming to swell the workforce of the cities. Today, faced with competition from other world suppliers and rising capital costs, traditional industries have declined, but new manufacturing skills have developed, particularly in the southeast, which is now the driving force of industrial change.

COUNTRIES IN THE REGION

Ireland, United Kingdom

INDUSTRIAL OUTPUT (US $ billion)

Total	Mining	Manufacturing	Average annual change since 1960
307.8	43.4	195.0	+1.8%

INDUSTRIAL WORKERS (millions)
(figures in brackets are percentages of total labor force)

Total	Mining	Manufacturing	Construction
7.8	0.5 (0.2%)	5.5 (18.5%)	1.75 (5.9%)

MAJOR PRODUCTS (figures in brackets are percentages of world production)

Energy and minerals	Output	Change since 1960
Coal (mill tonnes)	100.0 (2.1%)	-49.2%
Oil (mill barrels)	673.6 (3.0%)	N/A
Natural gas (billion cu. meters)	43.7 (2.4%)	N/A
Nuclear power (mill tonnes coal equiv.)	22.2 (3.3%)	No data
Peat (mill tonnes)	4.4 (21.8%)	+20%
Clay & Kaolin (mill tonnes)	26.9 (13.5%)	-17%
Chalk (mill tonnes)	14.5 (85.1%)	-10.8%
Manufactures		
Wool yarn (1,000 tonnes)	161.3 (7.2%)	-23%
Ladies' stockings (mill pairs)	629.7 (11.0%)	+38.2%
Synthetic rubber (1,000 tonnes)	312.8 (3.1%)	-4.3%
Steel (mill tonnes)	18.9 (2.6%)	-23.5%
Automobiles (mill)	1.9 (2.0%)	+4.0%
Televisions (mill)	3.1 (2.8%)	+44.8%
Beer (mill hecoliters)	60.2 (5.8%)	-4.6%
Confectionery (1,000 tonnes)	832.0 (12.3%)	+7%
Chocolate (1,000 tonnes)	480.1 (10.1%)	+12%

N/A means production had not begun in 1960

FUEL FOR GROWTH

The Industrial Revolution, the chain of events that transformed the largely agrarian and locally based economies of Europe and North America into the multinational industrial concerns of today, started more than 200 years ago in Britain. This blossoming of technology and entrepreneurial skill was made possible by Britain's large coal reserves that fueled new ironmaking processes and provided steampower to drive new machines.

Energy resources have remained the outstanding industrial asset of the British Isles. At the beginning of the 20th century, only the United States produced a greater quantity of coal than Britain. However, British production began to fall as easily worked seams became exhausted. Competition from other energy sources further accelerated the decline.

By 1990 Britain was only the eighth-ranked producer in the world. The coalfields of central England had increased their relative share of a declining national output, while those of northern England, Scotland and Wales – less economical to work – had cut back significantly. Many had ceased production altogether, and by the beginning of the 1990s the last few pits in Wales were closing.

Since the mid 1960s natural gas, especially from reserves beneath the shallower southern parts of the North Sea, has supplied a significant part of domestic energy needs. Soon after this, the search for new sources of fossil fuels revealed oil in its deeper northern waters, though frequent bad weather conditions there add to the difficulties of extraction. Estimated North Sea oil reserves amount to more than 4 billion tonnes. As a result, Britain has risen to become the world's fifth largest producer of energy, and it is the only major Western economy to have a trade surplus in the three main fossil fuels.

Ireland, by contrast, has only small energy resources. There is no coal. Traditionally peat (or turf) was dried and burned as fuel. Much of the center of the island is rich in this decomposed plant material, which is formed in boggy conditions. It is used as fuel for some electricity generation, but more efficient fuels are usually imported. In recent years, however, natural gas has been discovered offshore, and is now being exploited.

Declining assets

The British Isles are relatively poor in other minerals. Iron ore is the most significant of its metal deposits, and in the past metals were mined quite extensively. Gold and copper from Britain and Ireland formed the basis of trade with the Mediterranean at least 4,000 years ago, and many areas still bear the marks of former largescale mining. The abandoned wheelhouses of old copper and tin mines are dramatic features of the landscape in southwestern England, and lead was formerly mined in Scotland and the Pennines of England.

Today all such activity has ceased to be economic, and the metals for industry are imported, often from British-owned mines in many parts of the world. Although Britain's steelmaking industries grew up around its iron-ore producing areas, even here there has been a switch to the use of richer imported ores. The situation is rather different in Ireland. It was not a significant producer of metals until the late 1970s, when a major lead and zinc mine – Europe's largest – began operations in Navan in the east.

A handcrafted finish *(above)* Rolls Royce cars have a high international reputation. British manufacturing skills, which were at the forefront of the industrial revolution, are still in demand for quality goods.

Energy consumption and production *(below)* Forced economies have brought about a decline in British mining, and coal is now imported. Britain, however, has a surplus of fossil fuels and is the world's fifth largest producer of energy.

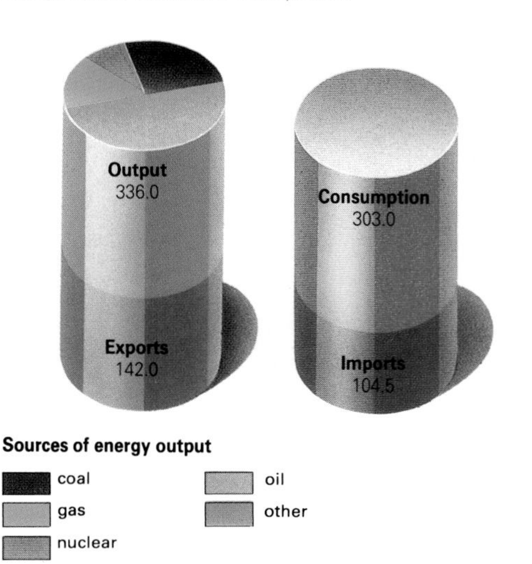

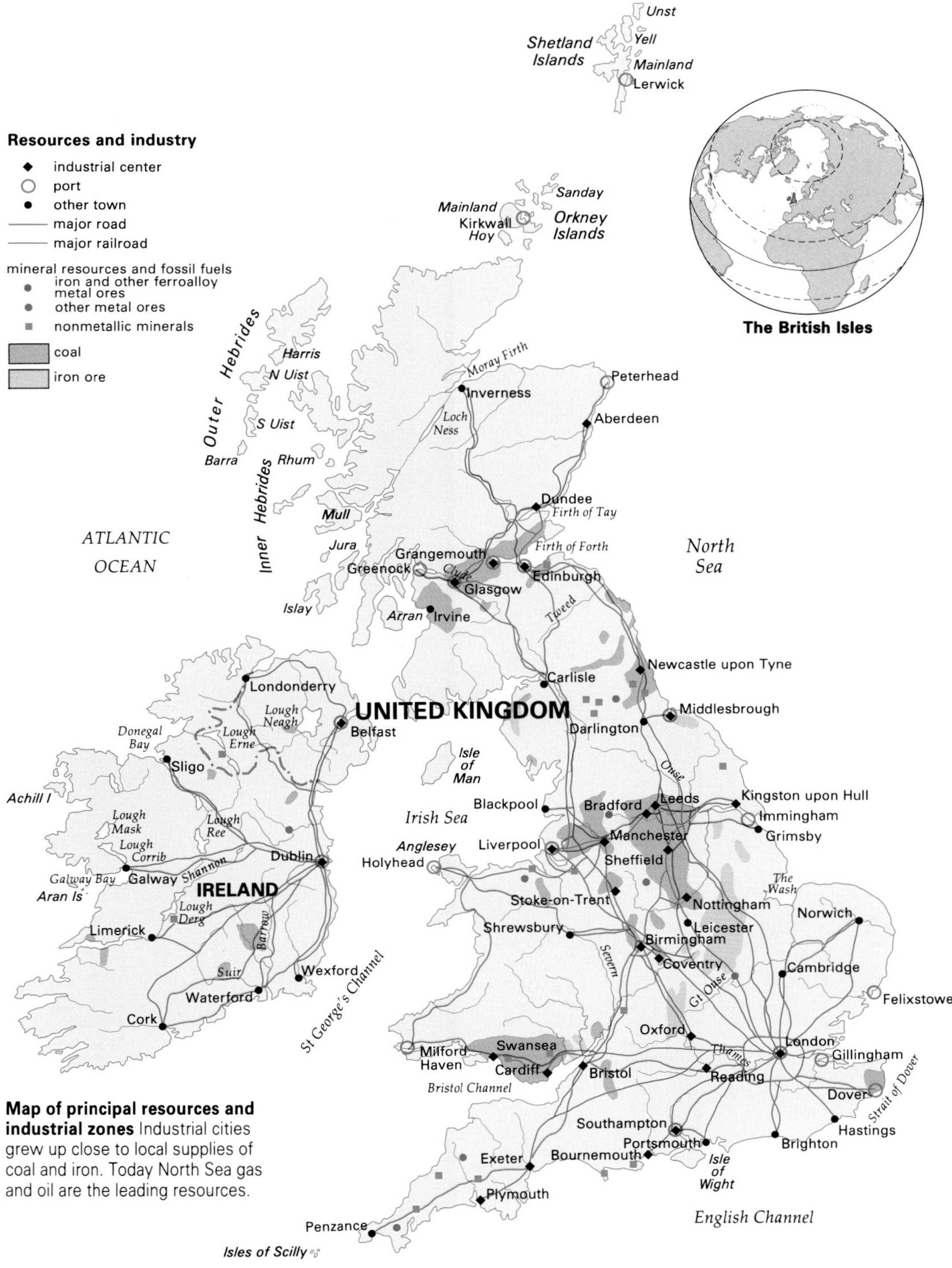

Map of principal resources and industrial zones Industrial cities grew up close to local supplies of coal and iron. Today North Sea gas and oil are the leading resources.

The British Isles' complex geological structure means that there is a wide range of stone for building and other purposes. In the past, slate was extensively quarried as a roofing material. Today this has virtually ceased, but the quarrying of other stones such as chalk and limestone remains important. The extraction of sand and gravel from sedimentary beds has increased dramatically in recent decades, often with unfortunate environmental effects. China clay (kaolin), a fine white clay that is used in ceramics, as an industrial filler or extender and in medicines, is quarried in the southwest.

CHANGING PATTERNS OF INDUSTRY

During the 19th century large industrial centers developed in central and northern England, southern Scotland, and south Wales, close to the major coalfields. There were particular regional specializations: cotton and woolen mills were concentrated in northern England; pottery and ceramics in the northwest Midlands; steelmaking in the northeast, central Scotland and south Wales. Britain became a leading manufacturing country, and its

numerous products were exported widely around the world.

In the years before World War II, however, the production of capital goods, such as ships and heavy machinery, began to decline as Britain lost its former competitive edge in a contracting world market. Cutbacks in one sector of an industry adversely affected others. Within the steel industry, for example, a decline in shipbuilding affected heavy engineering. Reductions in both lowered the demand for steel, which in turn led to a fall in the consumption of coal. A similar complex of interrelated trades existed in the textile industry. As production fell, it tended to become concentrated in fewer, larger plants using the best technology of the time.

The drift to the southeast

At the same time, population growth and changes in spending power were producing a rapid increase in the demand for consumer goods. New manufacturing industries developed in the larger, more affluent centers of the Midlands and the southeast, where the market for products such as radios, washing machines, cookers and refrigerators was greatest. Most important of all was the growth of the automobile industry.

In the years immediately after World War II industrial reconstruction and growth was bolstered by government policies for regional development, particularly targeted at the older industrial zones. Yet it proved impossible to halt completely industry's drift to the Midlands and southeast. Then, in the 1970s and 1980s, increased international competition from new suppliers, particularly in Japan and Southeast Asia, and a lack of capital investment forced a new round of cutbacks in the older industries.

Some, notably steel, emerged from this period of crisis and reconstruction as much more efficient operations, though

Bright lights (*above*) A huge petrochemical plant at Grangemouth, Scotland, lights up an otherwise gloomy prospect. In its industrial heyday, Scotland was a prosperous center of coal mining, shipbuilding, steel and engineering, but cutbacks have hit the area hard.

Warp and weft (*below*) Tweed, a flecked or patterned woolen cloth noted for its durability, is prized as a fashion fabric for medium to heavyweight suits. It is a thriving craft industry in parts of Scotland, Ireland and northern England.

with reduced output, fewer centers of production, and a far smaller workforce. Others, such as shipbuilding, shrank drastically to the point of virtual extinction. Even the petrochemicals and vehicle construction industries – both postwar success stories – were compelled to cut back on sites and workers.

By the late 20th century the proportion of the workforce involved in manufacturing had fallen to little more than one-fifth (in 1901 it was just over half, and in 1971 one-third). Most of these were employed in electronics, food processing, chemicals, industrial glass manufacture and pharmaceuticals rather than heavy engineering. At the same time service-based activities such as banking, insurance, retail selling, tourism, and the entertainments and leisure industries had become increasingly important as employers and creators of wealth.

Most of this expansion had taken place in the southeast. For example, in the 1980s high-tech electronics and computer-systems companies proliferated in the Thames valley corridor, close to Heathrow international airport and connected by expressways to London. In the last decade of the century, closer ties with the European Community and the physical link of the Channel Tunnel between France and southern England seemed set to act as a still more powerful magnet in attracting new industries to the southeastern part of the region.

THE SELLING INDUSTRY

One area of activity that has had spectacular success in Britain is advertising, brought into being by the rise of a larger, more affluent and better-educated population, ready to be informed and persuaded into buying an ever-wider range of consumer goods. British advertising is possibly the most sophisticated, witty and creative in the world, and a handful of British companies such as Saatchi and Saatchi are internationally acknowledged leaders in the field. The industry is a major employer of writers, designers, actors, directors and technicians, and also an important training ground for work in the media. Some creative people who began working in advertising have gone on to become household names as novelists or movie directors.

Promoting products is only part of the story. Companies may make conscious use of advertising to create a corporate image. Political parties promote their policies before elections, and government departments often spend lavishly on health education campaigns or on promoting shares in newly privatized industries.

The advertising industry, located mainly in London, is now very big business indeed. Major companies are prepared to spend vast sums on integrated campaigns (usually television commercials backed up by hoardings and newspaper or magazine advertisements) to launch a campaign and maintain a high profile. Independent radio and television channels are completely funded by advertising, and newspapers and magazines are heavily dependent on advertising revenue.

Industry in Ireland

Industrial development did not have the same momentum in 19th-century Ireland as it did in Britain, partly because of its position as an offshore island, and partly because of its lack of natural resources other than agriculture. About one-fifth of what industry there was at the beginning of this century was concentrated around Belfast, in what is today the province of Northern Ireland. Here there was access to cheap supplies of coal shipped across the narrow strait from western Scotland.

Linen making was the chief activity of many small towns in the north, since the damp conditions suited the growing of flax and there was a ready supply of soft water, used to soften the plant fibers for spinning. Textiles are still important in Northern Ireland today, and cotton (especially shirtmaking), wool (mainly for carpets) and artificial fibers are now as significant as linen. Shipbuilding, once Belfast's leading heavy industry, has declined, and engineering skills have now been diverted into aircraft and vehicle manufacture. Government policies have had some success in attracting multinational companies into the province.

In the Republic of Ireland, most industry was, until comparatively recently, based on agriculture. Food processing (including Guinness, the world-famous beer brewed in Dublin) remains important. The government has actively promoted industrial growth and manufacturing diversification by encouraging overseas investment, particularly into the computer and electronics industries. The recent expansion of mining has spawned a number of related processing and engineering industries. Traditional manufacturing crafts, such as crystal glassware, woven woolen cloth and ceramics, enjoy considerable success as offshoots of the island's prosperous tourist industry.

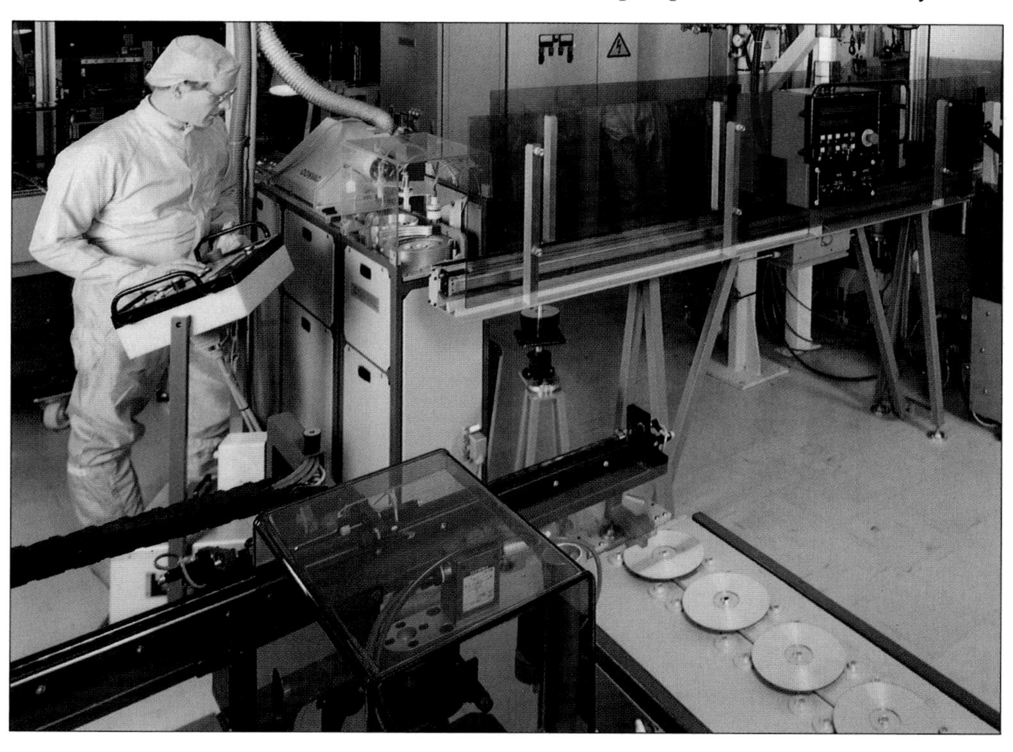

Disks for pleasure A company manufacturing compact disks typifies the new high-tech face of industry. Conditions are kept spotlessly clean and free from dust in this laboratory-cum-factory, and the workers are skilled technicians.

PEOPLE FOR THE JOB

The transformation of Britain from an agricultural to an industrial economy was made possible by a large reservoir of labor. At the beginning of the 19th century, population levels were rising, but most of the people, and the wealth of the country, were concentrated in the agricultural south, dotted with market towns and cathedral cities. Industrialization turned this pattern on its head. With the coming of the factories, people left the countryside in droves to find work in the new towns of the north, such as Manchester and Leeds. Later they migrated from stagnating or declining industrial centers to new "growth" areas offering more and better jobs.

The population continued to grow throughout the 19th century. But the increase in the north of England was almost three times that of the south and four times that of the central counties. By the second half of the 20th century the picture had changed yet again. The population of the former industrialized areas in the north of Britain had actually declined, while that of the southeast continued to grow.

In the postwar period, within its policies for industry, the government introduced a number of planning restrictions and financial inducements to attract new industries and employment to the old industrial centers. In the 1960s and 1970s several government administrative departments were moved from London to the regions, and major companies were encouraged to relocate their head offices outside the capital. Incentives, including exemptions from local taxes, and the provision of cheap loans, purpose-built premises and subsidized housing for key workers, were offered to firms willing to build factories in areas of industrial blight. New industrial estates were set up with easy access to the newly improved road network.

Despite these initiatives, the old industrial areas continued to suffer higher levels of unemployment than the rest of Britain. Even in the 1960s and early 1970s, boom years of nearly full employment, almost twice as many workers were unemployed in Scotland than in the south. These areas were consequently more vulnerable to the effects of economic recession when it occurred.

A changing workforce
In the old industrial communities, people lived close to their work. Sons (and often daughters, too, though generally married women did not work outside the home) followed fathers into jobs, building up pools of local skills in certain trades: cotton spinning in the mills of Lancashire in the northwest, cutlery in Sheffield, Yorkshire, and china in Staffordshire, for example. Much of this has changed. New industries are often located on greenfield sites away from the old factories and centers of population, and people usually travel several miles to work. Many traditional skills have become redundant, and new ones must be learnt.

In contrast to the past, few school-leavers now expect to spend all their working lives with one employer or even in the same industry, and some may have to retrain several times during their careers. The trend away from heavy industry means that far fewer jobs demand physical strength and more can be done on a part-time basis. Women are now an important part of the workforce, and many expect to work for most of their adult lives, whether or not they have children. Where once industry relied on a static, loyal workforce, the need now is for adaptable workers willing to accept new methods and new working practices.

Shop floor meeting Britain has a long history of trade union organization. However, recent legislation and the fear of job losses has curbed the power of the unions. Strike action has declined from its peak in the 1970s and 1980s.

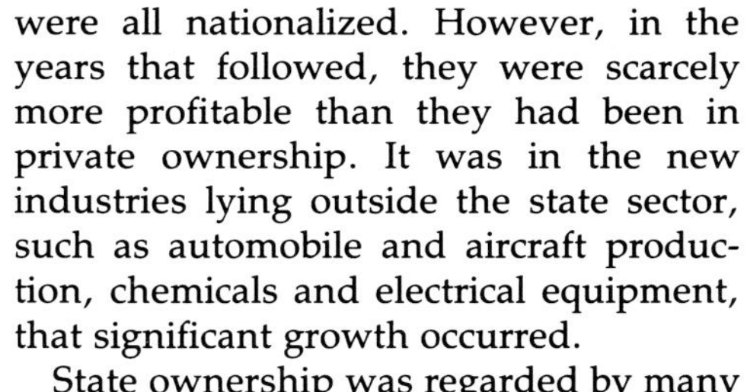

Keyboard control (*left*) Computers monitor operations in the hot metal mill at Britain's largest steel plant at Port Talbot, south Wales. The industry has undergone recent rationalization.

A new way of making glass (*below*) Innovation in industry tends to be greatest in the private sector, driven by the need to secure more orders and higher production. In 1959 Pilkington Glass, a British company, announced a new approach to making glass that saved on time, labor and money. The company unveiled a way of turning liquid glass out of the furnace directly onto a bath of molten tin. The glass floats on the tin while slowly cooling, allowing it to form clearly with no need for polishing or grinding. Rights to the process were later licensed to companies around the world. Although Pilkington retains a manufacturing branch in the northwest, in 1991 it moved its headquarters to Belgium.

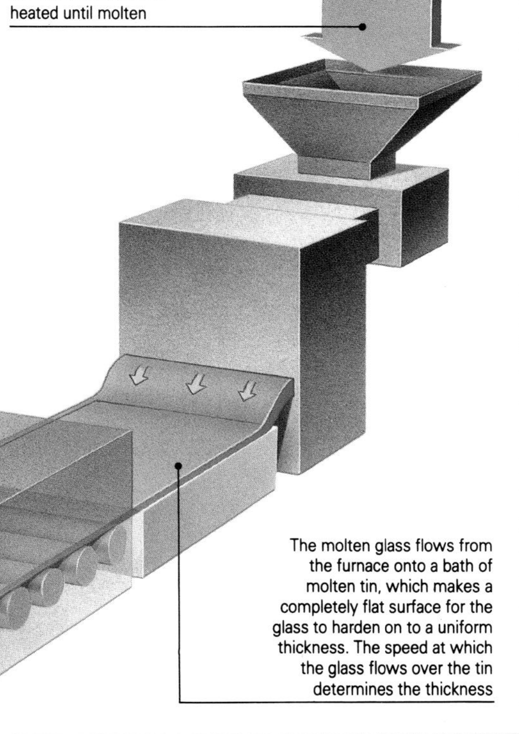

Dismantling state-owned industry

In the first general election after the end of World War II in 1945 a Labor government was returned to power. As part of its program to create a welfare state, it immediately placed many of Britain's major industries under state ownership. Coal, electricity, gas and steel, along with inland transportation (for example, railroads and inland waterways) and docks were all nationalized. However, in the years that followed, they were scarcely more profitable than they had been in private ownership. It was in the new industries lying outside the state sector, such as automobile and aircraft production, chemicals and electrical equipment, that significant growth occurred.

State ownership was regarded by many as a cause of low productivity, stagnation and inefficiency in British industry, and it became the target of mounting criticism from successive Conservative governments. Steel in particular suffered from swings in political opinion – nationalized in 1948, it was denationalized in 1953, and taken back into public ownership in 1967. The Conservative governments of the 1980s, led by Margaret Thatcher, were elected with the stated aim of freeing industry from government regulation and opening them up to market forces.

The coal-mining industry became the main target of government policies to curb union power and prune inefficiency in the state sector. Plans announced in 1984 to close up to 20 unprofitable pits, with the loss of thousands of jobs, provoked a strike that lasted for more than a year before the mineworkers voted to return to work. By 1991 many major industries had been privatized, including steel (for the second time), electricity, gas, water and telecommunications. The intention had been stated to denationalize all or part of the coal industry, the last of the energy-producing and distributing industries to remain in public ownership.

FROM SELF-HELP TO POLITICAL POWER

The trade (or labor) union movement came to birth in Britain during the 19th century as a reaction to the harsh life experienced by early factory workers. Local self-help societies were formed to combat the long hours, arbitrary dismissal, low wages and unhealthy conditions imposed by factory owners, and these associations then developed links with workers in other trades.

The movement grew rapidly, but did not acquire legality until the Trades Union Act of 1871. It played a vital role in establishing the political strength of working people around the principle of equal votes. At first its appeal was to workers in skilled trades, but by the end of the 19th century more than 2 million workers had joined the movement. Among the many thousands of unskilled workers it represented were young, unmarried women working in textile sweatshops, whose interests had previously been neglected.

In the early years of the 20th century, the movement's representative body, the Trades Union Congress, played a leading role in establishing the Labor Party as Britain's second political party. Links between the two remain, largely based on historical sentiment. Legislation since 1979 has greatly curbed the power of the unions (whose closed shops and other restrictive practices were often blamed for industrial stagnation) and the number of members has fallen as old industries and skills have been replaced by newer ones that have no established tradition of trade union membership.

Food processing: a growth industry

Food processing is now one of Britain's most important manufacturing sectors. Together with the beverage and tobacco trades, it employs four times as many people as the metal manufacturing industry and shows better profitability than textiles. The phenomenal growth that has taken place within the food-processing branch reflects significant changes in people's lifestyles.

In preindustrial Britain, as in most agricultural societies, commercial food processing was limited to the milling of flour and the brewing of beer. Granaries, flourmills, maltings and breweries were family owned and served local or regional markets. Gradually, with the development of machinery for manufacturing and the creation of an industrial workforce demanding a greater variety of foodstuffs, food production became more diversified, concentrated in the hands of a number of companies operating on an ever-larger scale.

Today most of the major groups within the British food industry are internationally based companies such as Cadbury-Schweppes, Allied-Lyons and Rank Hovis McDougall. Most processing and packaging is carried out in huge, centralized plants. Some are located close to sources of raw materials; for example, whisky distilling – long-established in parts of Ireland and in Scotland – is influenced by the quality of the local water, which adds a characteristic peaty flavor to the product. Quick-freeze plants are located in lowland parts of England where vegetables are grown, while canning and preserving plants are found in fruit-growing areas.

1 The raw ingredients of whiskey distilling are barley and yeast

2 The barley is steeped in water

3 The barley is "malted" by allowing it to germinate in warm, moist conditions. It is then dried

4 A mixture of malted and unmalted barley is milled to form "grist"

Yeast

6 Yeast is added and fermentation, which converts the sugar into alcohol, begins

The whisky industry *(above and right)* Whisky (spelt this way in Scotland, but whiskey in Ireland) has been made in both countries for centuries, and today supports a thriving export industry. Two resources found in abundance in the region are crucial in its manufacture: barley and fresh water. The word whisky comes from a Celtic expression meaning "the water of life," though it is believed that a knowledge of distilling was gained from the Middle East in the Middle Ages. Scotch whisky is distilled twice after being dried over the peat fires that give it a smoky flavor. Irish whiskey is distilled three times, making it exceptionally smooth. By law, Irish whiskey must age three years in the cask; most is at least 5 to 7 years old.

From the foodstore to the plate *(below)* Marks and Spencer, one of Britain's largest and most successful retail chains, derives more than a third of its income from the sale of food. It specializes in luxury and convenience foods that can be served to family or guests without apology.

A taste revolution As in other western countries, Britain has seen a marked trend toward chemical-free, "healthier" foods. At first, small specialist foodstores were alone able to meet the demand, but now many sectors of the industry have profited by developing product lines that promote a healthy lifestyle.

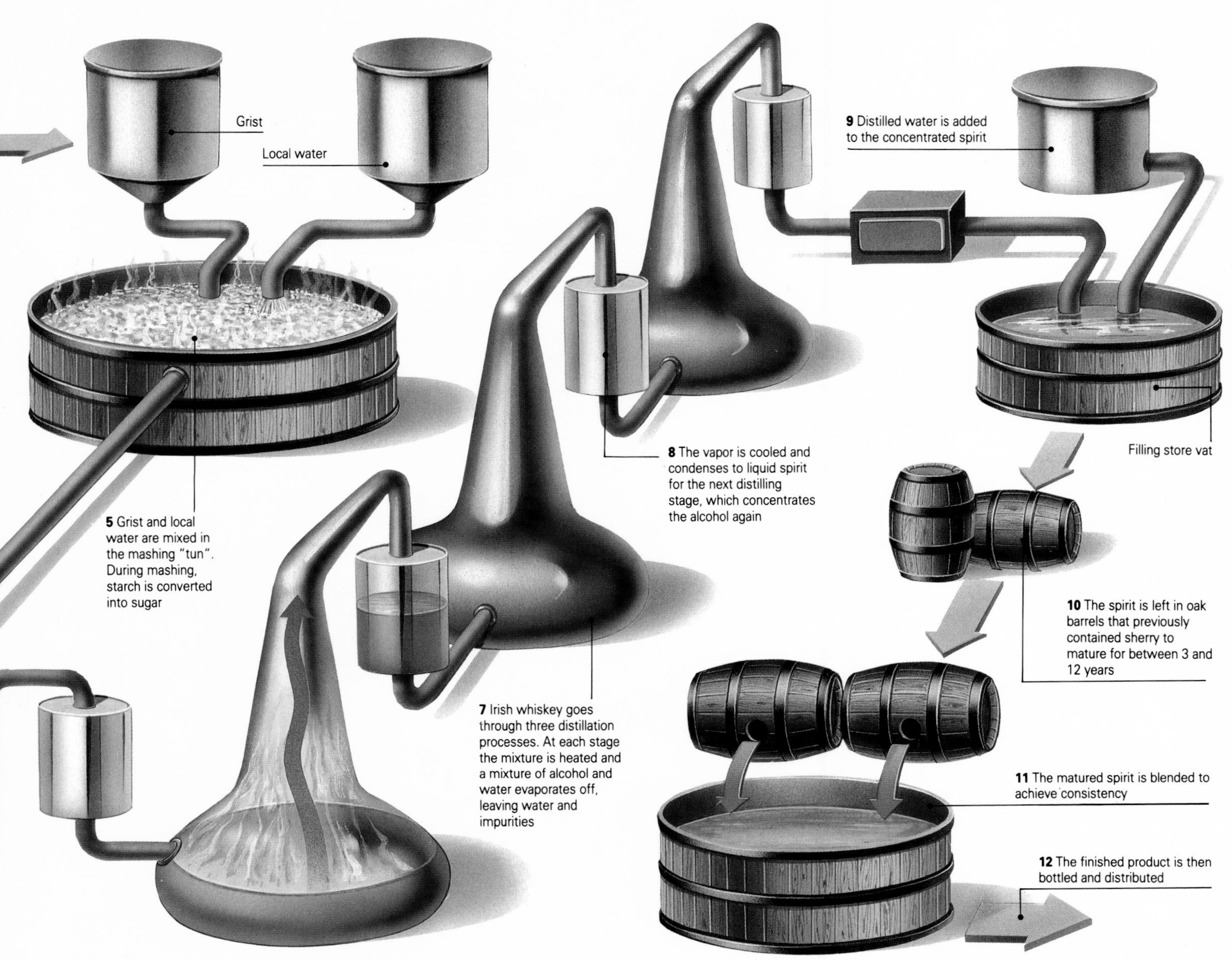

For well over a hundred years, from the beginning of the 19th century, British grain production declined. Increasing amounts were imported, and the milling of white, refined flours became a factory-based operation, mostly located near the ports. Even after the massive postwar increase in cereal farming, milling still remained concentrated in the hands of a few large companies.

In recent years, however, public awareness of dietary health issues has increased, along with a growing revulsion against blandness of flavor. Both these factors have helped to create a market for a greater variety of bread types made with unrefined flour containing more of the husk of the grain. As a result, there has been a significant revival in the fortunes of the small, local mills of rural Britain that have continued to use traditional milling methods, including using wind or water power to grind the grain.

Changing tastes

This is only one small example of the way eating habits and diet have been revolutionized in Britain. Relative affluence, ease of foreign travel and, to some extent, the influence of immigrant communities such as Pakistanis, West Indians, Chinese and Greek-Cypriots, have helped develop a popular taste for foods from many different countries. Changes in lifestyle have led to a demand for food that is quick and easy to prepare. This is typical of trends throughout the world wherever standards of living have risen and the purchasing power of the population has greatly increased.

Exotic foodstuffs are now imported in large quantities. Some, such as fruit and vegetables, wine, cooked meats and cheeses, are processed in the country of origin, but may need special handling, such as refrigerated storage and distribution facilities, before making their way to retail outlets. Other produce, whether imported or grown domestically, is processed (by heating, preserving, freezing or drying) and then packaged in cans, jars, bottles and containers before being passed on to the consumer.

One important area of growth, made possible by increasing ownership of refrigerators and freezers, has been the preparation of cooked chilled or frozen dishes, ranging from meat pies and precut chips (french fries) to pizzas and Tandoori dishes, supplied to the major supermarket chains. Such products are expensive to prepare, but consumers have shown themselves willing to pay high prices for conveniently packaged and attractively presented foods. Despite some worries about food safety, this trend is likely to continue as more and more women work fulltime and food marketing becomes a weekly rather than a daily routine.

Oil and water

The North Sea is rich in oil and natural gas fields, with huge untapped reserves. However, ever since the Danish government granted the first oil-drilling concession in 1963, the five North Sea oil producing countries have found extracting the oil both difficult and expensive. Subarctic winds and temperatures, the remoteness of the area and the depths to which the rigs have to drill are major factors in North Sea drilling. One of the most demanding operations, however, is providing accommodation and food for the crew.

Working shifts of 12 hours on a week-on, week-off basis, the 200-strong crew of a North Sea rig make the most of their spartan living accommodation. Food is flown in daily, and in an average week 73 kg (160 lb) of butter, 2,545 liters (560 gallons) of milk, 455 kg (1,000 lb) of vegetables and 910 kg (2,000 lb) of meat will be eaten. The crew are also offered a wide choice of leisure activities ranging from on-board cinemas to games and sports facilities.

In return, the crew perform very complicated and dangerous tasks. A major industrial accident occurs almost yearly, with over 500 deaths reported in 20 years. More than 30 percent of the workers on oil rigs have had a serious work-related accident.

A vivid reminder of the hazards of off-shore drilling occurred on 6 July 1988 when the Piper Alpha rig blew up in the North Sea, killing 167 crew members.

Braving the North Sea weather, these rigs are sitting atop one of the richest oil and natural gas reserves in the world.

A NEW INDUSTRIAL REVOLUTION

DEPLETION OF RESOURCES · NEW DIRECTIONS · MANAGEMENT AND LABOR

France has become a major industrial power in a remarkably short space of time. During the three decades following World War II the region's mineral and energy resources (particularly coal and iron ore) formed the basis of a hugely prosperous iron and steel industry. However, as these resources have become depleted, competitive foreign imports have become more and more important. Fueled by these imports, high-tech industries such as aerospace and telecommunications have experienced dynamic new growth, and multinational corporations play an increasingly significant part in the country's industrial progress. Wine making – one of the few remaining family-run industries – is big business, as is the food industry in general. Tourism throughout the whole country is also a major source of revenue.

COUNTRIES IN THE REGION

Andorra, France, Monaco

INDUSTRIAL OUTPUT (US $ billion)

Total	Mining and Manufacturing	Average annual change since 1960
304.9	191.2	+3.1%

INDUSTRIAL WORKERS (millions)
(figures in brackets are percentages of total labor force)

Total	Mining and Manufacturing	Construction
5.7	4.1 (16.9%)	1.3 (5.3%)

MAJOR PRODUCTS (figures in brackets are percentages of world production)

Energy and minerals	Output	Change since 1960
Coal (mill tonnes)	13.5 (0.3%)	-76.8%
Oil (mill barrels)	27.1 (0.1%)	No data
Natural gas (billion cu. meters)	2.9 (0.1%)	-62%
Nuclear power (mill tonnes coal equiv.)	43.1 (13.3%)	N/A
Potash (mill tonnes)	1.6 (5.5%)	-24%
Bauxite (mill tonnes)	0.6 (0.7%)	-70.2%
Manufactures		
Cement (mill tonnes)	30.9 (2.8%)	+115%
Steel (mill tonnes)	18.9 (2.6%)	+9.4%
Automobiles (mill)	3.8 (8.0%)	+274.7%
Plastics and resins (mill tonnes)	3.4 (6.6%)	+57%
Fertilizers (mill tonnes)	5.8 (3.6%)	+57.6%
Vacuum cleaners (mill)	2.3 (6.4%)	+51.3%
Telecommunications equipment (US $ billion)	12.0 (12.9%)	No data

N/A means production had not begun in 1960

Built for speed (*above*) The TGV (Train à Grande Vitesse) was introduced in 1981 on the Paris to Lyon line. The train's sleek design allows it to have a cruising speed of 265 kph (165 mph) and a top speed of 378 kph (235 mph).

Energy production and consumption (*below*) France is poor in fossil fuels. Nuclear power, based on substantial uranium deposits, accounts for more than half the country's energy production, and industry is dependent on large amounts of imported oil.

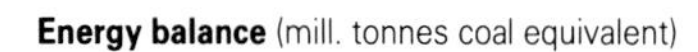

Energy balance (mill. tonnes coal equivalent)

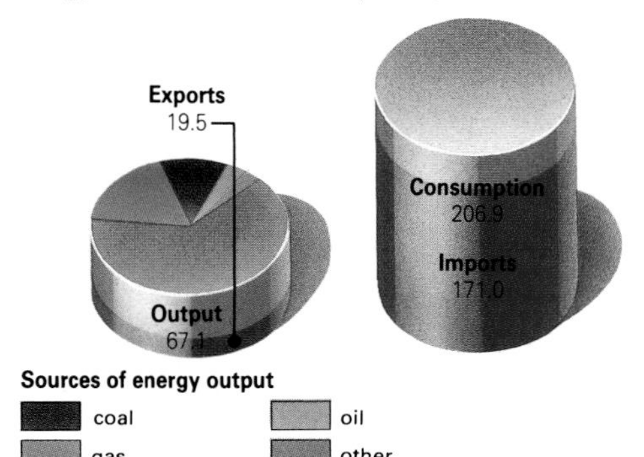

DEPLETION OF RESOURCES

France has a wide range of mineral resources, but many of the deposits are too small or low in quality for large-scale commercial exploitation. The principal exceptions are iron ore, bauxite, uranium, potash and sodium chloride (which is found both as common and as sea salt). Iron ore is found mainly in the eastern province of Lorraine, which once contained the largest iron-ore deposits in Europe. Despite the large reserves, intensive mining has taken a heavy toll, and output fell from 60 million tonnes in the early 1960s to 10 million tonnes in the early 1990s. Although relatively easy to extract, the ore's low metal content has recently rendered it less valuable than high-grade imports from other leading producer countries.

France was once one of the world's foremost producers of bauxite (rock containing aluminum ore), which takes its

Map of principal resources and industrial zones In the past, heavy industry grew up close to local supplies of coal and iron in the north, east and southeast, but the postwar decline in mineral extraction has seen centers shift to other parts of the country.

name from the village of Les Baux, near Arles, in the southwest. Large deposits were first mined here in 1882, but reserves are now very low and nearly half of France's aluminum needs are met by imports. Many mineral resources, such as lead, zinc, gold, silver, copper, tin, antimony, tungsten and vanadium (used to form alloys such as vanadium steel), are mined, but only in very small quantities.

Large deposits of common salt are found in Lorraine and along the edge of the Jura mountains. On the Atlantic and Mediterranean coasts sea salt is produced by a process of evaporation. The chemical and fertilizer industry is supplied with potash from Alsace, while sulfur, another key resource of the chemical industry, is produced mostly from petroleum refining and natural gas.

From coal to nuclear power

Coke-burning blast furnaces for the iron industry were established as early as 1785, but it was not until the 19th century that coal mining took place on a large scale around the Massif Central. By the turn of the century, the northern coalfields of Nord and Pas-de-Calais had become the main centers of production. Output increased after World War II when the industry was modernized, but fell again from the late 1950s onward, replaced by more competitive foreign imports. French coal, lying in thin seams deep underground, is costly to extract.

France's principal natural gas fields – near Lacq at the foot of the Pyrenees – were first exploited in the 1950s but by the 1970s supplies were nearing exhaustion. Almost 90 percent of natural gas is now imported from the North Sea, Algeria and the former Soviet Union. France's reserves of oil are even more limited. Virtually all oil is imported from the Middle East and North Africa.

The region's growing dependence on imported fuels has accelerated the search for alternative energy supplies. Hydroelectric power is important in the Alps, Pyrenees, Massif Central and along the Rhône valley. Nuclear power is another major French industry; France is one of the world leaders in nuclear technology. Uranium deposits in the west and Massif Central supply almost half of the industry's annual consumption.

Food-processing plants are concentrated mainly near the large urban markets or in the agricultural west. Wine, one of the country's leading exports, is produced in greatest quantities around Bordeaux and in Languedoc–Roussillon in the southwest. Dairy products, including cheeses, yoghurt and butter for the international market, are processed in large modern plants in Normandy in northern France.

NEW DIRECTIONS

After World War II French industry underwent a spectacular transformation, helped by massive foreign investment particularly from the United States. France's general postwar recovery was coordinated by an organization called the Commissariat General au Plan. Under the guidance of Jean Monnet (1888–1979), it devised a series of five-year plans, the first of which gave priority to the reconstruction of heavy industry in the region. The results, in terms of overall productivity and competitiveness, were impressive. Industrial production increased by almost 7 percent a year, until the economic recession of the late 1970s and early 1980s – caused by the world oil price rises – called a halt to the period of sustained growth. However, industry has emerged from the recession leaner but fitter; employment – after several years of severe job losses – has stabilized and productivity has increased again.

Winners and losers

Not all sectors of industry benefited equally from the postwar developments. Falling demand, foreign competition and the world recession took a heavy toll on traditional industries such as coal, iron, steel and shipbuilding, with the industrial basins of the north in particular being badly affected.

A large chemistry set *(above)* The Total refinery in the Bouches du Rhône area of southern France refines and processes oil and petroleum into many different types of chemicals and solvents. France is one of the leading chemical manufacturers.

The smell of success *(below)* Perfume is big but expensive business in France. Six million rose heads, pressed and crushed, are needed for each kilo of pure rose essence. 1 kg (2.2 lb) of French-grown Rose de Mai essence costs $16,000 to make.

The textile and clothing industries have also contracted in the face of competition from low-cost foreign manufacturers. The industry is most concentrated in the north, where factories produce wool and weave linen from local Flanders flax, as well as milling jute and cotton. In the south, Lyon has been the center of silk production since the 15th century, now supplemented by the manufacture of rayon and nylon. Foreign competition has stimulated technological improvements but has also resulted in closures, high unemployment, rationalization and a number of mergers in the industry.

The decline of traditional industries has been counterbalanced by a rapid expansion in others, causing a fundamental shift away from heavy industry toward light industry. The construction industry was one of the fastest to expand: the number of houses, hotels and roads that were built between the 1960s and 1980s

soared dramatically, stimulated in part by the growth in tourism. Passenger travel and the transportation of freight were both transformed by the building of the expressway, or autoroute, between Paris and Lyon, and by the success of high-speed trains (Trains à Grande Vitesse or TGV) – the pride of the nationalized railroad industry.

France's participation in setting up the European Community (EC) in 1957 was a major factor in the postwar manufacturing boom. As customs duties on foreign imports were lowered, consumers, given a wider choice of goods, were no longer obliged to buy just French products. Spurred by foreign competition, manufacturers began to produce goods to appeal not only to the home market but also to their European neighbors.

Among the most dynamic and successful branches of French industry today are mechanical and electrical engineering, electronics, vehicles, aerospace and nuclear energy, which were all at the forefront of the postwar technological revolution. Because these industries require high investment, their development has been accompanied by the emergence of large multinational corporations, tempted into the region by state planning and government grants.

The chemical industry, for example, has come to be dominated by a few large corporations such as Rhône–Poulenc or the chemical divisions of oil companies such as Shell or Total. The all-important automobile industry is now controlled almost exclusively by the two giants, Peugeot–Citroën and Renault.

Industrial relocation

Following the recession of the 1980s, the French government issued a range of restrictions and incentives aimed at encouraging industry to relocate from the congested area of Paris to the underdeveloped south and west. The vehicle industry was among those that moved to more spacious locations in the provinces, with room for expansion and a better working environment. Increasingly, production facilities are located in provincial sites, while head offices, research and technical facilities remain near the capital, which is still at the center of decision-making and innovation. Meanwhile areas with high rural unemployment gain from new industrial plants.

Brittany, in the northwest, for example has benefited not only from the arrival of Citroën, but also from the relocation of CNET (the research center of the telecommunications industry). This in turn has attracted numerous smaller companies specializing in electronics and telecommunications. The Pays de Loire, Provence and the Cote d'Azur in the west and south also enjoyed increased prosperity and employment as the result of industrial relocation.

Decentralization has undoubtedly reduced – but not removed – the imbalance between the industrialized center and the underdeveloped regions. However, many companies, rather than moving to more distant areas, have preferred to accept lower government subsidies by relocating on the fringes of the Ile-de-France, the prosperous heart of the country that has Paris at its center.

High flying industry *(right)* One of France's growing aerospace companies, the aircraft engine builders Turbomeca, has delivered more than 27,000 engines to customers in 120 countries. The company's four plants employ about 4,260 workers.

THE AEROSPACE INDUSTRY

The development of Toulouse in the southwest as the national center for the French aerospace and aviation industries has been a particularly notable success. The area now houses important research facilities, such as the National Center for Space Studies (CNES) as well as several civil and military aircraft manufacturers that have participated in prestigious international projects. These include the planning of the spacecraft Hermes; the satellite launcher, Ariane; and the supersonic airliner, Concorde.

Thousands of suppliers are generally involved in an aerospace production program, providing a pool from which skills, facilities and finance can be drawn. The aeronautics industry has attracted numerous small and medium-sized firms, most of them within the high-tech sectors of computing, electronics and robotics.

The French government played a crucial role in the industry's growth. Two of the most important companies, Aerospatiale (aerospace and missiles) and SNECMA (aero engines), are nationalized, while the government is a majority shareholder in others, such as Dassault Aviation, the military aircraft manufacturer.

International cooperation has become increasingly important in an industry where the research and development costs are extremely high and the prospect of commercial returns often distant and difficult to predict. Aerospatiale, for example, was instrumental in setting up the consortium of French, British, German and Spanish companies manufacturing the Airbus family of airliners. Other companies, such as Matra, the missile manufacturer, are also actively seeking cooperation within Europe and with the United States, Europe's principal competitor.

MANAGEMENT AND LABOR

Before World War II French industry was characterized by very many small, family-owned firms employing only a few people. These existed side by side with far larger enterprises owned and run by great industrial dynasties such as Schneiders (metal and iron working), Michelins (tires and rubber) and Pechineys (aluminum and chemicals).

The dominance of these industrial barons resulted in a form of industrial paternalism in which the relationship between owners and workers rested on a tacit pact of mutual assistance. In return for a lifetime of work devoted to the firm, company employees and their families could expect decent conditions of employment and other benefits such as subsidized housing.

Industry's social structure

From the mid 1940s to the early 1970s the French government took the line of biggest is best, encouraging mergers of small and medium-sized companies and offering incentives to diversify or increase the size of plant. This growth process, known as industrial concentration, had a profound effect on the social structure of industry. By the 1960s the great owner-managers had mostly been superseded by state-appointed managing directors who were steeped in the most up-to-date business philosophies and methods.

It was not until the economic crisis of the 1970s – which forced large companies to make cutbacks – that smaller firms and one-person craft enterprises were given the chance to flourish again. During the recessionary period, companies employing fewer than 20 people showed the highest annual growth rate of almost 3 percent, while companies with over 100 employees showed negative growth. However, these trends did not affect all sectors of industry to the same degree. The textile and clothing sectors continue to have a high proportion of small companies, while the chemical industry is overwhelmingly dominated by large companies employing over 500 people.

The sweeping social and economic changes that took place during the postwar period were reflected in structural

The sounds of silence (*left*) Testing sensitive recording equipment and musical instruments is part of an expanding high-tech acoustics industry in which France is building an international reputation. Thousands of highly absorbent baffles from ceiling to floor cut out internal echoes as well as protecting this research laboratory from all outside noise.

Boring work (*right*) The Channel Tunnel is probably the oldest construction project in the world. The emperor Napoleon (1769–1821) drew up the first plans for a submarine link with Britain. After 27 false starts, the tunnel link was due to be completed in 1993. Almost 10,000 French workers have worked on the complicated and often dangerous project.

A cold blast for the steel industry (*below*) French steel workers, used to working with temperatures of up to 1,000 C (1,500 F), are facing the industry's biggest production cool-down. France has a long steel-making tradition, and there are large reserves of coal and iron ore, but foreign competition is reducing orders and jobs.

SILICON VALLEY OF THE RIVIERA

Ever since the 1950s, the policy of the French government has been to encourage scientific and technological research. It has also funded product development in a range of industries with the aim of establishing regional centers of high-tech manufacturing that can compete with the best in the rest of the world.

The "Mediterranean Corridor" – the densely built-up coastal strip linking northern Spain and Italy through France – has become just such a center. The industrial and research estate of Sophia-Antipolis near Nice is a symbol of the area's dynamic development. The estate, which came into being in 1970, has an area of 2,300 ha (5,685 acres) – a quarter of the size of the city of Paris – and houses hundreds of French and multinational companies, mainly specializing in areas such as computing, electronics and robotics, as well as several research laboratories. Sophia-Antipolis provides employment for over 5,500 people – many of them highly-qualified experts from all over the world who have been attracted not only by the prestige of the estate but also by the pleasures of the Mediterranean climate and lifestyle.

The success of Sophia-Antipolis has led to plans to double its size, and there are a growing number of projects for similar science estates in other locations, particularly near cities in the south that have already developed a base of high-tech activity.

changes to the labor force. The devastation of two world wars and slow population growth in the century before World War II created a serious shortage of labor during the period of industrial and economic expansion that followed it. Industrial demand for labor was further heightened by a growing tendency for school-leavers to pursue higher education, and consequently to seek white-collar jobs rather than manual employment in industry.

An influx of immigrants

Large numbers of foreign workers, particularly from North Africa and Portugal, were encouraged to emigrate to France to compensate for the domestic shortfall. By the 1970s, immigrant workers accounted for 10 percent of the industrial workforce, with most of them employed in poorly skilled jobs in the construction, metalworking or mining industries. The increased demand for labor was also a major contributory factor in encouraging growing numbers of women to enter the labor market. Although female employment grew most rapidly within the service sector, there was also a marked increase in light industries such as electronics.

Between 1954 and 1975, the industrial workforce rose by 1.3 million. After 1975, however, employment suffered badly from the effects of the recession. Immigrant workers, who generally lacked adequate legal protection, were particularly badly affected. Between 1974 and 1978 alone, over 1 million jobs were lost. Employees of small companies were the primary victims, but those working for large enterprises were by no means safeguarded either. This was clearly demonstrated by massive redundancies in the steel industry and the collapse of a textile giant, Boussac, in 1978.

The old industrialized regions of the east were the areas most badly affected by unemployment during the recessionary period. By the mid 1980s Lorraine still had one of the highest levels of unemployment in the whole of France. Redundancies were caused by a combination of factors, including labor-saving technological improvements to production methods and the movement of companies away from the heavily industrialized areas to the south and west.

Industrial investors had been attracted to these areas by the large number of people – frequently unskilled – who were in need of jobs, and were ready to accept less generous working conditions than the more unionized workforce of the industrial urban areas. There are three major trade unions in France – the biggest being the communist-led General Labor Confederation (Conféderation Générale du Travail: CGT) – as well as a number of smaller specialized bodies for teachers and managers. However, it is the state, not the unions, that has been most influential in improving conditions for workers. As a result, union membership is very low by European standards: in 1980 only 25 percent of French employees belonged to a union, in contrast to an average of about 40 percent in other Western European countries.

Nuclear power

A nuclear reactor (*above*) at Paluel, on the north coast. Large deposits of naturally occurring uranium have been a major factor in the country's concentration on nuclear power as an energy source. Uranium is milled and concentrated (the ore contains only about 1 percent of uranium oxide). It is then enriched and converted into uranium hexafluoride, the fuel for the reactors.

Nuclear power now supplies almost 75 percent of French electricity – the highest proportion in the world. By comparison, fossil fuels – mainly coal, oil and gas – currently supply about 80 percent of the world's power. Once reserves of fossil fuels are exhausted they cannot be replaced, and the fear of depleting stocks too quickly, combined with increasing demand for domestic and industrial electricity, has prompted many Western governments to develop nuclear power.

It is certainly a more fuel-efficient form of energy – 1 kg of uranium produces as much energy as 3 million kg of coal. Under normal conditions experts claim that it is less damaging to the environment than conventional power sources. However, despite stringent safety precautions, the potential for disastrous accidents remains, as the explosion in 1986 at Chernobyl in the then Soviet Union demonstrated, when radioactive gas was released over a huge area of northern Europe. Methods for disposing of radioactive fuel waste are also a cause for concern to environmentalists.

There are two main types of nuclear power reactors: thermal reactors and fast-breeder reactors. Both types harness the explosive energy released by splitting the nuclei of heavy elements such as uranium or plutonium (a process known as nuclear fission). Essentially nuclear power stations work on much the same principles as power stations that are run on fossil fuels. Heat, in this case from the energy released through fission, is used to boil water, which produces steam. The steam drives a turbine which in turn generates electricity. The basic fuel for thermal reactors, natural uranium, consists of over 99 percent of the isotope V-238 (the form that decays to make plutonium) and less than 1 percent of the isotope V-235 (the form that creates heat as it decays). Many thermal reactors, therefore, use enriched uranium fuel, in which the amount of V-235 has been artificially increased to between 2 and 3 percent.

Over half the reactors in the world, and most of those under construction, are a specific type of thermal reactor that are known as Pressurized Water Reactors (PWRs). Alone, PWRs supply 70 percent of France's nuclear power. As the reactor core heats up, the water circulating through it has to be kept under pressure to prevent it from turning to steam. The pressurized water then passes through a heat exchanger where it transfers its heat to a second water system, and this produces steam to drive the turbines.

Fast-breeder reactors are fueled by a mixture of uranium and plutonium. Plutonium is a radioactive chemical element that is synthetically produced from uranium during fission. In a fast-breeder reactor the uranium and plutonium fuel inside the core is surrounded by uranium to create, or "breed", more plutonium.

The French nuclear program

France was one of the first countries to pursue a nuclear power policy. The determination to proceed with nuclear power was formed as early as 1945, prompted by the country's diminishing stocks of coal and by the expense of foreign oil. It has continued, despite opposition and technical problems over the years.

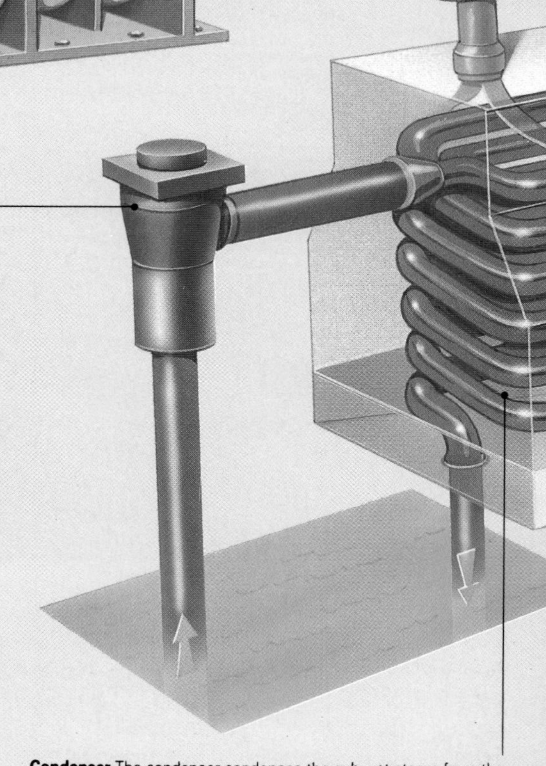

Generator The spinning turbine runs the generator, which produces the electricity

Water source A large water source, a lake or the sea, is needed for the condenser to cool the steam from the turbine. The water circulates in the condenser and is returned to the source

Condenser The condenser condenses the exhaust steam from the turbine into water, which is then pumped back to the steam generators to be converted into steam again. The water in this sealed system is warm and therefore does not need as much energy to convert it to steam

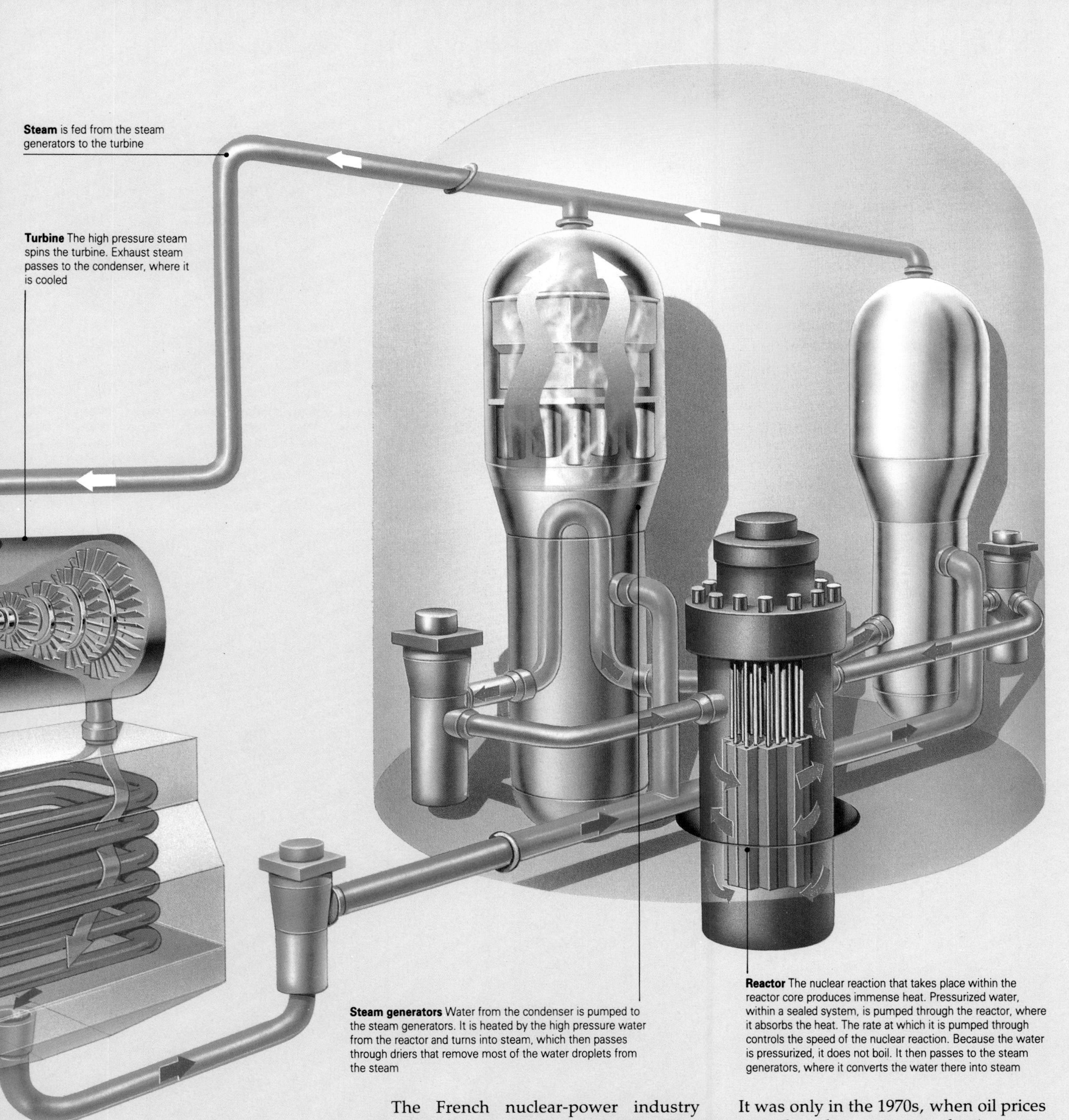

Inside a reactor The uranium fuel is sealed into rods that are arranged in clusters. Control rods between the clusters can either be lowered or raised to regulate the reaction and the power output. The heat is removed from the reactor by a liquid or gas coolant. This heat is then used in a heat exchanger to create steam, which drives a turbine, creating electricity. The whole of the reactor is enclosed in a thick shield of concrete or steel to prevent the radioactivity produced by the reaction from escaping.

The French nuclear-power industry grew fairly slowly at the beginning, mainly because of the extremely high investment and development costs and the experimental nature of the technology. The first gas-cooled, graphite-moderated reactor, located at Marcoule in the southeast, became operative in 1959. However, later reactors of this type failed to live up to the expectations created by the Marcoule experiment.

It was only in the 1970s, when oil prices rose and nuclear-power production costs had fallen, that France's nuclear program accelerated. Increased investment was aimed primarily at PWRs, and the first was built at Fessenheim in eastern France in 1978. The world's first commercial fast-breeder reactor, Super Phénix, was opened at Creys-Malville in the southeast in 1986, although technical difficulties delayed its coming into full operation.

THE INDUSTRIAL GATEWAY TO EUROPE

THE SHIFT FROM COAL TO GAS · CRISIS IN TRADITIONAL INDUSTRIES · HOUSING THE MULTINATIONALS

Although not richly endowed with natural resources, the Low Countries have exploited them efficiently and their manufacturing and service industries today make them prosperous members of the European Community (EC). Coal and iron ore are the traditional mainstays of heavy industry in the region, particularly in Belgium and Luxembourg. However, natural gas, discovered in the Netherlands in the mid 20th century, has since become essential to manufacturing industries across the whole region. Since the 1960s the country has more than compensated for the decline of its iron and steel works by soaring to prominence as an efficient administrative base for multinational companies. Since the establishment of the EC, the Low Countries have effectively become the industrial gateway to Europe.

THE SHIFT FROM COAL TO GAS

Known natural resources and the way they have been exploited have changed dramatically during the recent history of the Low Countries. In the 19th century industrialization was most advanced in Belgium, which mined its rich deposits of coal and iron ore, mostly in the Sambre-Meuse valley and in the southern region bordering on France. Luxembourg also mined deposits of iron ore in the southwest of the country, but little else. Only the Netherlands was believed to have no resources at all, and its economy was based on agriculture and overseas trade.

In 1902 coal was discovered in the South Limburg area of the Netherlands, and state-owned mines (Staatsmijnen)

Energy balance (mill. tonnes coal equivalent)

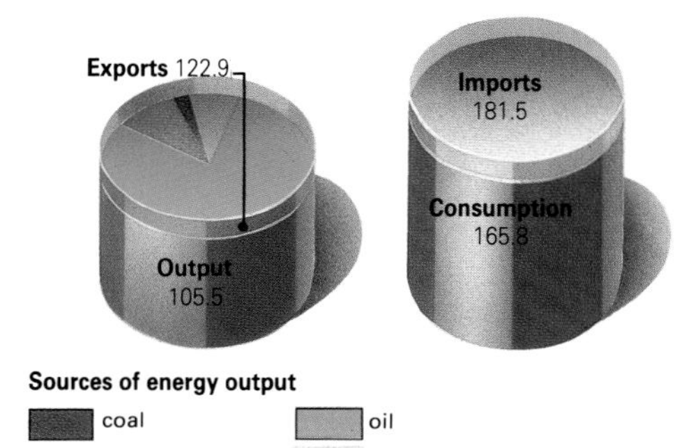

Sources of energy output

coal
oil
gas
nuclear

Energy production and consumption (*above*) Natural gas output is more than 10 times greater than output of any other fuel. It has replaced coal and oil in industrial use, though these continue to be imported. Luxembourg produces significant hydroelectric power, but must import other fuels. Across the region, imports exceed consumption and exports exceed output.

COUNTRIES IN THE REGION

Belgium, Luxembourg, Netherlands

INDUSTRIAL OUTPUT (US $ billion)

Total	Mining	Manufacturing	Average annual change since 1960
123.6	13.7	94.3	+3.0%

INDUSTRIAL WORKERS (millions)
(figures in brackets are percentages of total labor force)

Total	Mining	Manufacturing	Construction
2.7	0.1 (0.1%)	1.9 (18.2%)	0.6 (5.5%)

MAJOR PRODUCTS (figures in brackets are percentages of world production)

Energy and minerals	Output	Change since 1960
Coal (mill tonnes)	3.0 (0.06%)	-88%
Natural gas (billion cu. meters)	60.5 (3.1%)	N/A
Nuclear power (mill tonnes coal equiv.)	15.2 (2.2%)	N/A
Manufactures		
Steel (mill tonnes)	16.2 (2.2%)	+18.2%
Petroleum (mill tonnes)	78.5 (2.8%)	+95.9%
Plastics and resins (mill tonnes)	5.2 (10.3%)	+340%
Dyestuffs (metric tonnes)	498.4 (39.8%)	No data
Carpets (1,000 sq. meters)	382.1 (17.1%)	+18.3%
Cigars (billions)	1.9 (21.6%)	-16%
Chocolates (1,000 tonnes)	350.9 (7.3%)	+48.5%

N/A means production had not begun in 1960

were set up to exploit it on a large scale. A few years later large deposits of salt were found in the east, followed in 1945 by small quantities of crude oil. Finally the discovery in 1959 of what was then the largest known gas field in Europe brought about a revolution in the Netherlands' energy supply and encouraged an increase in manufacturing.

By the mid 1980s the reserves of natural gas in the region were estimated to be the second largest in western Europe. They provided more than 50 percent of domestic energy requirements, and improved the trade deficit. Following the installation of a 10,000 km (6,200 mi) pipeline network supplying power stations, large factories and neighboring countries, natural gas has become the chief energy resource in the region.

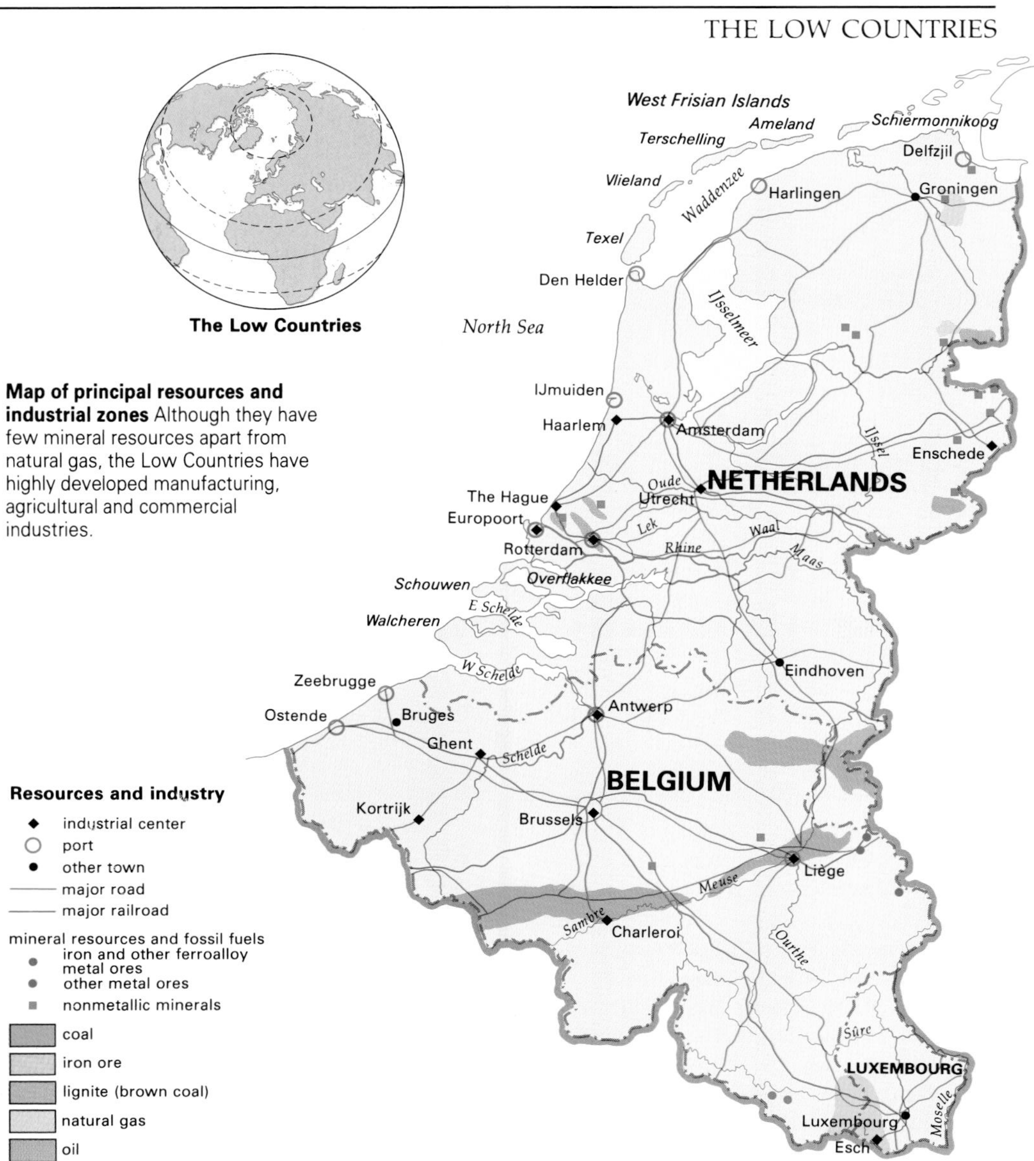

Map of principal resources and industrial zones Although they have few mineral resources apart from natural gas, the Low Countries have highly developed manufacturing, agricultural and commercial industries.

The thriving port of Rotterdam is one of the great gateways to Europe. Rotterdam handles more tonnage than any other harbor in the world, and is the most modern port in the Netherlands. It is also a major oil refining center.

The gas-fired revolution

In 1990 the Dutch company, Gasunie – in which the multinational oil companies Shell and Esso each have a quarter share – produced 75 US billion cu m (2.6 UK billion cu ft) of gas, which was almost equally divided between the home and export markets. The largest amounts of natural gas come from the Slochteren field in the northeast. Gasunie keeps reserves high by adding gas from the Norwegian Ekofisk field.

Increasing dependence on natural gas across the whole region led to the gradual phasing out of coal mining. The coal deposits, once the basis of Belgium's wealth, were, in any case, diminishing steadily. In 1927 yearly output was 25 million tonnes compared with 6 million tonnes in 1983. The Dutch were the first to close their mines followed by the Belgians who shut their last mine in 1983.

Minerals and building stone

Once coal stocks had been depleted, Belgium had few other mineral resources to fall back on. There are small deposits of lead, zinc, copper and antimony in the Ardennes, and the country has significant amounts of building stone, including granite, sandstone and marble. In the absence of local deposits of minerals for commercial use, mineral processing has become important in Belgium, which imports copper ore from Zaire and zinc and lead ores from Peru, Sweden and Zaire. The Netherlands also exploits a surface mineral, Maastricht chalk, whose high lime content makes it excellent for the production of cement. It can also be used for the manufacture of fertilizers. In addition, the Netherlands has some petroleum of its own, but it imports much more; petroleum is the second most important fuel for energy across the region. Most is imported through Rotterdam, Europe's leading port, where a complex of storage facilities, oil refineries and chemical works handles the majority of the petroleum imported for western Europe.

CRISIS IN TRADITIONAL INDUSTRIES

In the latter half of the 20th century the traditional manufacturing industries of the Low Countries began to founder. During the 1960s and 1970s Luxembourg was a significant European steel producer. The industry received government support dominated by Acieries Reunies de Burbach-Eich-Dudelange (ARBED). Luxembourg depended on steel-based heavy industry, international trade and banking for most of its wealth, with approximately one-third of the workforce employed in some kind of manufacturing industry. The year of the world oil crisis, 1973, marked the beginning of a radical change of emphasis. Local deposits of iron ore were running out, and foundries in southern Luxembourg became increasingly dependent on iron ore imported from France. Coke and petroleum, equally important to manufacturing processes, also had to be imported.

Rising prices made heavy industries less competitive, and long-established companies unused to change were unable to compete with the lower production costs of newly industrialized countries. Between 1974 and 1981 industrial production declined by more than one-fifth, mostly accounted for by cutbacks in ARBED. Luxembourg's steel industry came to an end in 1981 when its reserves of iron ore virtually ran out.

Manufacturing industry in Belgium went through a similar crisis in the 1970s and 1980s. The southern region bordering France had become a thriving industrial area on the strength of its iron and steel industry, which until 1974 was expanding and producing a surplus. Belgium produced the most steel per head of population in Europe but it had a narrow home market. In the mid 1970s competition from Third World countries, a stagnant domestic demand and falling productivity due partly to outdated equipment created a slump that affected production, profits and employment. Belgium's largest steel company, Cockerill – one of the biggest manufacturers in the world – was forced to merge with Hainault-Sambre, the government controlling 80 percent of the new company. An expensive restructuring plan was set up to lower capacity, and at the cost of 26,000 jobs it broke even in 1985.

Industry and transportation in Antwerp (*above*) Petroleum is one of Belgium's chief imports, used both for energy and as a raw material to feed the textile and chemical industries. Most oil refineries are located in or near Antwerp. Cobalt, uranium and radium are also important in Belgium's mineral-processing industry.

Fabrique National, Belgium (*right*) Arms manufacture began in Belgium in the 14th century, and in 1889 Fabrique National was set up to supply the Belgian armed forces. Later it supplied arms to NATO, and moved into joint ventures involving American and European aeronautics companies, as well as making tennis rackets and golf clubs.

The oil-importing boom

As traditional sectors declined throughout the region, there was new growth in oil refining, petrochemicals production and manufacturing using imported raw materials. The Netherlands, in particular, has a huge concentration of petrochemicals refineries around Rotterdam. It imports crude oil from the Middle East, particularly Saudi Arabia, and exports refined petroleum products to the European hinterland by means of a vast network of inland waterways. In Belgium too, oil refining and petrochemicals have begun to take the place of coal mining, iron and steel. New manufacturers established plants around the port of Antwerp and surrounding coastal regions of the

Flemish north, well outside the traditional French-speaking industrial zone, which has gone into a relative decline.

By the 1990s Antwerp and Rotterdam were the most important refining centers in western Europe. The storage and refining capacity in Belgium and the Netherlands is much higher than in other EC countries and the region has attracted leading world names to build branches there. Belgium's flourishing chemicals industry, a lucrative offshoot, is dominated by the Petrofina company. Antwerp is the main manufacturing center, but heavy chemicals are also produced in the Sambre-Meuse valley. Two-thirds of the industry's output is exported.

Spreading the manufacturing base

The region is also prominent in food processing (including tobacco), vehicle assembly, electronics, finance and banking. Traditionally Belgium had a thriving textile industry, Flanders (in the west of present-day Belgium) was an important center for woolen cloth in medieval Europe, and the area is famous for its lacemaking, still a leading craft industry. The mainstream textile industry, however, has been overtaken by competition from certain Third World countries where workers are paid lower wages and overheads. The region lags behind its European neighbors France, Italy and Germany in the design and production of high-fashion garments.

New industries, for example, telecommunications, biotechnology, genetic engineering, computer software and robotics are becoming established, taking advantage of the region's central and accessible position in Europe. Although new industries in Luxembourg are on a smaller scale, the country manufactures artificial textiles, plastics, pneumatic tires and fertilizers. At the beginning of the 1990s the government created 5,000 new jobs in Luxembourg through promoting diversification into metalmaking, aluminum and glass production, together with making a significant investment in the chemicals industry.

ARMING THE WORLD

The Low Countries, notably Belgium and Luxembourg, have a long association with the arms industry, which developed alongside production of iron and steel. Luxembourg produces gunpowder and explosives, while Belgium is one of Europe's leading manufacturers and traders of weapons.

Liège, the provincial capital in the east, is acknowledged as the arms center of the world. Fabrique National d'Armes de Guerre, one of the world's most aggressive manufacturers and exporters of small arms, rifles, Browning pistols and machine guns, has its headquarters there. The company makes more than 90 percent of Belgian small arms production exported each year to areas of conflict: the Middle East, Africa and Central America.

The city of Liège has historical links with the industry dating back to the Middle Ages. In the 15th century, Charles the Bold of Burgundy (1433–77) forbade the inhabitants to produce weapons, but they ignored his command, so the city was burned to the ground and its people slaughtered. Liège never had qualms about supporting both sides in a battle. When Fernando Alvarez, the third Duke of Alva (1507–82), attacked the Low Countries in 1567 both the defenders and invaders used weapons made in the city.

HOUSING THE MULTINATIONALS

Extensive involvement in the EC combined with the powerful positions of Rotterdam and Antwerp as ports, and highly developed transportation and banking facilities have made the Low Countries a comfortable base for a number of international companies. The Goodyear Tire Company established a manufacturing base in Luxembourg in 1951, and quickly became the second most important employer after the steel giant ARBED. Shortly after, Goodyear created a European technical research center there, and built a factory in Bissen to produce steel cables for covering tires.

The large oil companies Esso, Petrofina, Tamoil and Nayna/Universal are well established in Belgium, with refineries close to the harbor in Antwerp. In the Netherlands Shell, Esso, British Petroleum and Kuwait Petroleum have refineries around Rotterdam harbor where the Netherlands Refining Company is also based. A French-owned Total refinery in the Dutch seaport of Flushing receives oil by pipeline from Rotterdam and delivers products to the American-owned Dow chemical plant located across the river Schelde at Terneuzen.

Four of the world's largest companies, Royal Dutch Shell, Unilever, Philips and AKZO, also have offices in the Netherlands. AKZO, originally a Dutch company, leads the world in manufacturing synthetic fibers, which accounts for about one-third of its business. Chemicals, paints, plastics, adhesives, pharmaceuticals and hospital supplies make up the remainder. Philips specializes in electrical and electronic equipment, and Unilever, the holding company for more than 500 smaller companies across the world, manufactures and markets soaps, foods and other products for household consumption. The Royal Dutch Petroleum Company is one of two parent companies of the Royal Dutch Shell Group and has offices in the Hague (the other is based in London). The group is one of the world's largest corporations, and has companies in more than 100 countries. As well as their primary involvement in oil, petrochemicals and associated industries, they also invest heavily in metals and coal, and have interests in solar energy, forestry and consumer products.

Wrapping flowers for export (*above*) Less than 5 percent of the Dutch workforce is employed in agriculture, which accounts for 24 percent of its exports. Flowers are often raised in high-tech greenhouses in the southwest, but are cut and boxed by hand for shipment by air to international markets. Germany, France and Britain together purchase nearly 70 percent of the output.

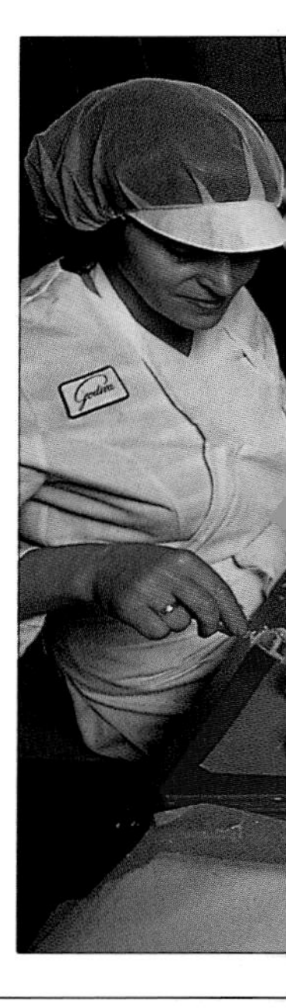

Luxury Belgian chocolate (*right*) From Brussels, hand-dipped Godiva chocolates find their way into gourmet shops all over the world, where they have an affluent and discriminating following. The right to use the word "chocolate" caused great controversy in the European Community when makers of real milk chocolate vehemently advocated that the lower-priced product made with vegetable oil should be called vegolate.

BELGIAN CHOCOLATES

Some of the finest chocolates in the world are made in Belgium. A Belgian chocolate is traditionally known as a praline, a smooth fine paste created when crushed hazelnuts, usually from Italy, are rolled together with sugar so that the natural oils are extracted from the nuts. In part, the quality of the chocolate depends on the smoothness of the paste, and to achieve this a large part of combining the filling with the chocolate covering is still done by hand. The chocolate itself, which comes in dark, milk or white varieties, contains a high concentration of cocoa mass, making it richer and darker than other types of chocolate.

The industry is characteristic of the region in that it uses mostly imported raw materials, often from former colonies, and the product is chiefly exported. Godiva, a Belgian company and one of the most prestigious in the business, uses high-quality cocoa beans from South America and Africa, combining them with thick cream and dairy butter, not vegetable oil. Godiva was established in 1929 in Brussels and is now world famous; it creates a new chocolate collection every six weeks for the luxury gift market. It makes 50 different types of filling, including whipped cream, whole nuts, cherries, grapes and liqueurs, sometimes in distinctive shapes such as the heart-shaped "coeur de Bruxelles." Belgian chocolate firms also pay great attention to presentation, using beautiful boxes and handmade fabric wrappings.

Europe's support network

The new economic growth that followed the collapse of iron, steel and coal has encouraged a thriving import–export business in the region, and developed a sophisticated range of supporting service industries. In the Netherlands, the Randstad area that includes The Hague, Rotterdam, Delft and Amsterdam with its airport, Schiphol, is a new economic center, dense with banking and freight industries as well as manufacturing. Brussels, which has an outstanding transportation network of expressways, railways and waterways has also created a new economic center around its airport, Zaventem. Together with the transportation, storage and administrative facilities located around the ports of Rotterdam and Antwerp, the area forms an inviting "gateway to Europe" that is increasingly popular with foreign companies. American and British oil and petrochemical companies such as Dow, ICI, Hercules and Union Carbide settled there in the 1960s, and two decades later they were followed by Japanese companies.

In 1970, 30 firms accounted for 53 percent of Dutch employment in industry and the country's resident multinationals employed a quarter of the industrial workforce. Service industries provide more than 60 percent of employment in the Low Countries, but these are often closely involved with big manufacturing concerns. The electronics giant Philips, for example, uses transportation and catering services, which are themselves significant employers.

Coping with unemployment

The Low Countries have a generally high standard of living, but in recent years unemployment has become a problem. This is due mainly to the switch from labor-intensive manufacturing industries to highly mechanized and computerized activities. Limburg, a southeastern province of the Netherlands, was designated a "restructuralization region" in 1973 under a policy aimed at establishing new industries, providing retraining for the workforce and an expansion of service sector jobs. This was achieved partly by transferring government departments from the western Netherlands to Heerlen. In spite of the Dutch government's efforts, however, unemployment in South Limburg is running at twice the national average. Labor-intensive cottage and craft industries, such as textiles, have declined significantly, and even in agriculture redundancies have been caused by the trend, increasingly experienced in the West, toward fewer, larger but highly mechanized farms.

Dutch blue and white china Delft pottery is a traditional manufacturing industry in the Netherlands, and is part of a long artistic heritage. The city of Delft, just outside the port of Rotterdam, is well placed to export its traditional blue and white pottery to collectors all over the world.

Belgium is one of the most heavily unionized countries in the West with 70 percent of the 4 million workforce involved, although "closed shop" policies are illegal. Unions are more likely to be organized in major industrial sectors than in the professions or crafts industries and they exert a heavy influence on Belgian politics. Job-sharing and part-time working schemes alleviated unemployment slightly, but in 1983 Belgium and the Netherlands had the highest level in Europe. Strikes during the 1980s were particularly prevalent, but the trend was for demands to move from higher pay toward more job security.

In Luxembourg the workforce was reduced when ARBED, the company dominating the ailing steel industry, had its capacity cut by the government in 1983. Early retirement and the diversion of labor to community programs lessened the severity of redundancies.

Philips – spreading light

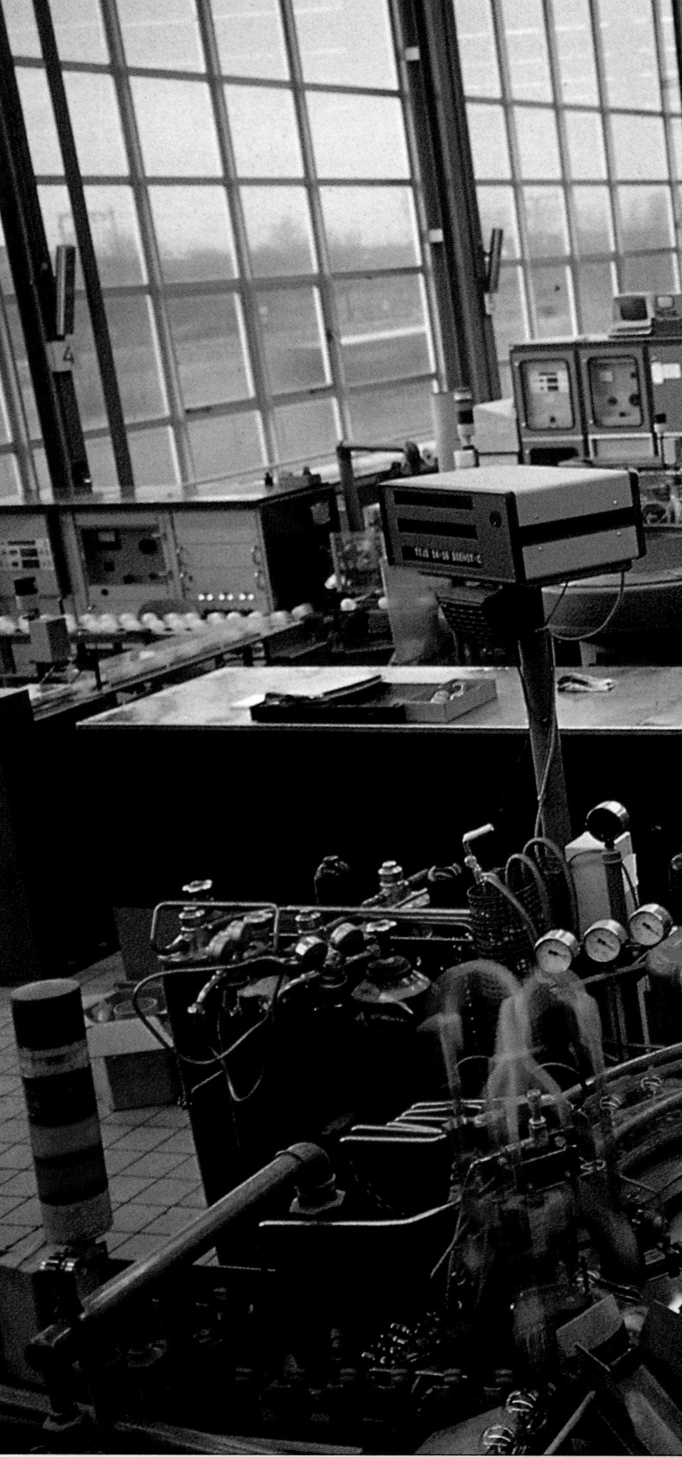

Eindhoven, a commercial and industrial commune in the southern Netherlands, is the home of Philips' Gloelampenfabrieken NV (Philips' Incandescent Lamp Works). The town is unusual in that it is dominated by and is famous for a single industry: its landscape includes Philips' buildings dating from before and after World War II. Offices in the town itself house the company's headquarters, while further divisions are located in towns to the north, at Hilversum, Apeldoorn and Best, as well as in other parts of Europe and the United States.

Philips began in 1891 when Gerard Philips built a light-bulb factory with a start-up capital of 75,000 guilders. By 1895 the company was making its first moves toward a European-wide sales organization for incandescent lamps, and by 1910 it had an enormous workshop where hundreds of female workers mounted tungsten filament inside the new-style tungsten filament lamps. Approximately a century after its creation the company had invented and marketed the highly successful compact disk and was active in almost every part of the electronics industry, supplying schools, the health sector and other industries, as well as manufacturing a wide range of domestic appliances for the private consumer.

State-of-the-art assembly *(above)* As an aid to producing high-tech electrical equipment, Philips supplies its factories with the latest in automated production and quality-control systems. This fully automated assembly line produces power electrolytic capacitors in a factory in Zwolle, in the Netherlands.

It started with light bulbs *(right)* In 1986, when this picture was taken, Philips had been manufacturing light bulbs for 95 years, and the production line was turning out 1.5 billion of them annually. Philips Lighting has sales divisions in 60 countries and its products can be found everywhere, from halogen headlights to the floodlights that illuminate major buildings in cities all around the world.

The constant movement in research and development at Philips' laboratories in the Netherlands has kept the region at the forefront of the electronics industry in Europe, though more recently Japan has dominated the world stage. In 1914 the laboratories began research into incandescent lamps and gas discharges that led to the manufacture of X-ray tubes, then radio and television tubes, and eventually electronic circuits. The first experiments with television began in 1925, and there were public demonstrations of the new technology three years later. It was not until 1950, however, that the first black and white television set reached the market, the same year as largescreen projection in the cinema. The next few years saw the introduction of the long-playing record (and new record players), washing machines, refrigerators irons and vacuum cleaners. Microphones and film sound equipment were the next areas of development, and Philips was involved in producing new designs in all these fields.

A talent for innovation

The company claims to have launched the first electric razor in 1939, the first video cassette recorder in 1971 and the development of an electronic pacemaker and heart defibrillator four years later. The 1980s saw the revolutionary compact disk and the portable liquid crystal display television set, followed in the 1990s by recordable CDs, desktop electron microscopes and digital compact cassettes.

A strong research and development division in Eindhoven employs half of Philips' research staff worldwide and is considered vital to all product divisions. Experience gained in any one project can often be applied in many fields and the company uses it to decide if a new invention is technically and economically feasible. Sometimes research is done in conjunction with other companies, universities and government organizations. An example of such cooperation was the Eureka project for High Definition Television (HDTV) which led to Philips being able to launch its own HDTV set on to the market in 1991. A trend toward the miniaturization of components, products and systems has led to increasing demand for small, portable, battery-powered appliances such as lamps and shavers, and has encouraged the development of more complex sound and vision systems. The company is constantly working on the integration of audio and video equipment, computers, fax machines and telephones, as well as developing systems for electronic banking and shopping, security alarms and teletext.

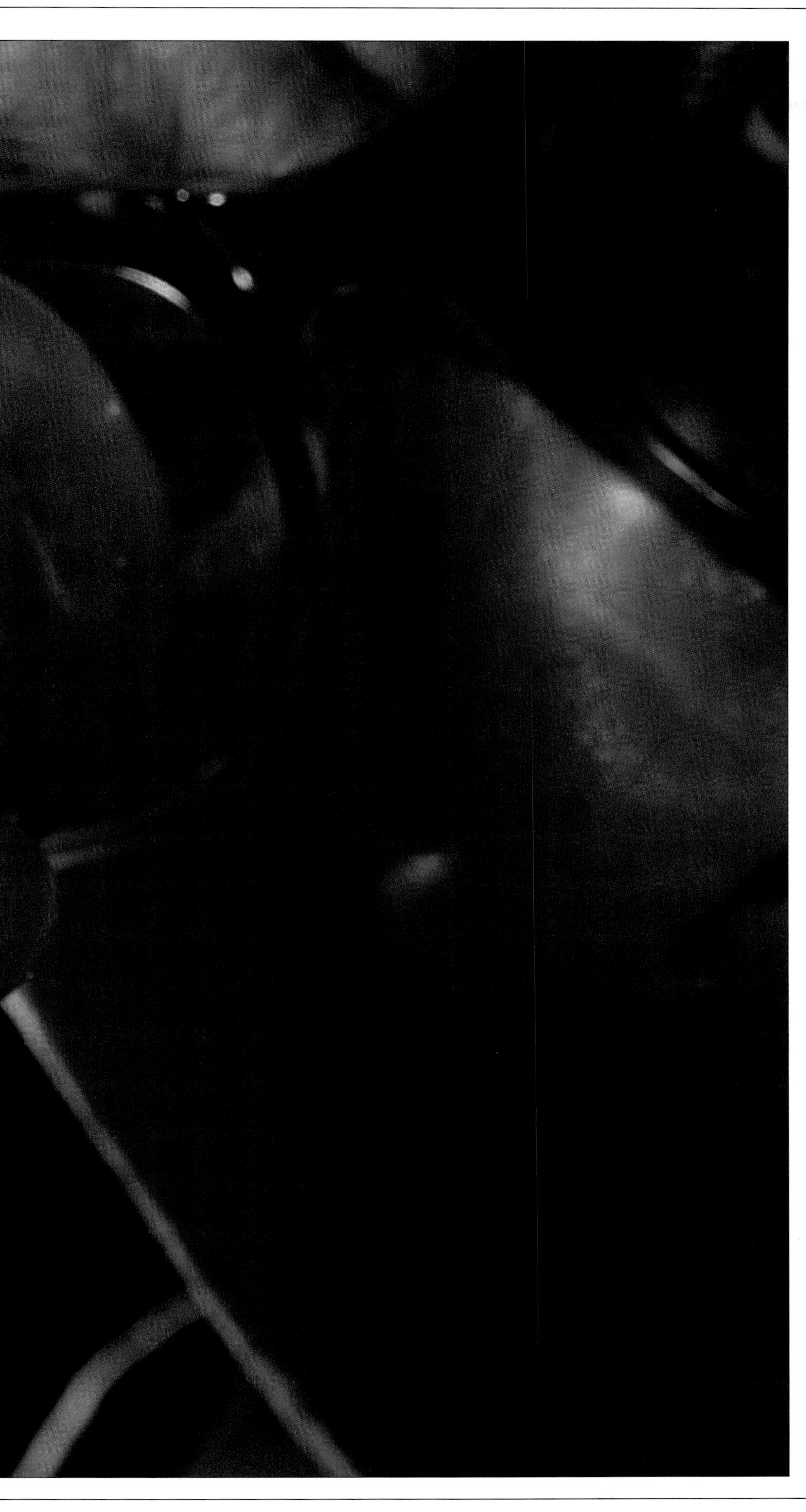

The dirty work of diamond cutting

Antwerp in Belgium is the leading diamond cutting and finishing center in the world. This industry has its roots in Belgium's history as a colonial power in Africa, particularly South Africa, where the majority of diamonds are mined.

Diamonds and other gems are classified and valued on their size, and on the way they refract or bend light from its original path through the air. The way a diamond is cut, and the number of facets (or faces) it has, give the stone its luster – the amount of light reflected from the surface of the gem – and greatly add to its value.

There are five steps involved in making a finished, cut diamond. The first is to mark an uncut diamond in a way that maximizes the number of cut diamonds that can be made from the original stone. Cleavers and sawers split the diamond along the marked lines. Cleaving involves cutting a groove in the diamond and using a chisel and mallet to split the stone: it can take up to eight hours to cut through a large diamond. Before the diamond is sent for faceting it is girdled, or rounded out. Faceting usually requires the work of two people: one to cut the first 18 faces and the other – a more senior brillianteer – to cut the remaining 40 of a standard 58-facet diamond. It is a dirty business, and diamond dust, lead and oil from the sawing and faceting stages cover the brillianteers' hands and faces.

A brillianteer examines a rough cut diamond The two most popular cuts are the brilliant cut (58 facets) and the 18-facet simple cut. Any other type is called a fancy cut.

WINDS OF CHANGE

A LAND OF LIMITED RESOURCES · LATE STARTERS · CHANGING MARKETS AND A GROWING WORKFORCE

The industrial resources of the Iberian peninsula are modest, given its land area and the size of the population. Spain possesses a wide range of minerals, including iron pyrites, mercury, zinc, silver, lead, gold and copper. Energy resources in both countries are limited. Portugal depends on its rivers for hydroelectricity, while Spain exploits local coal and has nuclear power. Industrialization in the mid 20th century was stimulated by state intervention. In Spain this resulted in one of the fastest rates of industrial growth in the world. In 1986 both countries gained full membership of the European Community (EC), and this encouraged significant updating of outmoded manufacturing processes. Craft industries are widespread, and both Spain and Portugal have a thriving tourist industry.

COUNTRIES IN THE REGION

Portugal, Spain

INDUSTRIAL OUTPUT (US $ billion)

Total	Mining and Manufacturing	Average annual change since 1960
143.2	126.0	4.2%

INDUSTRIAL WORKERS (millions)
(figures in brackets are percentages of total labor force)

Total	Mining	Manufacturing	Construction
5.8	0.1 (0.6%)	4.1 (21.6%)	1.6 (8.2%)

MAJOR PRODUCTS (figures in brackets are percentages of world production)

Energy and minerals	Output	Change since 1960
Coal (mill tonnes)	37.3 (0.8%)	+95%
Nuclear power (mill tonnes coal equiv.)	8.8 (2.8%)	N/A
Iron pyrites (mill tonnes)	2.0 (9.2%)	-8%
Mercury (mill tonnes)	1.5 (28.2%)	+30%
Marble (1,000 cu. meters)	0.95 (10%)	+187%
Manufactures		
Cement (mill tonnes)	30.8 (2.8%)	+370%
Steel (mill tonnes)	11.9 (1.6%)	+513%
Sulfuric acid (mill tonnes)	3.7 (2.5%)	+60%
Washing powders (mill tonnes)	1.3 (8.5%)	+115%
Olive oil (1,000 tonnes)	409.0 (25.5%)	-25.6%
Wine (mill hectoliters)	33.6 (11.7%)	-36.3%
Canned fruits (1,000 tonnes)	298.4 (5.8%)	+29%
Footwear (mill pairs)	176.4 (3.3%)	-16.5%

N/A means production had not begun in 1960

A LAND OF LIMITED RESOURCES

Although Spain and Portugal are, in world terms, relatively poor in mineral resources, metals have been mined there for thousands of years. Copper was being mined some 3,000 years ago in southwestern Spain at Minas de Río Tinto, the oldest known copper mining area in the world. It is no longer extracted there, but in Portugal a new mine yielding extraordinarily rich ores was opened in 1988 at Neves Corvo. The world's richest source of mercury is to be found at Almadén in southern Spain, where one mine has been in operation since 400 BC. In 1989 its output ranked second in the world (after the Soviet Union), and reserves are expected to last for another two centuries.

Spain's production of high-grade minerals, such as iron pyrites (for sulfuric acid production), zinc and lead, is high by world standards. However, the extraction of less valuable, low-grade ores – such as antimony, tin, tungsten (which is widely distributed) and iron ore – fell dramatically after Spain and Portugal became full EC members and state protection of these internationally uncompetitive metals was lifted.

Spain has small reserves of gold and silver, which are exploited chiefly for making jewelry and crafts for the tourist trade. Quarrying of rocks is important, especially granite, marble and alabaster for use in the construction industry. The country also has good supplies of potash, used by the chemical industry and in the manufacture of fertilizers.

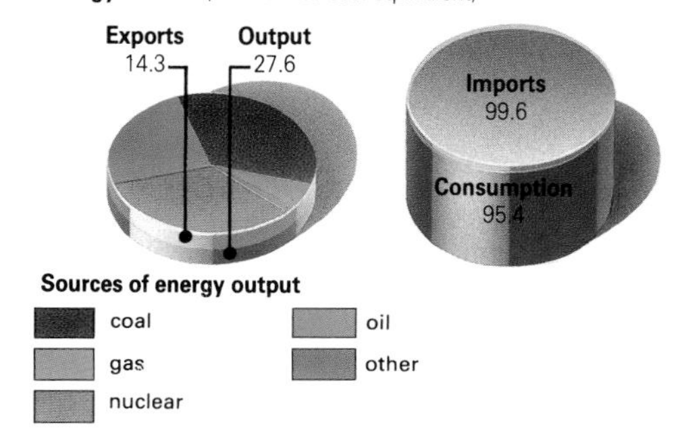

Energy production and consumption (*above*) Since Spain and Portugal have no significant reserves of oil or gas, they import close to 70 percent of their energy needs. Coal, nuclear power and hydroelectricity supply the rest. When imports exceed consumption they re-export the surplus.

Reserves of energy

The region's energy resources have come under ever-increasing pressure following the massive growth in industry since the 1960s. Lacking any significant oil, Spain depends on local coal, hydroelectricity, nuclear power and imported petroleum. Its coal reserves of 700 million tonnes – only 0.1 percent of known world resources – are divided roughly equally between brown coal (lignite) and hard coal (anthracite). At one time Spain had one of the largest coal-mining industries in western Europe, but the production of both types of coal declined in the 1980s, mainly as a result of increased competition from cheaper imported fuels from the United States.

Nuclear power provides one-third of Spain's electricity needs. The nuclear power program, which was accelerated in the 1970s when international oil prices rose steeply, exploits the substantial out-

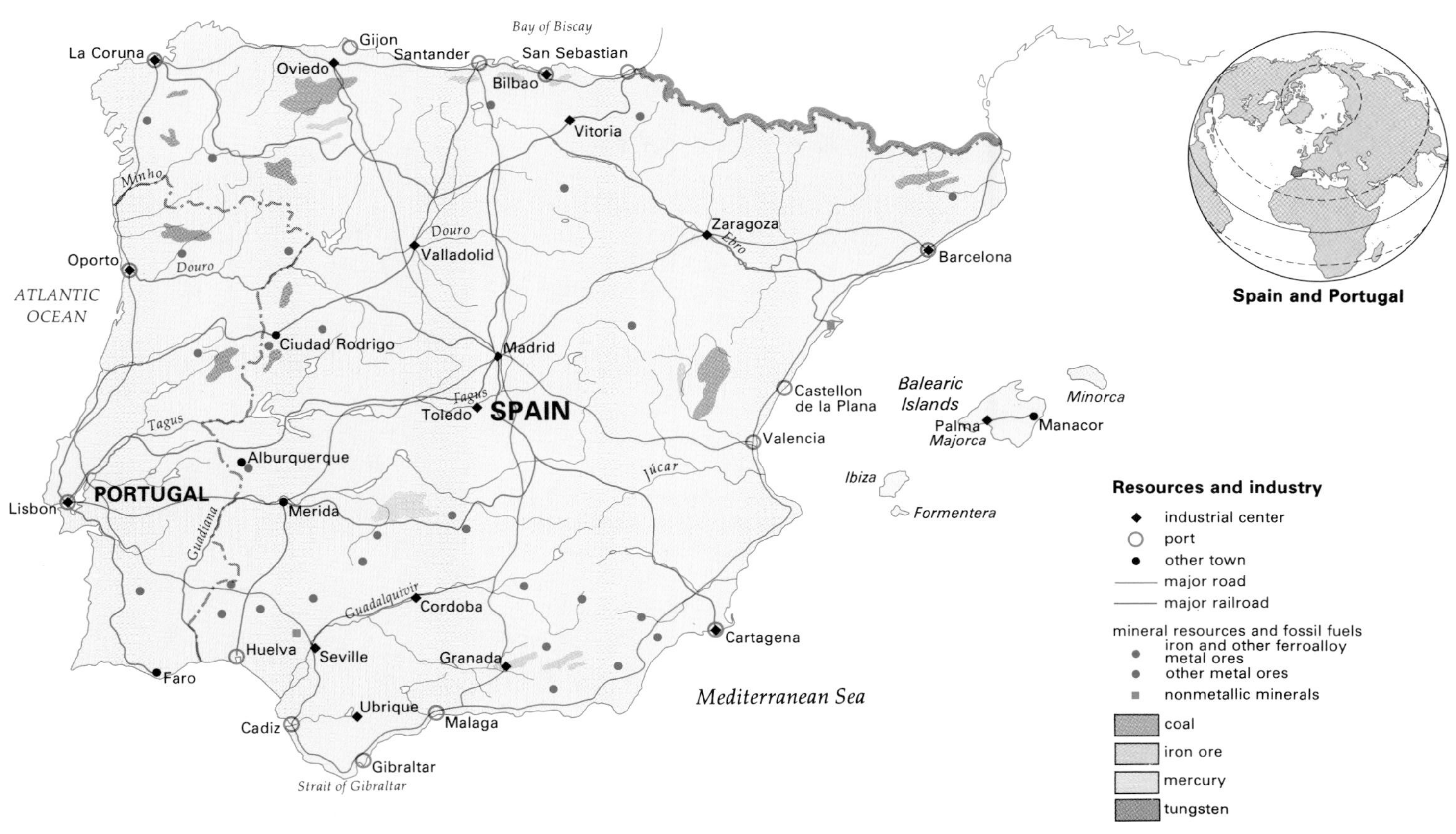

Map of principal resources and industrial zones (*above*) Manufacturing has always been on the move in the Iberian Peninsula. Madrid and the center of Spain attracted industry away from the port cities during the early 20th century, and now provincial towns such as Zaragoza are becoming industrialized.

Mined to exhaustion (*left*) Although the Río Tinto mine was one of the oldest copper mines in the world, it was not until 1868 when the mines were opened up to foreign investment that it became an important source. The mine was formerly Europe's greatest producer of copper, but it is now closed.

put of uranium on the Spain–Portugal border between Ciudad Rodrigo and Alburquerque. Spain previously imported nuclear fuel from the United States, but now concentrated ore is sent for enrichment to the Eurodif consortium plant at Tricastin in France, as Spain lacks a uranium enrichment plant.

Portugal, lacking fossil fuel reserves, harnesses its rivers to generate hydroelectricity. This supplies 46 percent of its energy needs. Spain is less reliant on hydroelectric power, which is severely constrained by the lack of regular rainfall in the peninsula, and by the increasing demands of irrigated farming in the Ebro and Guadalquivir river valleys. In 1989 drought in Spain caused an alarming 53 percent decline in hydroelectric power output. The Spanish and Portuguese governments are beginning to investigate the use of solar power as a source of electricity, as the southern areas of both countries receive constant sunshine.

LATE STARTERS

In 1990 Spain and Portugal ranked between sixth and tenth in world output of a wide range of manufactures, including chemicals, textiles, leather and rubber footwear, clothing, tires, automotive products and cement. High output in products such as coffee extracts, cigars and cigarettes was explained by the continuing influence of trading links with former colonies in Africa, India, Central and South America and the Philippines. Food processing – for example, sardine and tuna canning, olive-oil production, wine (including port and sherry) and brewing – was also important.

However, manufacturing growth had been slow in coming. In common with most of southern Europe, Spain and Portugal entered the industrial era later than their northern neighbors. In the past, both countries relied on cheap imports of raw materials from their overseas empires, rather than develop their own resources for industry. The rigid social structure that concentrated wealth in the hands of a landowning elite stifled initiative and technological innovation, and tied up money in property.

Historical manufacturing centers

Although some manufacturing industry had developed by the late 19th century, it was confined to coastal valleys with access to ports for overseas trade. Textile manufacturing, for example, generally centered around Barcelona and Manresa in northeastern Spain, where imported cotton and Iberian merino wool were processed, served by dyeing and textile machine works. Lisbon, Portugal's principal port and capital, also had some textile manufacturing. Steel, metalworking (including shipbuilding) and papermaking – based on locally produced woodpulp from the Pyrenees – developed around Bilbao in northern Spain, and leather and shoe production was traditionally based around Alicante and Valencia in the southeast.

It was the state, led by General Franco (1892–1975), that provided the impetus for Spain's industrialization in the mid 20th century. Following his assumption of power in 1939, the aim was to achieve military self-sufficiency by supporting the development of national steel, truck, aircraft, weapons, shipyard and chemicals industries. In 1941 Franco founded the National Institute for Industry (Instituto Nacional de Industria or INI) – a state-owned holding corporation – to foster and finance the nation's industries.

Growth and reconstruction

The 1960s and 1970s saw a rapid expansion in manufacturing, bolstered by new export markets, an increase in tourism that boosted local demand for manufactured goods, and rising foreign investment in the region's industries. However, it was dependent on the import of crude oil for local refining, and the sharp rise in international oil prices in 1979 abruptly curtailed the boom. Manufacturing in the five major industrial areas of Madrid, Barcelona, Valencia, Bilbao and Lisbon declined significantly. With the world markets in recession, Spain and Portugal's steel, chemical and shipbuilding industries suffered cutbacks in output. The shipyards had based their success on the construction of tankers; a market that collapsed spectacularly with the oil crisis.

Further rationalization followed when Spain and Portugal joined the EC in 1986, necessitating the removal of import tariffs and other measures to protect domestic manufacturing. In this sharper competitive climate many companies were shown to be inefficient and reliant on outmoded machinery. Industry began to move in a new direction, becoming increasingly dominated by foreign-owned multinationals.

These companies have tended to concentrate their activity in particular areas of the region, depending on the accessibility of resources and transportation routes. Lying at the center of the peninsula's radial road network, Madrid offers easy access to the rest of the region for the internal distribution of goods. There are also extensive international air routes for both passengers and freight. American and French firms predominate among the foreign-owned companies in Castile,

The fruits of the sea *(above)* The well-stocked fishing grounds of the Atlantic have created a thriving industry in Spain and Portugal. Anchovies, sardines and tuna are canned fresh as they are unloaded from the boats.

Shipshape in Cadiz *(right)* Spain and Portugal share long maritime histories and shipbuilding is a well-established industry. Cadiz has one of the few European shipyards capable of building the latest breed of international supertankers.

IBERIA'S BUILDING BOOM

Spain is Europe's fastest growing construction market, with annual growth rates of over 8 percent since 1987. The combination of tourist-related expansion – including major projects such as the 1992 Olympics in Barcelona and the World's Fair in Seville, as well as massive public works programs such as the completion of a national motorway network – has created a record demand for building materials.

Both Spain and Portugal possess plentiful supplies of raw materials, particularly limestone, marble, granite, slate, aggregates (usually sand mixed with cement to make concrete) and clays. Many of Spain's cement works (in which powdered alumina, silica, lime, iron oxide and magnesia are burned together in kilns and pulverized) are found along the coast. This gives easy access to export markets in West Africa, Italy and France. Around one-third of its annual output of about 30 million tonnes is exported, while slate (of which Spain is the world's largest producer), is also exported to the United States and Germany. A quarter of Portugal's slate, most of which comes from around Porto, is also exported.

The begrimed face of industry A coalminer in Asturias, northern Spain. During the Franco regime, trade unions were outlawed in Spain. Although democratization brought greater freedom for workers, it removed the safety network of state paternalism that protected jobs in mining and heavy industry, and unemployment rose steadily throughout the 1980s.

which are chiefly involved in making electrical and electronics products (AT&T, IBM, ITT, Thomson), photographic products (Kodak), vehicles (Citroën, Renault, Deere) and tires (Michelin).

Companies producing chemicals and artificial fibers predominate in Catalonia, northeastern Spain, attracted by its traditional textile and dyeing industry and the plentiful supplies of freshwater (lacking around Madrid) that come from the Pyrenees. Predominant among the multinationals operating there are German and Swiss chemical companies such as Agfa, BASF, Bayer, Henkel, Hoechst, Ciba–Geigy and Sandoz, supported by ICI and Courtaulds (British), Du Pont (American) and Pirelli (Italian).

CHANGING MARKETS AND A GROWING WORKFORCE

One of the region's key resources is its people. During the massive industrial expansion of the 1950s and 1960s, increasing numbers of people, freed from work on the land by the spread of mechanized methods of farming, migrated from rural areas to the newly industrializing towns. A new class of workers was created on an unprecedented scale: skilled and semi-skilled workers, technicians, managers, administrators and salesmen.

However, most local manufacturers were unable to exploit this growing home market for consumer goods because they lacked the capital, new technology or motivation for expansion. The way was therefore open to foreign companies, quick to take advantage of the cheap, growing workforce of young people, and of the advantageous currency exchange, to invest in the region's industries.

One field of manufacturing that has been particularly affected by foreign investment is the automotive industry. In the 1950s, Madrid and Barcelona were the main centers. However, since 1970, the multinational companies chose to locate their manufacturing and assembly plants in a number of different medium-sized provincial cities with sufficiently large skilled or semi-skilled workforces, so that they would not have to compete with each other for workers. Renault's main factory is in Vallodolid (central Spain), Peugeot's in Vigo (northwestern Spain), Ford's in Valencia (eastern Spain) and General Motors near Zaragoza (northeastern Spain).

United States investment *(above right)* In 1907 the first Ford automobile was sold in Spain. In 1920 the company built Spain's first car assembly line and in 1989 a new $125 million plant was opened. Most of the Fiestas, Orions and Escorts made at Ford's Valencia plant are destined for other countries in Europe, with only 150,000 Fords being sold annually in Spain itself.

Salt of the eath *(left)* The marshy terrain of Las Marismas near Cadiz in southwestern Spain is an unlikely place for industrial enterprise. Salt-laden air from the nearby Atlantic Ocean becomes trapped in the swampy valley, evaporates in the heat of the sun and leaves a rich deposit of salt for the local workers to collect using age-old methods.

The removal of tariff barriers that followed Spain and Portugal's admission into the EC altered the nature of the multinationals' operations. Traditionally they had been geared to meeting the needs of local markets, and output was diversified over a whole range of products. With unrestricted access for imported goods narrowing these markets, the multinationals began to develop more specialized production plants to serve the larger European market.

State initiatives

In the past, government intervention in industry, through INI, concentrated on establishing and controlling large capital-intensive state-owned enterprises: the Ensidesa integrated steelworks at Avilés in northwestern Spain, Casa aerospace at Madrid, and the Enusa uranium mining company in western Spain.

INI has had to be more competitive

since Spain joined the EC, but privatization and modernization are proving difficult. In the absence of any large non-INI Spanish firms with the capital for expansion, cooperation has been sought from foreign firms. Casa, for example, set up cooperation agreements with Aerospatiale in France for the Airbus project, and with Rolls-Royce (Britain), Saab (Sweden) and Messerschmidt Bolkow-Blohm (Germany) to assist in the European Fighter Aircraft program.

A more recent development by INI has been to promote industries, such as electronics, that will eventually function independently in the private sector. By 1990 INI owned 52 companies, indirectly controlled 126 and had shareholdings in a further 450. The most important INI enterprises are to be found in uranium, oil and coal production; in the shipbuilding, metals, vehicles, cellulose, paper and power industries; and in electronics, aerospace and weapons technology.

The workforce

Under the Franco regime workers' interests were represented by state-run "syndicates" that also represented the interests of the employers and the state. In 1971 a new law allowed managers, technicians and workers to set up independent associations, but workers' rights remained severely restricted. Militancy among workers increased in the 1970s, especially in the larger cities, and strikes were frequent.

In 1977 independent trade unions were officially recognized by the new democratic government. Although this allowed greater freedoms to workers, they remained dependent on the strength of the union and the size of the company they worked for, and job security became more vulnerable to economic market forces. Under Franco's regime, the state had protected workers' jobs, buying up bankrupt private firms such as Asturias coal mines, shipyards in Bilbao and Cadiz, and truck and bus factories in Barcelona and Madrid.

In the competitive climate of the 1980s, unemployment rose. Some 20 percent of the peninsula's workforce was unemployed in 1990, with half of all youth out of work. Jobs for skilled female workers were particularly scarce. To combat this, more effort was being made to create jobs by stimulating local state initiatives and private enterprise.

FOSTERING LOCAL ENTERPRISE

The intention of General Franco's centralized industrial policy, as exercised through INI, was to stimulate industrial activity in areas where individual entrepreneurship had previously been lacking. The result, however, was to stifle local initiative. Yet in some parts of the country, especially in the north in the Basque Country and Catalonia, where there was a strong tradition of local solidarity and self-reliance, state intervention was successfully resisted. Networks of traders, craftspeople and manufacturers that were operating before the Franco era were in fact able to expand their productivity in the 1950s and 1960s.

The Basque initiative to establish workers' industrial cooperatives was particularly successful. The largest cooperative, the Caja Laboral de Mondragon, began in a small way in the 1940s, centered around Vitoria, San Sebastian and Bilbao in northern Spain. Membership grew from 395 workers and four enterprises in the 1960s to 19,000 workers in 110 enterprises in the early 1990s. Some of the cooperative's enterprises are linked; for example, in the manufacture of parts for, and the assembly of, refrigerators and household appliances. Other enterprises, such as a supermarket chain, are completely separate.

Following Franco's death in 1975, substantial power was transferred from central government to 17 autonomous regions in Spain, including the Basque Country. Many of these regions are now introducing policies, backed by explicit or hidden financial incentives, to foster local entrepreneurs and support and encourage active competition for multinational business.

Traditional handicrafts

Spain and Portugal are renowned for their rich diversity of traditional handicrafts, or *artesania*. Most crafts were originally functional, making handmade tools, farming implements, utensils, furniture and clothing for a local market. Today the artefacts they produce are much more specialized and usually decorative.

Textile handicrafts are widespread. High-fashion garments are produced in Catalonia and Lisbon; Alentejo in southern Portugal specializes in distinctive red-and-green stocking caps for local cattlemen, and in Avila and Granada craftswomen weave fine blankets with regional motifs from local merino wool. Lacemaking is popular in country areas in central and southern Castile.

Leatherwork is generally carried out by small firms with fewer than 10 employees, who produce bags, belts, clothes and shoes. In the small town of Ubrique in southern Spain networks of craftspeople produce high-quality leather goods for prestigious international fashion houses such as Cartier, Givenchy, Gucci, Hermes and Yves Saint-Laurent.

Pottery, ceramics, glass and alabaster craft workshops are found throughout the region. Both Portugal and Spain are well-known for their brightly painted glazed tiles (*azulejos*). Many cities are renowned for their ceramic plates, bowls and vases with distinctive motifs. In some cases – as in Granada, Malaga and Seville – such handicrafts date back many centuries. Metalworking in iron, copper, brass, gold and silver is another age-old tradition, especially in the ore-producing district around Toledo. Here swordmaking – a highly skilled craft – dates back to the 1st century BC, though the elaborately decorated blades are no longer made for warfare, but for fencing and as curios for tourists. Wrought-iron screens are the speciality of Salamanca and around Granada traditional crafts include copper utensils, lanterns and metal candelabra.

Woodcrafts are less widespread but very distinctive, especially in the south. Perhaps the most outstanding products are the classical guitars made in Granada and the finely crafted fans of Seville.

Survivors from the rural past

A combination of factors account for the survival of the skills needed to produce these diverse traditional artifacts. Of fundamental importance is the geographic isolation of many communities. Thousands of villages and small towns in the vast rural interior are cut off from major

Chair man (*left*) Although Spain has a long history in furniture making, it never developed a specific school or recognizable hallmark of its own. Furniture in Spain always appeared in foreign form, though rendered in a more masculine and vigorous style. Spanish craftsmen did develop two ornamental flourishes: the baroque Spanish Colonial and patterned Arabesque are both used by carpenters today.

Dinner service industry *(above)* The distinctive porcelain and ceramic plates of the peninsula originated during the Moorish occupation in the 13th and 14th centuries. The southern ports of Malaga and Granada still produce traditional patterns mainly for tourists.

Beating steel into swords *(left)* This blacksmith in Toledo still crafts swords in the traditional way. Although modern weaponry has limited his customers, bullfighting matadors and ceremonial sword enthusiasts continue the tradition.

transportation routes or the newly industrialized towns where cultural change has been great.

Since the era of mass standardized production dawned much later in Spain and Portugal than the rest of Europe, the handicraft tradition withstood the pressure to modernize more successfully. Added to this, the craft industries were protected by state policies that concentrated on developing the energy, metals, chemicals and engineering industries. In both Spain and Portugal distinctive cultural traditions such as bullfighting, local fiestas and religious festivals create a large market for specific custom-made articles. Traditional costumes and handmade musical instruments such as the guitar are still necessary accompaniments of *flamenco* dancing, for example.

But it is the phenomenal growth of tourism in both Spain and Portugal that has given the greatest lift to the handicraft industries. The spread of tourism to even the most remote interior villages has provided new market outlets for many declining communities. Along the holiday coasts, traditional pottery wares are piled high on roadside stalls. And in the cities, tiny workshops give added color and character to many old quarters, providing an extra attraction for tourists. In Spain, government protection for the craft industry has been strengthened. Subsidies from the EC have allowed some of the autonomous communities, such as Andalucia, to support traditional industries by offering rural communities easier access to capital, as well as information on production techniques, organization and marketing. Increasing opportunities are being found to sell products abroad by establishing links with retail groups in the rest of Europe.

Age-old processes

Like champagne, sherry has a long history and a jealously guarded name. Grown in the chalky soil of Jerez de la Frontera, the palomino grape flavors the brandy-fortified sherry with its characteristic nutty, sweet but dry taste. Only sherry from the Jerez region, near Cadiz in southwestern Spain, can be called sherry. Sherry's official controlling body, the Consejo Regulador, can force other producers outside the region to classify their wines as sherry-type or regional sherry wines.

Two unique and ancient processes give sherry its distinctive character. Vintners use a strain of yeast that promotes a growth of mildewlike bacteria called *flor*, which ferments all the sugars in the wine into alcohol, making it very dry. Young sherry is sorted into groups according to its type of *flor* growth.

Unlike other wines, sherry has no vintage. This is due to a special blending process called the *solera* system. Once sorted, the sherry is placed into barrels stacked one on top of the other, up to eight high. The barrel at the bottom is the oldest sherry, and is used for bottling. When the bottom barrel is one-third empty, it is filled up by the barrel above it, which in turn is filled by the one above it, and so on. This enables the old sherry to educate and refine the young sherry, ensuring that the same quality wine is produced year after year.

The Consejo strictly limits the amount of sherry released each year. No more than 40 percent of a vineyard's stocks may be sold, keeping production and the amount of wine exported constant over the years.

An ancient craft A cellarman still uses time-honored testing methods to ensure his sherry retains its characteristic taste and quality.

LATE DEVELOPERS

LEAN RESOURCES · MANUFACTURING DIVERSITY · RISING TO NEW CHALLENGES

Few regions of Europe are as poor in minerals and fossil fuels as Italy and Greece. This hampered their industrial development until well into the 20th century, though the intellectual climate in Italy since the Renaissance has encouraged entrepreneurial flair and innovation. Greece, by contrast, suffered 400 years of stagnation under the Turkish occupation, and industrialization did not really begin to take off until the 1950s. Both nations rely heavily on imported fuels and materials. Greece has natural resources of coal and bauxite, while Italy has gas and hydroelectric power. People are important assets – Italy and Greece are rich in cheap labor and skilled artisans. The beauty of the region and the attractive climate have helped create a significant tourist industry, which is a major employer in both countries.

COUNTRIES IN THE REGION

Cyprus, Greece, Italy, Malta, San Marino, Vatican City

INDUSTRIAL OUTPUT (US $ billion)

Total	Mining	Manufacturing	Average annual change since 1960
301.9	0.9	256.0	+3.3%

INDUSTRIAL WORKERS (millions)
(figures in brackets are percentages of total labor force)

Total	Mining	Manufacturing	Construction
7.9	0.03* (0.0%)	5.7 (20.5%)**	2.1 (7.4%)

** Mining in Greece only ** Includes mining in Italy*

MAJOR PRODUCTS (figures in brackets are percentages of world production)

Energy and minerals	Output	Change since 1960
Coal (mill tonnes)	47.4 (1.0%)	+1758%
Natural gas (billion cu. meters)	17.0 (0.9%)	N/A
Bauxite (mill tonnes)	2.5 (2.6%)	-11.2%
Abrasives (mill tonnes)	7.3 (60.3%)	-15%
Marble (1,000 cu. meters)	3.6 (38%)	+50%
Magnesite (mill tonnes)	1.1 (6.3%)	No data
Manufactures		
Steel (mill tonnes)	24.7 (3.4%)	+300%
Automobiles (mill)	2.1 (4.5%)	+353%
Refrigerators & washing machines (mill)	8.5 (8.4%)	+119%
Footwear (mill pairs)	371 (8.3%)	-25.3%
Leather goods (mill sq. meters)	144.7 (22.5%)	-5%
Wine (mill hectoliters)	68.6 (24%)	-23.8%
Olive oil (1,000 tonnes)	758 (47.3%)	-3%

N/A means production had not begun in 1960

LEAN RESOURCES

The use of metals has a long history in the region. Knowledge of ironmaking technology, which contributed to the spread of the Greek and Roman civilizations around the Mediterranean, emerged about 3,000 years ago. Long before this weapons and tools had been forged from bronze, an alloy of copper and tin. The main source of copper in classical times was the island of Cyprus, and copper (*cuprum* in Latin) literally means Cyprian metal. Silver and gold were also important – the wealth of Athens in the 5th century BC was based on silver mines, worked by slaves. Later, the Romans improved methods of smelting iron, copper, lead and tin.

Nevertheless, neither Italy nor Greece can boast rich mineral resources, and those deposits that they have are often of poor quality, or uneconomic to extract. Italy has large metal-based industries, but relies almost entirely on imported raw materials, since the output of iron ore from the island of Elba, and of antimony, copper, mercury, lead, zinc and bauxite from small, scattered deposits has virtually ceased. Both countries mine limited quantities of silver and iron pyrites, and in addition Greece has deposits of chrome and nickel. Most Greek metals are exported, including high-grade iron ores from the Cyclades (where output had halved to 440,000 tonnes a year in the early 1990s), because of the lack of local processing or smelting facilities.

Bauxite, yielding aluminum, is the most valuable metal resource in Greece; reserves are estimated to exceed 500 million tonnes. Its current annual output of 2.5 million tonnes, from mines at Itea near Delphi on the Gulf of Corinth, rivals that of Yugoslavia and Hungary, Europe's other leading producers. Most bauxite is exported to the former Soviet Union in exchange for natural gas, but the French-owned firm of Pechiney is converting increasing amounts into aluminum at a nearby processing plant.

Energy questions

The region only has reserves of brown coal and these are quite small. Italy's mines, located in the upper Arno valley in the center of the peninsula, rarely yielded 2 million tonnes a year; today mining is virtually defunct. Greece is better placed with reserves estimated at 3 billion tonnes. Annual production in the early 1990s had risen to 50 million tonnes, spurred on by sharp rises in the price of imported oil.

Infinitesimal amounts of oil and natural gas are drilled in the Greek sector of the Aegean sea off Thasos. In Italy, however, oil production – mainly in Sicily – rose from 13 to 35 million barrels during the 1980s. Yet this covers only 5 percent of what is consumed by Italy's oil-refining industries, among the largest in Europe. The rest is imported, much of it from nearby reserves in the Middle East and Northern Africa. Natural gas output, from reserves in the Po valley, Adriatic Sea and Sicily, is more substantial, providing about 30 percent of requirements.

Hydroelectricity contributes 20 percent of Italy's electricity supplies. At first glance it is surprising that Greece, with its extensive mountains, derives only 10 percent of its power requirements from rivers. But the potential for hydroelectricity remains untapped because of the high risk of earthquake damage.This risk is why the Greek government canceled a nuclear-power project in the 1980s. Italy

Resources and industry

- industrial center
- port
- other town
- major road
- major railroad

mineral resources and fossil fuels

- iron and other ferroalloy metal ores
- other metal ores
- nonmetallic minerals
- bauxite
- mercury
- sulfur

Italy and Greece

Map of principal industrial and economic zones
Industry is concentrated in northern Italy, with very little manufacturing elsewhere. Bauxite, used to make aluminum, is a major resource in Greece.

Energy balance (mill. tonnes coal equivalent)

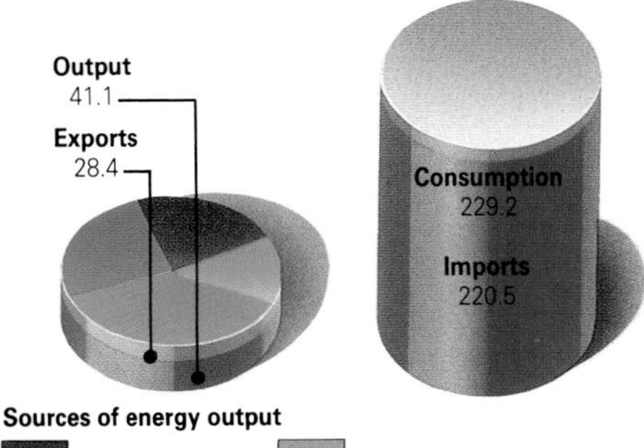

Sources of energy output

coal
gas
oil
other

Energy production and consumption *(above)* Italy and Greece produce little of the energy they use. Few natural resources and the threat of earthquake damage to dams or nuclear power stations mean that industry has to rely on imported energy.

Losing its marble *(left)* The quarries at Carrara, Italy, have always been active, but never so busy as in the last two decades. More marble has been mined in 1970–90 than in the previous 100 years. At this rate, there could be shortages early in the 21st century.

has also halted nuclear-power production, but on environmental grounds.

Resources from the land and farming

Marble is limestone that has been altered by heat and pressure from geological activity to take on a hard, crystalline form. It may be white, red, green, blue or black, or have colored veining, and can be polished. Both Italy and Greece contain rich outcrops of marble – the most famous are the pure white marbles of Carrara and Massa, in central Italy, which were used by the Renaissance sculptor Michelangelo (1475–1564) for many of his statues. It is still highly prized across the world. Since classical times marble has been a valued resource for building and remains so today – Italy leads the world, and Greece is fifth, in quarrying marble. The industry enjoyed a boom in the 1980s as international business corporations around the world wanted to use it as a building material for prestigious new headquarters.

About 60 percent of world supplies of pozzolan (a volcanic slag used to strengthen concrete), and of pumice and cinder (used for polishing and smoothing) are quarried in the region. In addition, 3.6

million tonnes of rock salt are mined each year in Sicily for use in the chemicals industries, though exploitation of the island's sulfur deposits is declining. The Greek island of Euboea supplies 1.1 million tonnes of magnesite – the principal source of magnesium, used in the making of heat-resistant ceramics.

Well-developed agriculture supports a flourishing food industry. Together Italy and Greece are the world's biggest producers of wine, and are second only to Spain and Portugal as producers of olive oil. Pastas, biscuits, ice cream, canned fruits and vegetables, especially tomatoes, and (from Greece) dried fruits and prepared tobacco leaf are all exported.

MANUFACTURING DIVERSITY

An economic chasm divides the region. Although industrial output exceeds US$300 billion, placing the region sixth after the United States, Japan, Germany, France and the Soviet Republics, it is concentrated in northern Italy. The area is the home of the Renaissance, that period of intense artistic and intellectual activity in the 14th and 15th centuries which had its roots in the culture of Italy's city-states. Today, however, northern Italy is a dense network of urban centers and transportation systems that connect the area with major European markets. In these conditions modern manufacturing has flourished, and northern Italy is a world leader in high-quality, fashion-designed and skill-intensive products in a number of fields.

Southern Italy (known as the Mezzogiorno) and Greece are poor by contrast. Fragmented by mountains and islands, they lie remote from markets and still suffer from the rigid social structures of the past. Most industrial activity is located in or around sprawling ports and cities such as Athens, Piraeus, Salonika, Bari or Naples. Greek industry is dominated by family-owned workshops, and consequently processing techniques are much simpler than in Italy.

The Greek government has subsidized some industrial expansion, such as energy production and chemicals in the port areas of Athens and Salonika. There is a Japanese-owned steel plant in Salonika making rolled goods and galvanized sheets. In the engineering field, a number of foreign-owned companies assemble a relatively small number of products, mainly from imported parts.

Italy has a large chemicals industry centered in the Po valley. Originally established in the 19th century to make copper sulphate sprays for vines from locally mined copper, it now manufactures fertilizers, synthetic fibers, rubber, plastics and other products from imported oil. These processes are assisted by nearby reserves of natural gas. In recent years, generous regional aid has persuaded private and state firms to

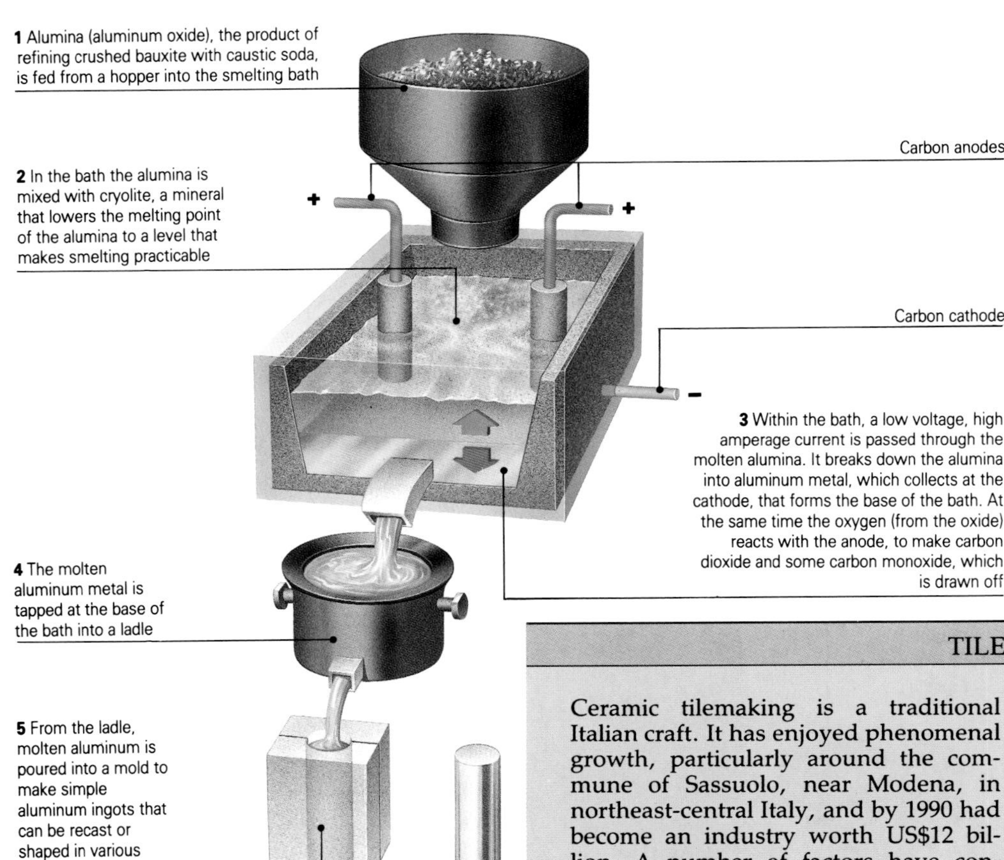

Aluminum smelting The third most abundant element, aluminum never occurs naturally in metallic form. It has to be refined from other ores, and the most important source is bauxite. Greece is Europe's biggest producer of bauxite. Until recently the high energy costs of producing aluminum forced Greece to export most of its bauxite, but it now has several aluminum smelters near cheap local power sources. In the smelting process, bauxite ore is first refined by crushing to form alumina. This is heated, and a high electric current passed through the molten metal to convert it to aluminum. Lightweight, resistant to corrosion, tough, malleable and an excellent conductor of heat, aluminum has countless industrial uses.

TILE STYLE

Ceramic tilemaking is a traditional Italian craft. It has enjoyed phenomenal growth, particularly around the commune of Sassuolo, near Modena, in northeast-central Italy, and by 1990 had become an industry worth US$12 billion. A number of factors have contributed to this success. A rapid rise in construction in the 1960s escalated demand for tiles, and wealthy local entrepreneurs were encouraged to invest capital in the industry.

There has been sustained innovation in process technology, often adapting ideas from glassmaking. For example, the introduction of a rapid single-firing process using roller kilns shortened the batch-production time from 16–20 hours to 55 minutes, slashing heating costs. Local natural gas is used to fire the kilns, and improved technology also allows cheaper local red clays to be used. Great emphasis is placed on design. Motifs range from reproductions of the paintings of the great Italian masters such as Leonardo da Vinci (1425–1519) to post-modernist designs.

Today a multitude of small firms in and around Sassuolo are responsible for about 25 percent of world decorative ceramic floor and wall tile production, and for about 50 percent of world exports. This typifies the recent pattern of industrialization, in which the balance is shifting to the "Third Italy" – that area of the central northeast that lies between the traditional industrial heartland of the northwest, now partly declining, partly revitalizing, and the stagnating Mezzogiorno.

The science of appliances *(above)* Famous for quality and durability, Italy's domestic appliance industry places great emphasis also on style and design. The development section at Zanussi, which makes washing machines, refrigerators and freezers, employs as many designers as engineers.

Sole men *(below)* Florence is the capital of the Italian shoe industry. Smaller factories and shops build up a loyal clientele, and craftsmen keep the shoe-shaped blocks (or lasts) of individual customers. Italian shoe styles are usually very narrow.

locate massive chemical-industry complexes in the close vicinity of ports in the Mezzogiorno, Sardinia and Sicily.

Steelmaking, which has grown rapidly since the 1950s, is also found near major ports. Genoa and Leghorn on the northwest coast, and Naples and Taranto in the south are able to take advantage of cheap imports of high grade iron ore from Northern Africa and coking coal from the United States. Low manufacturing costs have sustained the competitiveness of Italy's steel-using industries, including shipbuilding in Genoa and Trieste on the north Adriatic coast.

Engineering with passion and flair

Italy's rise to global eminence in the field of engineering was a product of Italian ingenuity and design, inspired by a national passion for sleek automobiles and domestic gadgets. Even before 1914 Italian manufacturers specialized in luxury and sports automobiles – an interest that companies such as Ferrari, Lamborghini and Maserati keep alive.

The automobile industry today – at present ranked fifth in the world – is dominated by FIAT (Fabbrica Italiani Automobili Torino; the Italian Automobile Factory of Turin). It makes elegant, compact, fuel-efficient cars for the popular market, as well as the more expensive Alfa Romeos and Lancias, and buses, trucks and tractors. It also produces machine tools and robots for automotive and other manufacturing applications. Nearly all of FIAT's plants are clustered around Turin, Milan and Modena in northern Italy. However, several branch assembly and component factories were opened in the 1970s in southern Italy and Sicily to tap cheap, unemployed labor and regional grants.

Since the 1950s the production of domestic products, such as refrigerators, washing machines, dryers, cookers and dishwashers, has developed rapidly. Companies such as Zanussi export electrical goods throughout Europe, though mostly now under the control of foreign-owned corporations such as Electrolux (Sweden) and Philips (the Netherlands). Olivetti, making electronic typewriters, calculators and computers, is likewise allied to a number of companies in the United States.

Fashion leaders

For centuries Italy has manufactured textiles, clothing, shoes and leather goods. Woolen yarn and fabric factories in Prato and leather tanneries at Santa Croce sull'Arno, both near Florence in central Italy, have been renowned since medieval times. Initially geared to local needs, production soared with the introduction of manufacturing machinery, making Italy a leading exporter of fabrics and clothing. Yet these industries remain the preserve of a myriad of small family firms situated in a number of towns, mainly on the northern edge of the Po valley.

The introduction of cheaper imports from Asia and Eastern Europe led to decline, but a shift into the manufacture of advanced computer-controlled textile and sewing machines has revived the fortunes of some companies. At the same time the production of cheap knitwear, clothing and footwear has moved to areas of northeastern and central Italy where there is a greater supply of labor, often employed as outworkers. Outworkers are also used in Greece, especially by foreign firms supplying department stores in Germany and the Netherlands.

RISING TO NEW CHALLENGES

Industrial development in Italy and Greece has had to overcome the handicap of niggardly natural resources. In addition, rugged terrain and – in Greece – numerous islands are a hindrance to transportation. These difficulties also add to the time and costs of moving goods and equipment. The success of northern Italy in meeting these limitations and challenges owes a great deal to individual skills, determination and entrepreneurial flair for business.

Italy's industrial wave

The fall of Constantinople, the capital of the Greek-speaking world, to the Turks in 1453 not only spelled an end to Greek commercial and industrial enterprise for 400 years, but also helped indirectly to bring about the conditions that fostered the modern industrial vitality of northern Italy. It was the removal of classical manuscripts from Constantinople to centers of learning in Italy to save them from destruction that helped in part to stimulate the climate of scientific inquiry and discovery that emerged with the Renaissance. During this time northern Italy came to lead the world in glassmaking and shipbuilding (Venice) and in leatherwork (Florence).

The lack of mineral resources, particularly coal and iron ore, stopped Italy from joining in the Industrial Revolution that swept through Western Europe in the 18th and 19th centuries. Italy entered modern manufacturing with the technologies of the "third industrial revolution", when the building of the first hydroelectric power stations in the north allowed it to join the race against the industrial giants of Britain and Germany in such areas as steelmaking, railroad engineering and shipbuilding. Three leading industrialists opened the way to Italy developing a particular expertise in the production of automobiles and electrical goods: Giovanni Agnelli (1866–1945), Giovanni Pirelli (1848–1932) and Camillo Olivetti (1868–1943).

Giovanni Agnelli – an industrialist who championed the automotive industry as a means of providing cheap transportation for the masses – founded FIAT in 1899. By 1938 its Turin plant was employing 30,000 workers making streetcars, trolleybuses, railroad engines, automobiles, tractors and airplanes. Giovanni Pirelli opened one of Europe's first rubber factories in Milan in 1872. By 1884 he had pioneered a method of producing insulated electric cables, and by 1899 he had introduced commercial tire manufacture. An electrical engineer by training, Camillo Olivetti began making typewriters in 1908 to speed office work. In the 1920s he revolutionized his manufacturing operation by applying American production-line methods and introducing aesthetically designed products, and so the company grew to become Europe's largest manufacturers of office equipment.

Today this spirit of entrepreneurialism and innovation is kept alive by a rising number of small, often family-owned businesses producing a variety of consumer goods, and gaining ever greater

A national passion *(above)* Pasta is an essential part of any Italian meal. It is made of a dough of durum wheat that is rolled or molded into flat ribbons, hollow tubes, or a variety of shapes, and then dried. Italy has several pasta regulatory bodies, and designs for new shapes must be accepted by each one. Engineering skills have even been applied to pasta design to ensure that the separate pieces collect an even coating of the sauce that usually accompanies a dish of pasta.

Southern migration *(left)* Northern Italy has traditionally been the country's industrial engine, but since the 1950s Italian governments have helped to finance new developments in the south, such as this electronics plant in Bari.

GREEK SHIPPING – A STORY OF OPPORTUNITY

For a small country, Greece has a very large merchant fleet – seafaring has always been fundamental to the livelihood of a state made up of so many islands. Without a fleet, trade could not have flourished in classical times between the Greek colonies that spread around the Mediterranean from Spain to Egypt, Syria and the eastern Black Sea. Since then, periods of maritime prosperity alternated with periods of eclipse by invaders and pirates, until the relocation of world trade to the Atlantic ports brought a longer decline.

The opening of the Suez Canal in 1869, linking the Mediterranean to the Red Sea and the Indian Ocean, brought new opportunities for expansion, and by 1938 the Greek merchant fleet had increased to 1.8 million tonnes; growth was rapid after World War II. Shipowning dynasties such as the Chandris, Livanos and Onassis families tapped the international boom in cargo, offering cheap carrying services between ports around the world. They were also quick to take advantage of increases in oil shipments from the Middle East and the rise in holiday cruise traffic.

The 1980s saw a definite slump. Many lines lacked the capital for new vessels and therefore bought secondhand in depressed markets. But this restricted their ability to supply the fast, scheduled container services that the international trade increasingly demanded. Real or perceived problems with government, labor and the infrastructure of industry in Greece encouraged many shipowners to register their fleet under flags of convenience. With the installation of a new telecommunications system in Piraeus this began to change in the 1990s.

presence in the export field. In the food industry, for example, the Campari, Cinzano, Martini-Rossi and Metaxa families have acquired worldwide reputations for wines and spirits, and the names of Buitoni and Lavazza are popularly associated with pasta and coffee.

A role for the state

Outside northern Italy, a common legacy of past foreign domination (Sicily and southern Italy were ruled by the Spanish Bourbon kings, and Greece by the Turks) snuffed out the entrepreneurial spirit. During the 20th century the state has therefore played a crucial role in both countries in providing money and initiatives to promote industrial growth. It was the economic recession of 1929–32, coupled with a political intention to discipline the workforce, that prompted the fascist government (1924–43) of Benito Mussolini (1883–1945) to involve itself in industrial policy making.

Leather crafts (*below*) Greece has a long tradition of leather-working going back to garments, armour and riding equipment in ancient times. Today, most of the handcrafted leather goods for sale are destined as souvenirs for tourists.

In 1933 Mussolini founded the Institution of Industrial Construction (Istituto Ricostruzione Industriale; IRI) to shape development and change. Firms were merged into separate subsidiaries, and it became the linchpin of state-run industry: today it employs more than 500,000 workers. The state is involved in as much as one-third of Italian business, either through direct control, or by holding shares. This far exceeds the degree of state interest in any other Western European country.

State initiatives have had four basic aims – to increase production of existing manufacturing industries by expanding local supplies of materials and energy; to support the development of new products and processes in electronics, aerospace and shipbuilding; to upgrade roads, airports, shipping lines and telecommunications; and to encourage industrial development in the Mezzogiorno and Italian islands through direct finance or subsidies to private companies.

State involvement in Greek industry was a condition of American aid to the country after World War II under the Marshall Plan, with the aim of overcoming the lack of local capital and combating the threat of a militant communist labor force. Electrification schemes were put into place to foster the development of key industrial sectors such as sugar refining, fertilizer production, bauxite and oil refining and shipbuilding.

Italy's "Second Renaissance"

Italy is an acknowledged world leader in design. Products from Italian manufacturers as wide ranging as tiles, office furniture, lighting, kitchenware, sportswear – not to speak of fashion – have a distinctive style and flair giving them a keen competitive edge in the international marketplace. Good design is a practical advantage in selling goods and services, ensuring that the product is aesthetically pleasing to the customer and satisfies their sense of personal style. Well-designed goods also feed a basic desire for social status – they show that the owner appreciates, and can afford to buy, beautiful things.

At first sight it is perhaps surprising that Italian design has acquired such a high reputation. Italian craftsmen were largely left out of the major European artistic developments embracing such influential movements as Art Deco, Art Nouveau, Surrealism and Bauhaus. During the 23 years of Mussolini's fascist dictatorship (1922–45), artistic talent was severely restricted and widespread poverty limited the consumption of upmarket goods only to the very rich.

After 1945, however, change was rapid. Design became a medium for Italians to affirm their democratic values, egalitarian ideals and individualism, and a means to restoring national pride. In the fervor of the times they rediscovered the great traditions and skills of the northern Italian Renaissance artists and craftsmen. A great number of underemployed intellectuals and architects, especially in Milan, channeled their talent into industrial and interior design.

To meet the challenges of making new products using innovatory materials and process technologies, close liaison was established between artists, craftspeople, architects, engineers, manufacturers and the public, and in this way an appreciation of design and its application was widely established. The patrons of Italy's "Second Renaissance" – unlike the first, which was limited to the wealthy few – were the millions of ordinary working men and women, many of them still poor, who were fervently catching up with the ideas and influences they had not previously experienced.

Soon, design permeated home, office, street and cafe. Realistically simple furniture appeared in 1945, as did the revolutionary Vespa (Lambretta) motor scooter. In 1950 came the striking La Pavoni Espresso coffee machine. The Milan Triennale (trade fair) of 1951 brought international recognition to Italy's pioneering designers and manufacturers, fostering and encouraging further good design and technical innovation throughout the decade.

The manufacturing boom of the 1960s, stimulated by Italy's membership of the European Community, escalated the demand at home and abroad for designer products. A mushroom-growth of government offices and large private companies encouraged higher standards of office furnishing. Tourism brought increasing numbers of wealthy, discerning buyers to shops and workshops. Some designers, like Magistretti or Zanuso, concentrated on furniture, others worked in a range of product areas such as rugs, fabrics, laminated plastics and lighting. The head of FIAT's design department was commissioned to design a new range of pastas.

Centers of design

In the new milieu of rising affluence and heightened aesthetic appreciation, small design-orientated firms proliferated, keen to apply their own design, artistic or craft virtuosity to making high-quality goods. Many were concentrated in the dense network of towns situated in the northeast and center of Italy, where close contact between designers, craftspeople, industrialists, engineers and retailers had become well established.

Top designers, however, gravitated to the fashion-conscious cities of Milan, Rome and Florence. Each of them has its own character. Milan is the economic capital of Italy, acting both as factory and showcase for Italian manufacturing. The city specializes in textile, clothing, shoe, household ware, lighting and furniture design, and increasingly in fashion and haute couture.

As the home of Italy's government, state corporations and bureaucracies, and of film and television, Rome concentrates on exclusive high-fashion clothing for an elite national and international clientele. Florence, retaining its medieval craft traditions, is the "leather capital" of Italy, pioneering soft and colored leather goods and footwear. It also specializes in embroidered linens and lingerie, and in the fine jewelry that has been made there since the time of the first Renaissance Medici rulers.

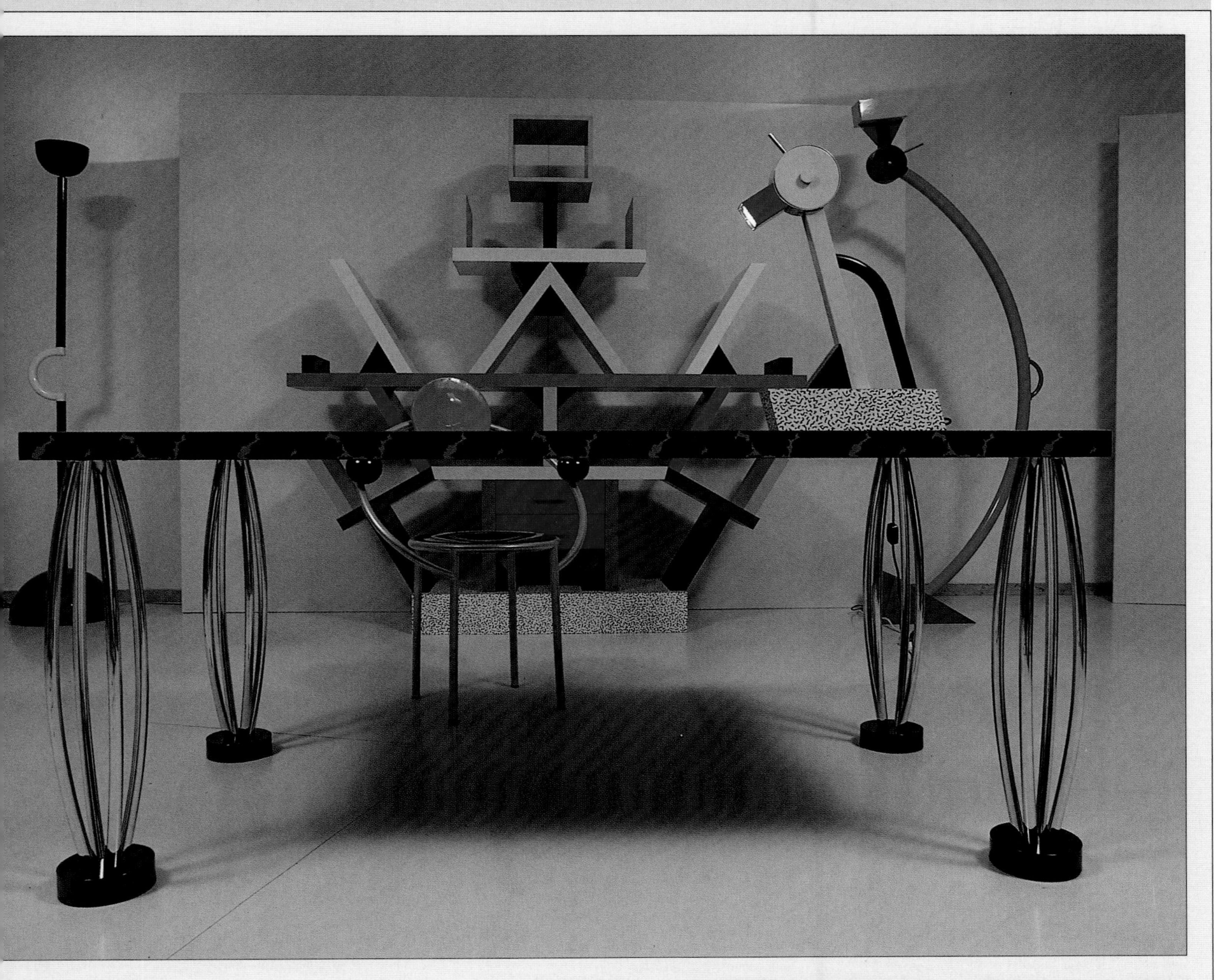

Furnishing ideas (*above*) Memphis is a group of designers making avant-garde furniture, fabrics and ceramics. Formed for the 1981 Milan Furniture Fair, the group produces limited-edition products that are heavily influenced by ancient Egyptian artifacts and modern pop art. Although Memphis caused controversy when it first appeared, it soon began to influence the more mainstream designers.

Horse power (*left*) The luxury sports car manufacturers, Ferrari, are a leading name in motor racing – one of Italy's national enthusiasms. The Modena-based company and Grand Prix racing team founded by Enzo Ferrari is now a part of the FIAT group.

Fashion city (*right*) Heavily industrialized Milan produces more than just engineering products. The city is one of the world's leading centers of haute couture fashion, an industry which was stimulated by local textile factories producing silk, cotton and hemp. International fashion empires from the city include Armani, Benetton and Valentino. The export of fine woolen goods made by other manufacturers still makes up a large part of Milan's economy

AN INDUSTRIAL HEARTLAND

THE CRUCIAL ROLE OF COAL · CENTERS OF EXCELLENCE · A TECHNICAL CULTURE

Central Europe is among the leading industrial regions of the world, particularly in its output of electronic products, automobiles, watches, chemicals and pharmaceuticals. This has been achieved despite the region's comparative shortage of natural resources. The driving forces behind its manufacturing growth after World War II were a large domestic market that encouraged manufacturing diversity, as well as largescale efficient production methods. Even more important in achieving exporting success is the willingness of its large and highly skilled population to adapt successfully to changing world conditions. Political stability, fostering cooperation between government, business and workers, has also helped. So, too, has the region's geographic location, which places it at the communications hub of Europe.

COUNTRIES IN THE REGION

Austria, Germany, Liechtenstein, Switzerland

INDUSTRIAL OUTPUT (US $ billion)

Total	Mining	Manufacturing	Average annual change since 1960
658.4	43.2*	558.2**	+2.9%

** Figure relates to West Germany (pre-unification in 1990)*
*** Figure includes mining in Austria, East Germany (pre-unification in 1990) and Switzerland*

INDUSTRIAL WORKERS (millions)
(figures in brackets are percentages of total labor force)

Total	Mining	Manufacturing	Construction
17.5	0.02 (0.0%)*	14.1 (30.6%)**	2.9 (6.4%)

** Figure relates only to Austria*
*** Figure includes mineworkers in Germany and Switzerland*

MAJOR PRODUCTS (figures in brackets are percentages of world production)

Energy and minerals	Output	Change since 1960
Coal (mill tonnes)	495.0 (10.4%)	+4.7%
Natural gas (billion cu. meters)	15.7 (0.8%)	N/A
Potash (mill tonnes)	2.9 (9.7%)	-13.3%
Salt (mill tonnes)	13.7 (7.5%)	-29.4%
Magnesite (mill tonnes)	1.1 (6.3%)	No change
Manufactures		
Steel (mill tonnes)	41.8 (5.7%)	+1.2%
Automobiles (mill)	4.85 (10.2%)	+221%
Papermaking/printing machines (1,000)	299.7 (86.8%)	No data
Clocks (mill)	26.1 (45%)	No data
Electric furnaces (1,000)	183.3 (83.8%)	No data
Transformers (mill)	3.0 (66.1%)	No data
Plastics and resins (mill tonnes)	9.3 (18.6%)	+131.8%
Cement (mill tonnes)	48.9 (4.5%)	+37%
Beer (mill hectoliters)	124.5 (12%)	+4%

N/A means production had not begun in 1960

Massive excavation at dusk *(above)* This open-cast mine at Grevenbroich in western Germany is typical of many throughout the region, supplying brown coal (lignite) to local power stations.

Energy production and consumption *(below)* Coal is the chief energy source, and is used to generate electricity. The region has few other fossil fuels and more than half the energy used is still imported. The continuing development of nuclear power should alter this balance in the future.

Energy balance (mill. tonnes coal equivalent)

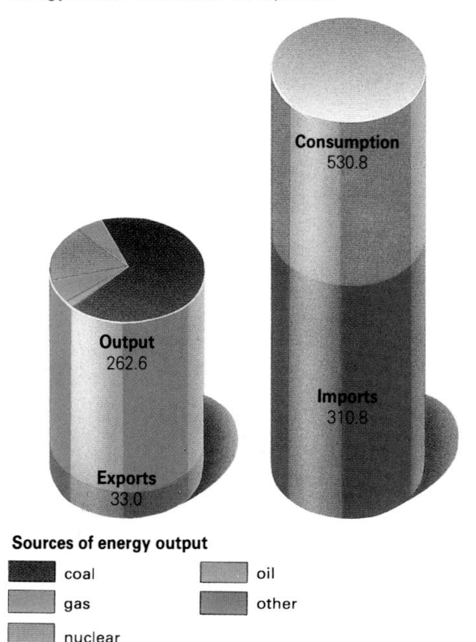

THE CRUCIAL ROLE OF COAL

Except for an abundance of coal, Central Europe is not rich in mineral deposits. Its modern industrial success has been achieved by harnessing its human resources to produce goods for export that pay for increasing amounts of imported energy, fuel and raw materials. In this, the region bears close resemblance to Japan since World War II.

Coal, however, particularly from the Ruhr and Saar fields of the lower Rhine, laid the foundations of Germany's industrial prosperity in the 19th century. Bituminous coke and coal supplied the booming steel, engineering and chemical industries, and brown coal (lignite) fueled production of electricity. Output soared, peaking at 190 million tonnes of bituminous and 253 million tonnes of brown coal in 1943.

After World War II, East Germany's communist leaders set the country on a course of rapid industrialization under state control. The expansion of its opencast mines made it a leading world producer of brown coal for fuel, power and synthetic oil, with production peaking in the mid 1980s. By contrast, West German coal production was highest in 1960. After

Map of principal resources and industrial zones *(right)* The region is heavily industrialized, particularly along the banks of the major waterways. Many manufacturing companies rely on imported materials, especially iron, because the local sources are not adequate. However, large deposits of potash are exploited by the chemical giants.

this date the demand for bituminous coal (particularly from the coalfields of the lower Rhine) fell as oil imported from the Middle East and natural gas, from local supplies or from the Netherlands, became cheaper.

Brown coal has a lower carbon content than bituminous coal, and therefore burns less efficiently, creating more pollution. Output has leveled in recent years as the government has turned to nuclear power in an effort to conserve the diminishing natural resources and to protect the environment. Germany today is estimated to have 80 billion tonnes of coal, equivalent to 7.4 percent of recoverable world coal reserves. Some 70 percent of this is brown coal.

Gas, oil and water

Coal aside, the region is deficient in fossil fuels, and most petroleum has to be imported. In the 1970s and 1980s a natural-gas field was found to extend beneath the North European Plain from the Netherlands into northern Germany, but this area supplies only 30 percent of domestic natural gas and 3 percent of domestic crude-oil needs. A small field in the Danube basin of eastern Austria yields 9 million barrels of oil.

Austria is also low in coal deposits, and Switzerland contains no fossil fuels. However, both these countries are able to take advantage of swift-flowing, glacier-fed Alpine rivers to produce hydroelectricity. Two of the world's highest dams are in southern Switzerland: Grand Dixence 285 m (935 ft) and Mauvoisin 237 m (780 ft). By the 1990s hydroelectricity provided 71 percent of Austrian and 60 percent of Swiss energy output, but virtually all Switzerland's hydroelectric potential had been harnessed. Nuclear power supplied the remaining 40 percent of Switzerland's energy output.

Metals and other minerals

The region's scant supplies of metal ores lie mostly in small, scattered deposits, many of them long since worked out or made uneconomic by competition from cheap imports. Production of iron-ore peaked at 20 million tonnes in 1960, but by 1990 had fallen to only 300,000 tonnes. Although copper, nickel, tin, silver and uranium were all being mined within what was formerly East Germany, output here seemed set to decline in the 1990s.

Within the region as a whole, the eastern Alps of Austria is the only place where iron ore is still mined in significant quantities, with output at more than 1 million tonnes. Austria is also the world's leading producer of natural magnesite, used extensively in the chemical industry. Sizable potash deposits, the basis of the local chemical industries, are found in northern Germany, and account for one-quarter of world potash production. There are large salt deposits in the Jura mountains of northern Switzerland.

Harnessing the icy waters This dam at lake Mooserboden in southwestern Austria is part of a major hydroelectric power complex. Neither Austria nor Switzerland is rich in fossil fuels, but water power provides a renewable source of energy.

CENTERS OF EXCELLENCE

Central Europe's successful manufacturing industries provide relatively more jobs (30–45 percent of employment) and higher incomes than almost anywhere else in the world. They generate 15 percent of visible world exports. Sophisticated, high-quality finished goods are typical of the region.

The industrialization of Germany

Mammoth industrial complexes first began to develop in Germany in the period of prosperity after unification in 1871. Price-fixing cartels between producers and suppliers helped to forge intricate production systems that slotted together the extraction and supply of raw materials, processing techniques, and the manufacturing of finished goods. These industrial complexes were concentrated in three main areas in the west and north of the country.

The most important center of heavy industry – which eventually grew to become Europe's largest – was based on the bituminous coalfields of the Ruhr and Lower Rhineland. These yielded coke for steel plants that in turn supplied metal to factories making tools, machinery and equipment. The Rhine and the Dortmund–Ems canal provided cheap and easy import routes for bulk supplies of raw materials, particularly iron ore.

A second industrial zone developed to the north of the Harz mountains around Hannover, Wolfsburg and Magdeburg, where local supplies of iron ore, salt and potash attracted steel, engineering, chemical and tire manufacturers. The area is also a rich agricultural district, and food processing flourished. The third industrial complex was in eastern Germany, around Halle, Leipzig and Stassfurt. Here huge chemical industries were supported by brown coal and potash.

Despite their preeminence, industry was not confined to these three zones. A good communications network and electricity supply meant that power, raw materials and finished goods could easily be transported to or from most places. Factories sprung up around the ports of the north, along the Rhine and within reach of Berlin, Germany's capital. In the center and south of the region, there had long existed a network of well-spaced, medium to large cities. In these, a strong spirit of communal pride was combined with craft traditions and well-established apprenticeship systems to produce the ideal conditions for local entrepreneurialism and industrial innovation.

The damage inflicted on the industrial capacity of the north and west during World War II was enormous. Although much was subsequently rebuilt, the north never recovered its preeminence. Several factors encouraged a shift southward. The division of Germany in 1945 led a number of key firms such as the optical manufacturer Zeiss to relocate from the eastern Soviet zone (later East Germany) to the southern American zone (later West Germany). Later, imports of Middle Eastern and North African oil via pipelines from Marseilles (France) and Genoa (Italy) offered cheaper fuel than Ruhr coal, and allowed giant chemicals corporations to diversify into petroleum products.

Being located at the center of European road and rail communications, manufacturers in central and southern Germany were well placed to supply neighboring markets in the European Community (EC).This advantage increased still more following the EC's enlargement to include new members in 1981 and 1986. An influx of immigrant workers (*gastarbeiter*) from southern Europe and Turkey during the 1960s and 1970s helped to keep labor costs low in Germany.

Germany today excels in the manufacture of automobiles and automotive components (including internationally renowned companies such as BMW, Mercedes-Benz and Volkswagen), printing materials and machinery, ceramics and porcelain, plastics, pharmaceuticals and instruments. The picture changes in the east where, during 40 years of communist rule, the emphasis was on the uncontrolled exploitation of raw materials to achieve self-sufficiency in industry, especially in chemicals and machinery, though specialization was limited. Old-fashioned and inefficient manufacturing plant will require large injections of capital before industry can modernize and diversify.

Alpine industries

Swiss industry has much in common with its northern neighbor, though on a reduced scale. Its long tradition of craft industries has given rise to a diversity of manufacturing skills with considerable specialization. It is, for example, a leading producer of high-precision goods such as watches, and optical and scientific equipment. Pharmaceuticals, machine tools, textiles and textile machinery are also important. Neutrality, political stability, privacy of accounts and a sound currency have long made Switzerland a safe banking haven – an attraction shared by neighboring Liechtenstein, which is also a tax haven. The Swiss developed insurance, especially of freight, to exploit their central position in Europe.

Austria industrialized much later than its neighbors. Before 1918, the major industrial areas of the Austro-Hungarian empire were in Bohemia, today part of Czechoslovakia. Austria itself was predominantly agricultural. Austrian industry expanded rapidly after World War II, particularly in the west. This was the area occupied by the United States after 1945 and was therefore the recipient of United States' investment aid. Industrialization is also widespread around Vienna and along the Danube corridor. Branch plants of foreign-owned (especially German) firms predominate, taking advantage of cheaper labor. Leading Austrian industries are steel, papermaking, chemicals, plastics, textiles, clothing, leather, shoe and electrical products.

Gradual opening up of relations with Eastern Europe in the 1980s enabled Austria to exploit its political neutrality and its geographic position to finance East–West trade. As a result of continuing political change in Eastern Europe, Austrian industrial growth seemed likely to be strengthened during the 1990s, particularly around Vienna and in the east and southeast in areas bordering Czechoslovakia and Hungary.

A BREAKTHROUGH IN STEELMAKING

In steelmaking, excess carbon has to be burned out of pig iron (the name given to the hot iron produced by smelting iron ore in a blast furnace). This removes the impurities that make iron brittle, producing refined steel. Today most steel is made in a basic oxygen furnace, also known as an L-D furnace, named after the towns of Linz and Donawitz in Austria, where the process was first developed.

The search for a method of using pure oxygen instead of air to refine steel had begun before World War II in Germany, Switzerland, Britain, Sweden and the United States. The breakthrough came in Austria in the 1940s when experimental scientists developed a pear-shaped vessel in which a supersonic jet of oxygen from a water-cooled lance, or pipe, was directed vertically down into a mixture of molten pig iron, steel scrap and lime. At the end of the process the molten steel and slag were separated.

The first commercial oxygen furnaces came into operation in Linz and Donawitz in 1952 and 1953. They had a capacity of only 39 tonnes, but today much larger vessels are in use, with capacities of 200 or 300 tonnes. The advantage of the L-D process over the open-hearth process previously in use is that it cuts the time taken to process the steel from about 10 hours to 40 minutes. It also allows greater use of steel scrap (up to 40 percent of the furnace charge in some cases). As a result, smaller quantities of coking coal and iron ore are required, and transportation costs are reduced.

Precision engineering (*above*) Swiss watches and clocks have a world-wide reputation for accuracy and attention to detail. Some 15 pieces of this jeweled movement are made by hand.

Factory-made German wurst (*right*) Though mass produced, the ingredients of this traditional delicacy are strictly controlled by tradition and by EC regulations. This Bruhwurst variety, which includes the original frankfurter, is made from finely minced pork or beef.

A TECHNICAL CULTURE

An aptitude for technical innovation, together with an imaginative entrepreneurial spirit, are traits that are associated with the people of Central Europe as far back as the Middle Ages. When Johann Gutenberg (c. 1390–1468), a craftsman from Mainz in central Germany, invented a printing press that used movable type, he created a technology that survived well into the 20th century. Germany is still recognized internationally for the manufacture of printing equipment and inks, and for papermaking.

Frequently, industrial innovation has come about as a response to the challenge of a particular situation or set of physical conditions. In his book *De re metallica* (1556), Georg Bauer (also known as Georgius Agricola, 1494–1555), mineralogist and scholar, has left a record of the equipment that was used to ventilate and to pump floodwater out of silver mines. The Swiss in the early 20th century were pioneers in the development of electric traction equipment, cable railroads and elevators to assist in transportation over mountainous terrain. They also led the development of tractor-plows to cultivate small, steeply sloping fields.

Having become leaders in the field of automobiles and aircraft, German scientists during the 1920s and 1930s devoted their attention to finding ways of converting local supplies of lignite into synthethic oil and carbonizing bituminous coal to obtain gasolene and diesel fuel in an attempt to overcome the country's lack of oil resources.

Many of the region's early industrialists united technological research and invention with entrepreneurial flair – a combination that has continued to sustain the large industrial corporations that they founded. Alfred Krupp (1812–87), for example, developed a method of making high-quality cast steel. In 1914 the company that bears his name made further important innovations in stainless steel. Bayer, the chemicals corporation founded by Friedrich Bayer (1825–80), who developed artificial dyes, went on to create polyurethane in 1937. Nestlé, established at Vevey in southwestern Switzerland to make a milk-based babyfood in 1866, invented instant coffee in 1937, and is now a leading brand name in all kinds of processed food.

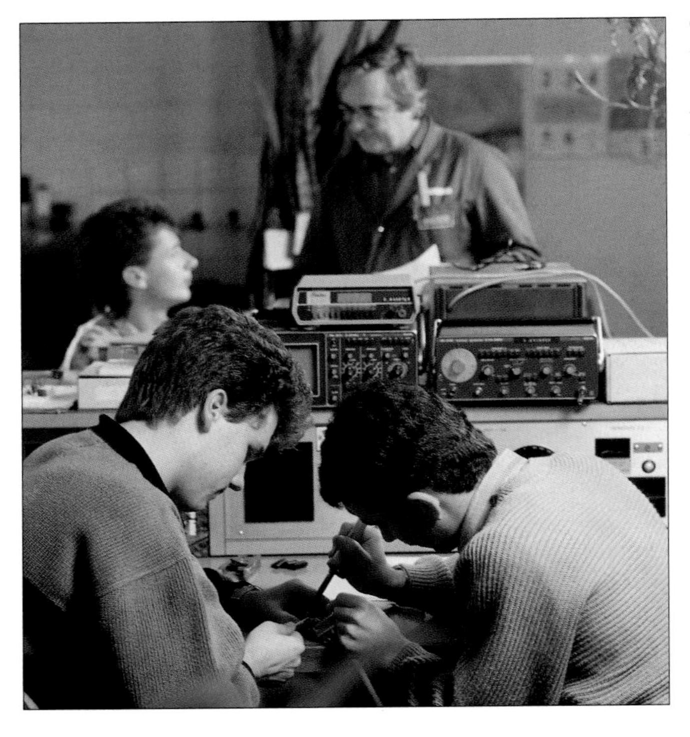

On-the-job-training *(left)* Part of the success story of manufacturing industry in the region is the efficiency of its highly trained workforce. Acquiring new skills has a very high profile in an environment where each company's working capital is the ingenuity of its workers and where several leading brand names compete on the basis of quality and speed.

Clean, precise and efficient *(right)* Robots in a factory in western Germany weld together this BMW chassis on an automated production line. Established in Germany for over 75 years, BMW now runs some of the most advanced manufacturing plants in the region. Automation is increasingly common in western German engineering; in the 1980s, for example, over 15,000 robots were installed in workshops throughout the area. Like their competitors, Audi, BMW produces automobiles that are highly regarded all over the world for their performance, safety standards and luxury design.

Supporting innovation

Numerous state-run institutes of technology have long been centers of technical research. In this way, a large pool of research scientists, engineers and technicians is continually being trained and replacing itself, and Central Europe has been kept at the forefront of innovation and invention. Since 1901, 40 Nobel prizes have been awarded to scientists from the region: 20 in chemistry, 14 in physics and 6 in medicine. Many of these prizes rewarded research that has subsequently had some major commercial application in industry. Among them were the prizes given to Wilhelm Röntgen (1845–1923) for the discovery of X-rays in 1901, to Karl Ziegler (1898–1973) for research on polymers (leading to great improvements in plastics) in 1963, and to joint German–Swiss teams for their work on electron microscopes and new superconducting materials in 1986 and 1987.

In Germany, initiatives taken both at national and at regional or local government level have worked to create a "social market economy" that combines a high degree of economic efficiency with social welfare programs. The legal framework exists to encourage very large corporations to achieve technical and organizational economies of scale while at the same time protecting family firms and workers' employment conditions. Banks actively participate in controlling and financing modernization in large and small firms alike, insulating them from stock market fluctuations.

Support is given to industry in many other ways. In Germany state governments and in Switzerland cantonal governments provide incentives for companies to locate in their region. They also encourage the building of science parks, and nurture small businesses generally. The regular upgrading of the highway, railroad and telecommunications links that connect the major cities is carried out by central governments.

Common goals

Both as producers and as consumers, the people of Central Europe are united by a strong commitment to high standards of quality. This has resulted in close cooperation between management and trade unions. Apprenticeship training schemes – originating in the medieval guild tradition – are a feature of the region, and have contributed significantly to the creation of an adaptable skilled workforce. Craftsmen's associations, which exist throughout the region, help to unite efficient standardized production with custom-made artistry.

There is a general concern to create clean, tidy, attractive living and working environments. Public awareness of environmental issues, particularly of the damage caused by industrial emissions,

FROM BEETLE TO PORSCHE

In terms of character and style, the gap between the Volkswagen Beetle and the classic Porsche is huge. Yet both these automobiles owe their existence to the skills of one man – Ferdinand Porsche (1875–1951). This Austrian automotive engineer worked for the German company of Daimler-Benz from 1916 until 1931, when he established his own firm designing and producing sports and racing cars in Stuttgart. However, it was not until after World War II, in 1950, that he designed the famous sports car that was to bear his name.

In 1934 Porsche was brought in to work on a plan devised by Adolf Hitler (1889–1945) for a cheap "people's car" (*volkswagen* in German). Working with his son, Porsche was responsible for the initial design of the automobile that was later dubbed the Beetle because of its small size and rounded shape. From 1937 the Beetle was mass-produced at the Volkswagenwerk in Wolfsburg. Left in ruins in 1945, the factory was reactivated under the direction of the British army and run by the state authorities from 1948.

The Beetle quickly became the means for Volkswagen's revival and rapid expansion during West Germany's "economic miracle" in the late 1940s and 1950s. Made popular by US servicemen returning home after tours of duty in West Germany, many Beetles were exported to the United States. In the 1960s and 1970s Volkswagen introduced production and assembly plants in North and South America, Africa, Southeast Asia and Australia, and all this time the Beetle's shape hardly changed from its original design. But it was gradually replaced by sportier models, and by the late 1980s was only produced in Volkswagen's Mexican plant.

came to the fore in the 1970s and 1980s, when the political success of the Green Party in Germany put environmental protection on the agenda of central government. Attempts are now being made to control pollution, to clean up the Rhine and other rivers, and to landscape exhausted quarries and mining waste tips.

The greatest challenge of the 1990s, however, is the upgrading and replacement of old-fashioned and worn out industrial plant in the east of Germany, and the integration of its former state-run industries into the economic and highly competitive structure of the West.

The pharmaceutical industry

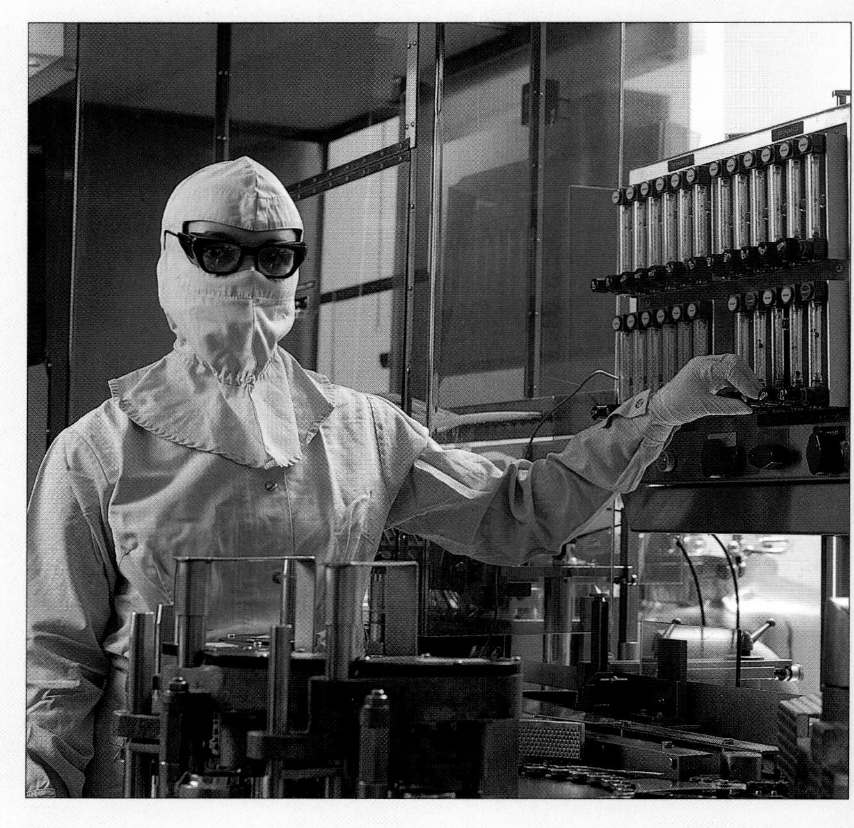

Exhaustive testing *(above)* is carried out according to strict guidelines before any product is offered to the public. Side effects in animals are carefully monitored, and constant control is exercised over the potency and purity of the drug.

Pharmaceuticals, or drugs, are used in the diagnosis, relief, treatment or prevention of disease in humans or animals and come in various manufactured forms. Some are produced as liquids or solutions, others as ointments or tablets. All draw on a wide variety of raw materials. Plants yield alkaloids such as quinine (from cinchona bark) and essential oils such as camphor, cinnamon or peppermint. Other ingredients come from animals, including fatty oils (cod liver oil) and insulin from hormones. Naturally occurring minerals, including borax, copper sulfate, iron and zinc oxide, are also used in the manufacturing process.

The pharmaceutical industry is closely linked with chemicals or petrochemicals production, which supplies products such as acetic acid, chlorine, sulfuric acid, ethylene or phenol for drugs manufacture. The actual production of drugs uses many processes used in chemical synthesis, and the manufacture of antibiotics or vaccines requires the cultivation of bacteria or molds under highly controlled laboratory conditions.

Central Europe has the largest pharmaceutical industry in the world outside the United States. The major centers of research, development and manufacturing all lie close to the Rhine in the south and west of the region. No fewer than 11 of Europe's biggest drugs manufacturers are located within three core areas: Frankfurt–Darmstadt and Ludwigshafen–Mannheim, both in central Germany, and Basel in northern Switzerland. One reason for this concentration is the close link between the pharmaceutical industry and medical research – several long-established universities (Heidelberg, Karlsruhe, Tübingen, Nuremberg) lie in this area. In the mid 16th century, the world's first pharmacopoeia – a book listing drugs, their uses and preparations – was in circulation in Nuremberg.

Giant corporations

In the 19th century a number of companies along the Rhine involved in the manufacture of chemical dyestuffs began to switch to modern pharmaceuticals. A riverside site was crucial because vast quantities of water are required to manufacture drugs, and the waterway provided easy access for raw materials and coal. Many of these companies remain world leaders and household names in the industry today.

Most are located in new, often mammoth, complexes producing both chemicals and pharmaceuticals. Ciba-Geigy and Sandoz, for example, are both based in Basel, and in the Rhineland of Germany are Bayer in Leverkusen, Hoechst in Frankfurt am Main and BASF (Badische Anilin und Soda Fabrik) in Ludwigshafen. Pharmaceuticals account for a varying proportion of their total business. While they are the chief product in Hoechst or Ciba-Geigy, in others (for example, BASF) they are only one of many manufacturing lines, alongside plastics, synthetic fibers, fertilizers, inks, dyes and paints.

All, however, have two features in common: a complex relationship between chemicals and pharmaceuticals in both research and manufacturing, and a need for a high-volume chemicals business to finance the costly pharmaceuticals and

Packaging a safe dosage *(left)* The active ingredients of any medication can often be made up in several forms. In capsules such as these the ingredients are more stable, an accurate dose is assured and storage is simple.

A riverside site *(right)* is a considerable advantage to this Ciba-Geigy pharmaceutical plant. Many of the leading names in the industry are based in the Rhine valley, where they use the water as a resource in processing as well as for cheap transportation.

medical research that may take years to complete. It is usual for the production of a new drug to take up to 12 years from presenting the initial concept to selling in the marketplace.

Only a handful of large corporations are able to afford the costs, take the risks, and have the facilities to manage the complex chemistry involved. For example, of the 147,000 patents granted to Bayer (the company that discovered aspirin in 1899) since 1863, some 39,000 are still in force and 33,000 are still pending. Despite this, the World Health Organization claims that, with present knowledge, only a third of 30,000 identified illnesses are as yet curable by pharmaceuticals.

Outside the area of research and development, a number of specialist manufacturing firms have developed to serve key niche markets in drugs. Many have particular expertise in converting the active substances produced by the big corporations into dosage forms. Some 400 of these are scattered throughout the major urban areas of Central Europe. However, many of them, including such well-known manufacturers as Roche of Basel, Merck and Wella of Darmstadt, and Boehringer of Mannheim, prefer to remain close to the major research, manufacturing and university centers in central Germany and northern Switzerland.

INDUSTRY IN TRANSITION

SCATTERED RESOURCES · DEVELOPMENT AND DECLINE · THE NEW REVOLUTION

With the exception of Poland, Eastern Europe's natural resources are scattered and fairly meager. Until 1945, industrial development in the region was limited and patchy, land being used primarily for agriculture. The postwar communist governments adopted the Soviet model of full state control of resources and industry, giving priority to heavy industry to supply capital and military equipment. Energy and raw materials were imported from the Soviet Union. Growth was rapid, particularly in mining zones, around big cities or in areas close to the Soviet Union. Initially the aim was to achieve national self-sufficiency in manufacturing, though some specialization occurred later. The political changes of 1989–90 exposed the acute need to restructure outmoded industry to meet market competition.

COUNTRIES IN THE REGION

Albania, Bosnia and Hercegovina, Bulgaria, Croatia, Czechoslovakia, Hungary, Macedonia, Poland, Romania, Slovenia, Yugoslavia

INDUSTRIAL OUTPUT (US $ billion)

Total	Mining and Manufacturing	Average annual change since 1960
260.6	260.6	+4.8%

INDUSTRIAL WORKERS (millions)
(figures in brackets are percentages of total labor force)

Total	Mining	Manufacturing	Construction
22.3	0.6 (1%)*	17.5 (30.8%)**	4.2 (7.4%)

* Figures refer only to Bulgaria and Poland
** Figures include mining equipment in all countries except Bulgaria and Poland

MAJOR PRODUCTS (figures in brackets are percentages of world production)

Energy and minerals	Output	Change since 1960
Coal (mill tonnes)	559.2 (11.9%)	+165%
Copper (mill tonnes)	0.65 (7.6%)	+116.5%
Lead (mill tonnes)	0.28 (8.3%)	-16.4%
Sulfur (mill tonnes)	5.05 (33.7%)	N/A
Manufactures		
Linen fabrics (mill meters)	201.2 (19.7%)	No data
Knitted sweaters (mill)	416.7 (35.7%)	No data
Men's and boys' suits (mill)	7.75 (11.7%)	No data
Footwear (mill pairs)	571.7 (12.9%)	No data
Nitric acid (mill tonnes)	4.95 (18.2%)	+149%
Cement (mill tonnes)	60.9 (5.5%)	+196.5%
Steel (mill tonnes)	52.6 (7.2%)	+132.5%
Buses (1,000)	39.3 (11.9%)	No data
Railway locomotives and rolling stock (1,000)	30.3 (16.5%)	No data

N/A means production had not begun in 1960

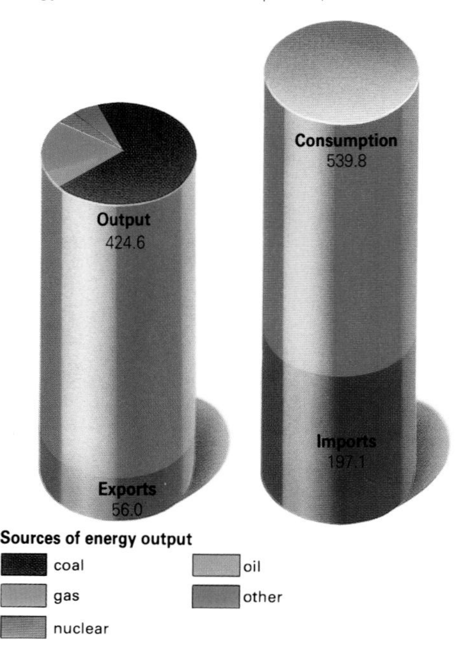

Energy production and consumption Low-quality brown coal fuels most of Eastern Europe's energy production. The region relies on imports of every other fossil fuel, particularly oil. Until the economic crisis of the 1980s the Soviet Union was the main supplier.

SCATTERED RESOURCES

Before World War II the only parts of Eastern Europe to have significant concentrations of industry were Bohemia-Moravia in western and central Czechoslovakia and Silesia, which is now mainly in south-central Poland. These centers, which had abundant and easily accessible deposits of coal and metal ore, were developed during the 18th and 19th centuries as the industrial heartlands of the old Austro-Hungarian and German empires. Elsewhere, mineral reserves were largely untapped and agriculture supplied the resources for the food processing, textiles and shoe industries that were characteristic of the region.

The years after 1945 brought enormous changes. Communist governments were installed and the countries of the region embarked on a program of massive industrialization to reduce reliance on imported goods. This created a pressing need to exploit national resources to the full. Existing mines were expanded, new ones were opened and prospecting was

Turning the screws (*left*) Although the region has an extensively developed industrial base, factories and plants are in desperate need of modernization. Outdated manual tasks are being performed at this Skoda engineering plant in Pilsen, Czechoslovakia.

intensified. Some sizable discoveries were made: copper, sulfur and brown coal (lignite) in Poland; lead and zinc in Bulgaria; chrome in Albania. But there were not enough raw materials to meet all the needs of industry, and the region became increasingly dependent on imports of coal, coke and iron ore. Some 75 percent of oil and 52 percent of natural gas requirements were supplied via the "friendship" and "brotherhood" pipelines from the Soviet Union.

Patchy energy reserves

Coal is Eastern Europe's prime resource. Reserves amount to 70 billion tonnes, or 6.4 percent of the world's total, equal to those in Western Europe. But their distribution is very uneven in quantity, still more so in quality. The Upper Silesian coalfields, which lie mainly in central-southern Poland but which extend into Moravia in central Czechoslovakia, produce most of the region's better-quality bituminous coal. Poland mines 69 percent of the region's coal in calorific heating value, but the coal basins found elsewhere are mostly small, containing poor quality brown coal.

Other energy resources are limited. Oil and natural gas are drilled in every country, but most is produced near Ploiesti in central Romania, where oil was first refined as early as 1856.

Albania and the countries of former Yugoslavia have harnessed their rivers and streams for hydroelectricity, which supplies 87 and 33 percent of their power respectively. However, their considerable hydroelectric potential has not yet been fully developed due to the threats posed by earthquakes.

The small quantities of uranium mined in the Ore Mountains of Czechoslovakia and the Villany area of southern Hungary are used in the generation of nuclear power. This supplies between 25 and 37 percent of electricity in Bulgaria, Hungary and Czechoslovakia, which are short of alternative energy sources.

Metals and other minerals

Most mineral deposits are either too small or too poor in quality to be significant. However, former Yugoslavia mines copper, zinc, silver and gold, and ranks among the world's top 10 producers of antimony, bauxite, lead and mercury. Poland is a producer of copper, zinc and silver, Czechoslovakia of antimony and mercury, Hungary of bauxite, Bulgaria of lead, and Albania of chrome.

The region's iron-ore reserves are very limited. Czechoslovakia's deposits in Moravia, though historically important, are now depleted. Small iron-ore mines were opened throughout the region to supply the expanding steel industry after 1950, but they are now closing, having become uneconomic to work in the emerging market economy. Former Yugoslavia mines most, though only 1.8 million tonnes.

Local industries draw on a wide range of other resources. Stones for building are widely available. There are world-class deposits of magnesite in Czechoslovakia and former Yugoslavia, sulfur and salt in Poland, and graphite in both Czechoslovakia and Romania. The long-established glass and porcelain craft industries of Czechoslovakia draw on local supplies of sand, potash, lime and china clay (kaolin).

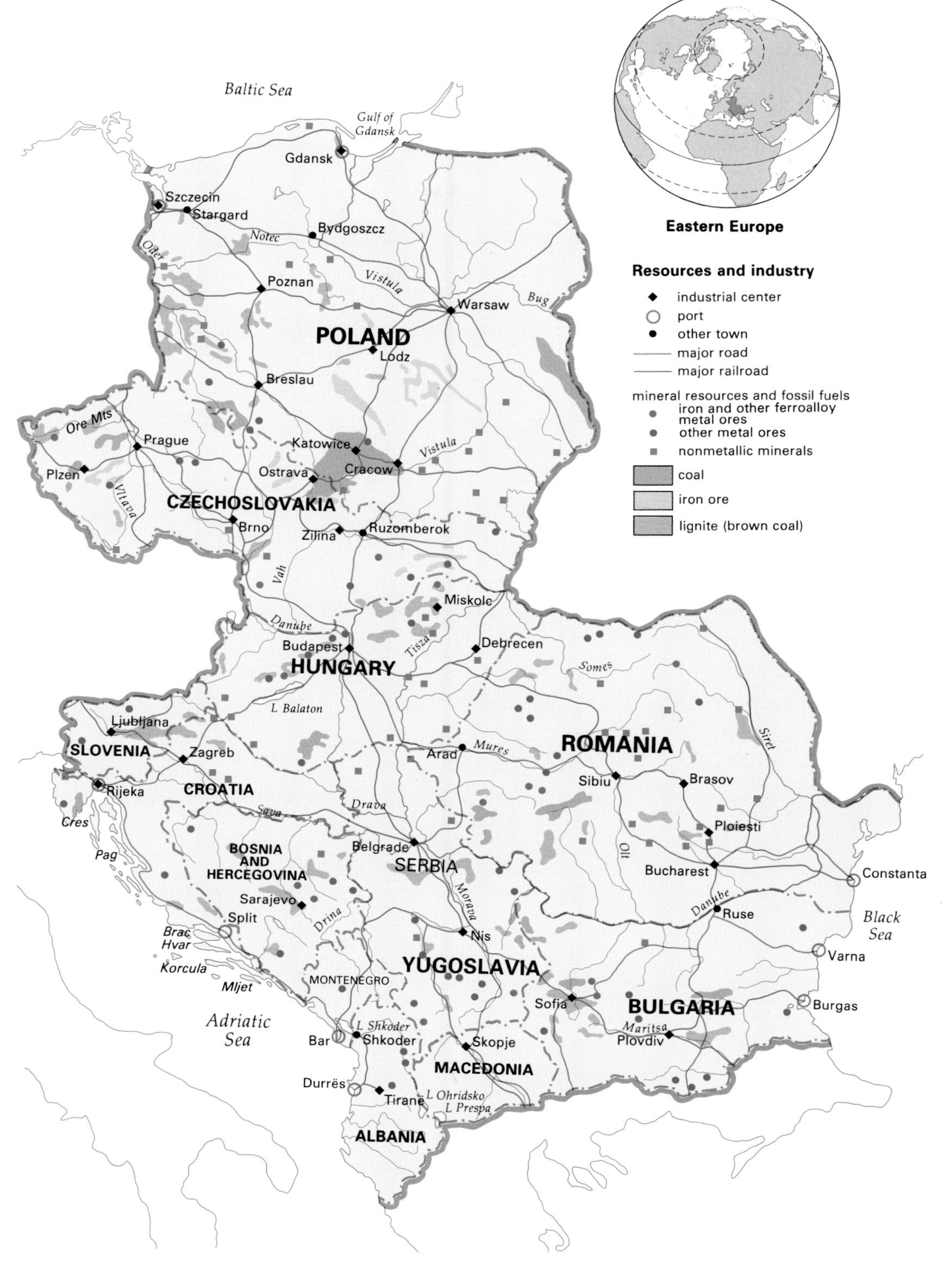

Map of principal resources and industrial zones Before 1945 there was relatively little industry except in Poland and Czechoslovakia, but during the next 40 years of communism natural resources were exploited to the full in the race to industrialize.

Reflection of a former glory *(left)* Glass was first produced in the Czechoslovakian region of Bohemia in about the 10th century. Local craftsmen traditionally specialized in decorative cut glass. Exploitation of the area's rich potash and lime deposits in the 18th century then led to the invention of a high-luster glass called Bohemian crystal. It became the leading glass in the world, only surpassed decades later by English lead crystal. Bohemian glass was highly prized until the early part of this century, but today the industry is a pale shadow of its former self.

Measuring decline *(right)* The Gdansk shipyard was the birthplace of Solidarity, the trade union and reform movement that was born in the industrial unrest of the 1970s and helped to bring an end to communism in Poland. However, the yard did not fare well in the harsh economic climate of the 1980s. A loss of over $5 million in 1987 prompted the government to announce that it was to close. After a national and international outcry, a Polish-American heiress, Barbara Johnson, agreed to buy the shipyard for several million dollars. But even this injection of cash could not assure the shipyard a future.

DEVELOPMENT AND DECLINE

Eastern Europe's industrialization after World War II was so rapid and spectacular that it became a model for development to be followed by the countries of the Third World and was admired even by non-communist regimes. Between 1950 and 1985 industrial growth increased more than tenfold – admittedly from a low base in the Balkan countries of Albania, Bulgaria, Romania and former Yugoslavia – and consistently exceeded world averages in every country.

Yet in 1945 Eastern Europe was devastated by war. Industry lay in ruins, and millions of people, many of prime working age, had perished. Reconstruction over the next five years concentrated on mobilizing labor and materials to rebuild prewar industries. Most heavy industries continued to be located in small areas of Czechoslovakia, Poland and Hungary. Elsewhere, industry followed the prewar pattern of the simple processing of food, textiles and wood products.

A plan for industry

The consolidation of communist power inaugurated a new era of largescale industrialization based on the prewar Soviet model. Natural resources and existing industries were brought under state control, and the location and character of future industrial development was centrally planned and directed.

Overwhelming emphasis was placed on an ambitious plan to develop heavy industry. This had many objectives. The local production of capital goods and technology would allow the region to catch up with the West, while heroic building projects would testify to communist superiority. The region's strong commitment to defense necessitated a large armaments industry; this need intensified as the Cold War between the Eastern and Western blocs developed into a spiraling arms race. There were also ideological motives: by creating industrial jobs and raising incomes, rural laborers would be attracted into the cities, creating a class of industrial workers in line with the Marxist theory of communist development.

The increase in coalmining and iron and steel output was to be accompanied by the construction of metal-processing plants and hydroelectric power stations. The aim was to provide the materials for growth and diversification in mechanical and electrical engineering, allowing the region to become self-sufficient in machinery and equipment. This would be followed by an extensive program of industrial modernization and expansion.

Hundreds of new plants and extensions were built. Fuel, power and metal-smelting facilities were located near sources of raw materials and engineering works near major cities or areas with exceptionally good transportation links, where existing plant and skilled labor – however limited – offered good development potential. Several key steel, metalworking, engineering and defense industries were located in strategic sites in less-developed areas of Poland, Czechoslovakia and Romania. This had the dual purpose of placing them away from the danger zones close to the borders with

A LOCAL INDUSTRY: SHIPBUILDING IN GDANSK

Gdansk in north central Poland, at the mouth of the Vistula river on the Baltic Sea, was an important trading port in medieval times. A shipbuilding industry was recorded here as early as 1572. It prospered over the next 200 years, but declined when the city – renamed Danzig – was absorbed into the German Prussian empire in 1772. It was only in 1945, when Poland regained her Baltic coastline, that the industry sprang back into life. Poland needed ships for overseas trade and fishing; the Soviet Union, whose own shipyards were full with orders for military vessels, wanted to take advantage of Gdansk's merchant shipbuilding capacity. Over the next few decades Gdansk became Eastern Europe's leading shipbuilding center, making Poland the world's eighth most important shipbuilding nation by 1985.

The Gdansk shipyards became the focus of world attention when widespread dissatisfaction over food shortages and inadequate living conditions provoked major strikes in 1970 and 1980. The protests became symbolic of all workers' movements in Eastern Europe that aimed to throw off Soviet-backed communist governments.

The achievement of this aim, ironically, contributed to the shipyards' subsequent demise. Suffering from a world shipping recession and plant inefficiency, and without the Soviet market to save them, they became redundant and uncompetitive. In the early 1990s, attempts to privatize the yards had still not found success.

western Europe, and of exploiting the plentiful rural workforce to be found in these previously underdeveloped areas. Yugoslavia, on the other hand, which had evolved its own independent "route to socialism", located its key industrial sites deep inside the Dinaric and Balkan mountain systems for fear of invasion by the Soviet Union.

Heading for crisis

Industry entered a new phase after 1960. Due to policy shifts in the Soviet Union, oil and gas were used to fuel the development of new industries. Production of consumer goods, food products and the road transportation sector were flagging. To bolster them plants processing petrochemicals, plastics, fertilizers and synthetic fibers were set up. For example, refineries were centrally located near Warsaw in Poland, Budapest in Hungary, and Ploiesti in Romania. Chemical and fiber factories were sited close to supplies of water, and rivers – especially the Danube and Vistula – became heavily polluted "chemical conduits".

An increase in the exchange of Soviet coal, electricity, oil, gas and iron ore for Eastern European manufactured goods encouraged specialization. Czechoslovakia gained renown for industrial machinery, Hungary for its Ikarus buses, Poland for railroad equipment and ships, and Romania for chemicals. During the 1970s improved relations with the West fostered trade links, and in order to acquire new technologies and skills, Western-owned companies were encouraged to set up plants in the region near the major cities. These included plants producing automobiles (Fiat, Renault and Volkswagen), color televisions, farm-machinery (Massey Ferguson), chemicals, synthetic fibers and tires.

Despite these and other attempts to diversify and modernize, a profound crisis hit industry during the 1980s. Centralized management had stifled innovation and initiative; equipment was old-fashioned and worn out; manufacturing and mining industries – once such a notable success – were now inefficient and obsolete. However, although the political upheavals of 1989 opened the way to a thoroughgoing rationalization and modernization of industry, the continuing economic crisis in Eastern Europe proved an effective stumbling block to any attempts at successful reform.

Bright new ideas Eastern Europe's tradition of glassmaking adapts to a modern industry in this Tungsram bulb factory in Budapest. Like many successful new projects it is a joint venture, half owned by the American company CIE General Electric.

THOMAS BÁTA, ENTREPRENEUR

Thomas Báta (1876–1932) was a Czech entrepreneur of the early 20th century. He opened his first shoemaking workshop in the Moravian village of Zlin, then part of the Austro-Hungarian empire, in 1894 and by World War I had risen to become bootmaker to the imperial army. Back in civilian life he returned to Czechoslovakia and rapidly built up a shoemaking business that soon grew to become one of the newly independent country's largest manufacturing firms. Using the latest mass-production techniques pioneered by the American automobile manufacturer Henry Ford (1863–1947), he was able to increase output and cut costs.

Báta marketed his shoes direct through his own range of shops, and advertised vigorously in newspapers and magazines. He based his enterprise in his native village of Zlin, building an integrated production plant and providing a good environment for his employees. By the 1920s, Zlin had grown to become a tree-lined industrial town, complete with workers' housing, schools, shops and other amenities.

During the 1930s, Báta opened several large factories abroad, and the company became one of the world's leading shoemaking multinationals, selling inexpensive, durable – but rarely fashionable – footwear. In 1948 the communist government nationalized Báta's Czechoslovakian factories, but preserved his unique integrated shoemaking facility. Control of the multinational company remained with the Báta family, and following the overthrow of the communist regime in 1989, they embarked on the attempt to regain ownership and control of their assets in Czechoslovakia.

THE NEW REVOLUTION

From the end of World War II until the period of economic reform initiated by President Mikhail Gorbachev in the 1980s, the policies of the Soviet Union exercised direct or indirect influence over most aspects of Eastern European industry, from resource exploitation to the eternal shortages of consumer goods. Even the location of key plants – particularly the steel, power and defense industries – was decided by Soviet officials, often with Red Army approval.

Independence and private enterprise

Not all the countries of Eastern Europe fell within the direct sphere of influence, however. Albania remained within the Soviet bloc until 1960, when it switched allegiance to communist China, then under the leadership of Mao Zedong (1893–1976). Albania's leaders emulated Mao's "Great Leap Forward" that combined big urban and small rural industry, but retained strict state control.

From 1948 Yugoslavia came under the guidance of President Tito (1892–1980) and evolved its own form of socialism. This marked a shift away from centralized control toward a greater degree of workers' self-management. Private businesses employing up to five people were permitted and became a particularly important part of the expanding building, transportation, handicraft and tourist industries.

Even before the momentous political events of the late 1980s, there were signs in the other countries of Eastern Europe that the state's stranglehold over all industrial activity was already weakening. Czechoslovakia and Poland, for example,

The original Budweiser (*right*) At this brewery in Czechoslovakia Budweiser Budvar was being made long before another similarly named beer appeared in the United States. Czechoslovakia's beer-making traditions are centuries old. In 1295 King Vaclac II gave the citizens of Pilsen the right to brew beer, leading to the first pilsner lager.

Bicycles built by two (*below*) Hungary, with one of the strongest economies of all the Eastern European countries, actively encourages joint projects with Western companies. This factory in Budapest has been set up by the United States' bicycle manufacturer Schwinn to produce products both for Hungary and for the United States.

tacitly encouraged small private workshops. But it was in Hungary that small-scale private enterprise was given the most leeway.

Black market economies had existed for years in all countries to provide goods and services that were in short supply. In Hungary, this unofficial economy became part of everyday life and was openly tolerated by the authorities. The system was known as *maszek*, a contraction of the Hungarian word for private sector. Enterprising factory workers, for example, might use the state's materials and equipment for unofficial business and earning money for themselves while satisfying the customers by providing their goods more quickly.

Semilegal *maszek* sowed the seeds that led to the official reintroduction of private enterprise: by the early 1980s, approximately 12,000 firms were licensed to operate privately. Hungary, like former Yugoslavia, also devolved more responsibility for production and marketing decisions to the managers of state enterprises. During the 1980s, similar changes spread to other countries in the region.

Toward the future

The reforms that swept through Eastern Europe in 1989 effectively removed the communist monopoly of power, paving the way for new attitudes to industry. The break up of the Soviet bloc – which had tried to project an image of unity – revealed marked national differences. The northern peoples of Poland, Czechoslovakia and Hungary were more or less committed to privatization. Those in Albania, Bulgaria, Romania and former Yugoslavia were divided or undecided over introducing sweeping changes.

Eastern European industry was entering a new era full of opportunity, but it had inherited deep-seated problems. The inevitable closure or rationalization of mines and factories that yielded surplus output or outmoded goods was bound to cause unemployment. There were fears that, even allowing for the growth of new service jobs, some areas would become industrial "badlands", the toxic wastelands of abandoned industries.

Years of unregulated industrial activity has left some parts of the region with a legacy of severe environmental damage. The excessive reliance on brown coal to generate power caused massive air pollution. The waste from metal works and chemical complexes resulted in degradation that amounted to an ecological disaster in the opinion of some experts.

Solving environmental problems, implementing technological modernization and developing new industries was likely to be a lengthy and expensive process. Despite the changed market conditions, many people were reluctant to enter a new spirit of entrepreneurialism. It was not easy to abandon the attitudes that regarded capitalism as a social and economic evil, which years of communist teaching had established. Yet internal efforts to revitalize industry, coupled with Western aid and investment in what could prove to be a lucrative new market, offered grounds for cautious optimism.

The brown coal of Eastern Europe

Coal, a primary fossil fuel, is a hard, carbon-rich material formed by the decomposition and chemical conversion of vast amounts of vegetable matter under conditions of great pressure and over millions of years. Most authorities agree that four major ranks of coal can be identified: anthracite, bituminous coal, sub-bituminous coal and lignite. The differences relate to hardness, and depend on how long ago coalification took place, and the degree of pressure to which the material has been subjected.

Anthracite is the oldest and the hardest coal, and is the best quality fuel. It contains between 80 and 95 percent carbon, burns with little smoke at great heat and leaves little ash. Bituminous coal, with a carbon content of up to 80 percent, is widely used in industry to produce steam and for coking. The poorer quality sub-bituminous coal, which contains less carbon and tends to throw off sparks during combustion, is frequently used to produce gas. Lignite contains the highest proportion of moisture, ash and volatiles: usually about 40 percent. It is therefore less heat efficient than other forms of coal, and gives off far greater amounts of air pollutants.

The coal most commonly found in Eastern Europe is usually known as brown coal. There is, however, no international agreement on the classification of brown coal. In the United States, the term applies to the poorest quality coal, worse than lignite. In Europe, brown coal is interpreted more broadly to include both lignite and sub-bituminous coal. The European definition is used here.

Quantity versus quality

Although brown coal is a low-quality fuel, it plays a critical role in Eastern Europe. Production rose rapidly after 1950, first to fuel the ambitious program of industrialization and – after 1974 – to cushion the effects of price rises or cutbacks in the supply of Soviet oil. The region's brown-coal reserves are currently estimated at 38.6 billion tonnes, or 3.5 percent of the world's reserves; in 1988, however, the region produced 350 million tonnes, or 23 percent of world brown-coal output. Czechoslovakia, Poland, Yugoslavia and Romania rank in the world's top 10 producers of brown coal, Bulgaria and Hungary in the top 15.

The reserves of brown coal are scattered among many deposits. Even so, most – especially the larger deposits lying in lowlands or basins – can be mined economically. It requires relatively few workers to operate the huge draglines that remove the surface overburden (material covering the coal seam). This can be up to 40 m (130 ft) thick. The technique is used at Konin and Turoszow in Poland, Tatabanya in Hungary and Kosovo Polje in Yugoslavia. Brown coal found in very hilly terrain is extracted via underground shaft mines or drift mines into hillsides; this method is common in the Dinaric mountains and at Kreka and Tuzla, all in Yugoslavia.

Most brown coal is used for electricity generation, chemicals manufacturing or is made into briquettes for heating. Because of its high waste content, it is converted to these uses on site to save transportation costs.

Since 1950, industrial complexes of varying degrees and types of integration have been developed in brown-coal-mining areas. The simplest of them comprise a thermal electric power (TEP) station, as at Turoszow in southwest Poland. Elsewhere, integration has gone further. At Konin in central Poland, the TEP plant provides electricity to an adjacent aluminum smelter. A complex at Kosovo in Yugoslavia produces chemicals, while at Tatabanya in northwest Hungary waste from mining and coal processing is also turned into building materials for the Budapest market. Attempts in Yugoslavia to manufacture coke from brown coal for use in steelmaking have failed.

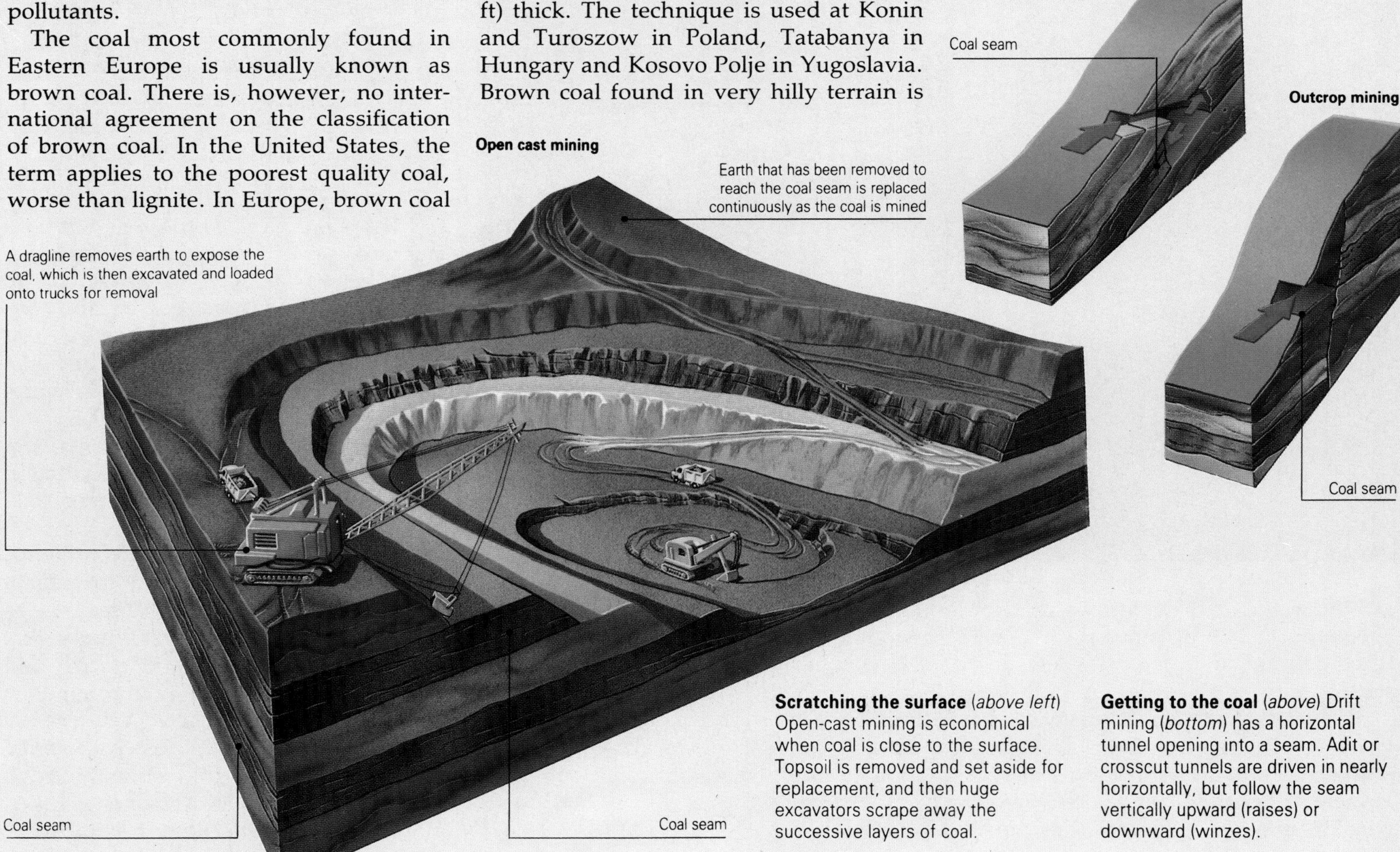

Scratching the surface (*above left*) Open-cast mining is economical when coal is close to the surface. Topsoil is removed and set aside for replacement, and then huge excavators scrape away the successive layers of coal.

Getting to the coal (*above*) Drift mining (*bottom*) has a horizontal tunnel opening into a seam. Adit or crosscut tunnels are driven in nearly horizontally, but follow the seam vertically upward (raises) or downward (winzes).

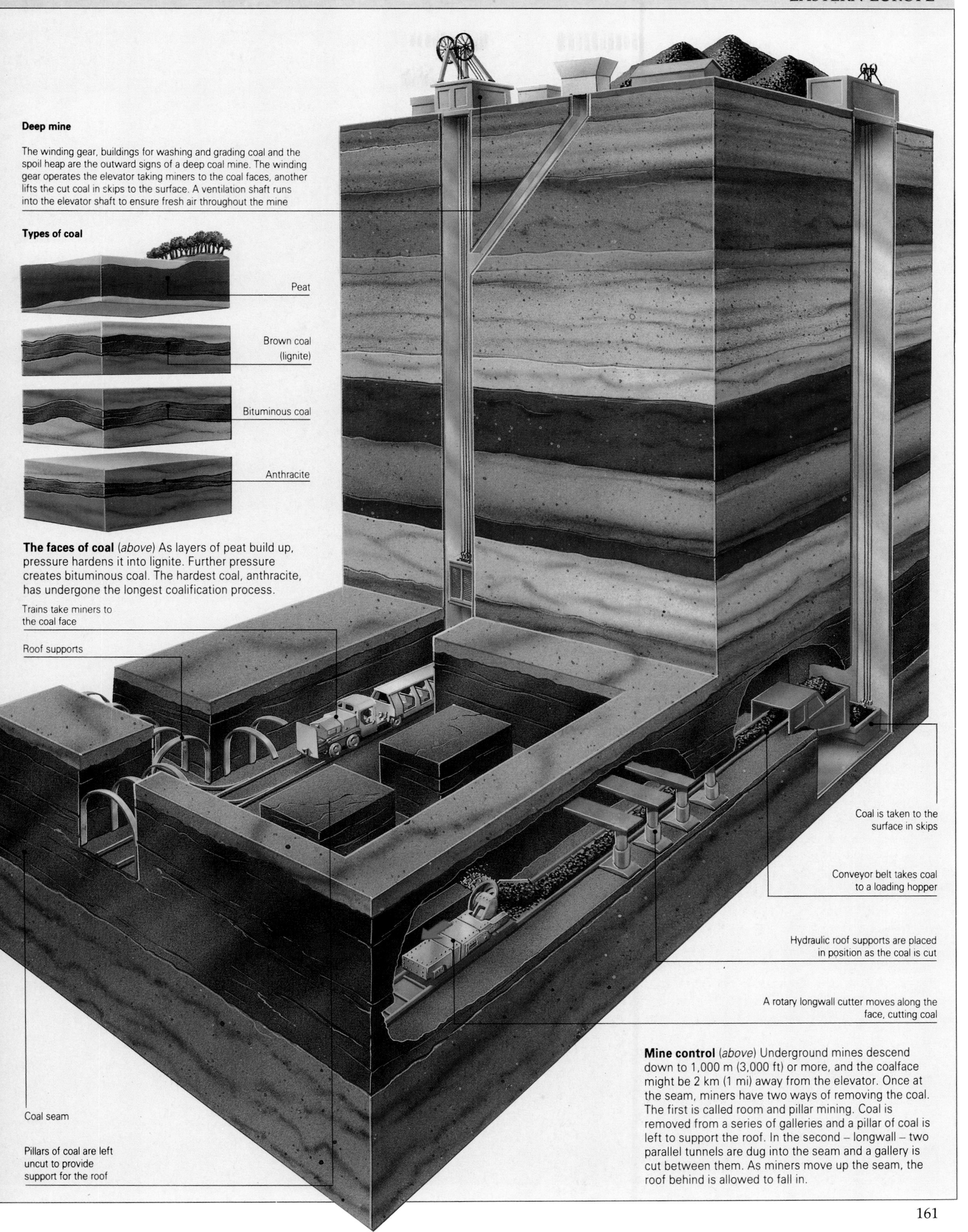

Deep mine

The winding gear, buildings for washing and grading coal and the spoil heap are the outward signs of a deep coal mine. The winding gear operates the elevator taking miners to the coal faces, another lifts the cut coal in skips to the surface. A ventilation shaft runs into the elevator shaft to ensure fresh air throughout the mine

The faces of coal (*above*) As layers of peat build up, pressure hardens it into lignite. Further pressure creates bituminous coal. The hardest coal, anthracite, has undergone the longest coalification process.

Mine control (*above*) Underground mines descend down to 1,000 m (3,000 ft) or more, and the coalface might be 2 km (1 mi) away from the elevator. Once at the seam, miners have two ways of removing the coal. The first is called room and pillar mining. Coal is removed from a series of galleries and a pillar of coal is left to support the roof. In the second – longwall – two parallel tunnels are dug into the seam and a gallery is cut between them. As miners move up the seam, the roof behind is allowed to fall in.

The dark side of industry

In the months that followed the downfall of communism across Eastern Europe in 1989–90, the true extent of the decline of the region's industrial plant was uncovered. The state of the factories would have an impact on the economies and environments of the region that would reverberate far into the future.

Modernization was almost an unknown concept in the industrial theory of communism. Factory managers were urged to rely on cheap human labor rather than on far more expensive machinery. Throughout Eastern Europe working conditions were primitive, and safety precautions rudimentary. The industrial death rate was far higher than in the West, and in the opinion of many Western industrialists who visited the region, most of the region's factories were lagging behind in technology by at least 25 to 50 years.

The environmental damage caused by uncontrolled industrial emissions and run-down, neglected plant and machinery is already enormous, and likely to grow still further. The advancing decay of the Soviet-built nuclear power stations in Czechoslovakia and Hungary for example, is provoking fears for the future. All the region's new governments face the problem of clearing up the mess. In September 1990 the new Czech government pledged to deal with "a wide range of environmental problems that had only come to light after the end of communist rule". The two areas of greatest concern were the high emissions of sulfur dioxide gas by heavy industries and the contamination of the water supply by chemicals leaked into the ground over a period of years. Czechoslovakia, Hungary and Poland all opened their state industries to Western investors to help get the factories out of the industrial dark ages.

Hazardous work A Bulgarian factory worker strips zinc from a plate used in zinc smelting. He is working without protection in a mist of sulfuric acid, making it almost impossible to breathe.

MASSIVE HEAVY INDUSTRY

RESOURCES TO FEED RAPID GROWTH · ACHIEVING INDUSTRIAL SELF-SUFFICIENCY · PEOPLE AS INDUSTRIAL FODDER

Northern Eurasia's diverse natural resources are among the richest and most plentiful in the world. The region, comprising the former Soviet Union and Mongolia, is a world leader in the production of natural gas, iron ore, coal, manganese, mercury and potash; and ranks second or third in the production of oil, copper, lead, zinc, diamonds and nickel. Yet the very diversity of terrain and climatic conditions that helped to produce such a wealth of resources also creates major difficulties in extracting them. In many cases resources are so remote that transportation problems make it uneconomic to exploit them. Communist domination of the region from 1917 until 1991 has shaped industrial development, encouraging production of engineering and the defense industries at the expense of consumer products.

COUNTRIES IN THE REGION

Armenia, Azerbaijan, Belorussia, Estonia, Georgia, Kazakhstan, Kirghiz, Latvia, Lithuania, Moldavia, Mongolia, Russia, Tadzhikistan, Turkmenistan, Ukraine, Uzbekistan

INDUSTRIAL OUTPUT (US $ billion)

Total	Mining and Manufacturing	Average annual change since 1960
752.6	752.6	+5.7%

INDUSTRIAL WORKERS (millions)
(figures in brackets are percentages of total labor force)

Total	Mining and Manufacturing	Construction
54.7	41.6 (31.8%)	13.1 (10%)

MAJOR PRODUCTS (figures in brackets are percentages of world production)

Energy and minerals	Output	Change since 1960
Coal (mill tonnes)	662.4 (14.1%)	+35.2%
Oil (mill barrels)	4452.9 (19.7%)	+311%
Natural gas (billion cu. meters)	711.7 (37.1%)	+1471%
Iron Ore (mill tonnes)	138.2 (24.3%)	+31%
Nickel (1,000 tonnes)	189.6 (23.5%)	+38.5%
Vanadium (1,000 tonnes)	9.6 (31.1%)	No data
Phosphate rock (mill tonnes)	12.0 (22.6%)	+83%
Manufactures		
Steel (mill tonnes)	163.0 (22.3%)	+149.6%
Cement (mill tonnes)	134.9 (12.3%)	+196.6%
Sulfuric acid (mill tonnes)	29.4 (20.5%)	+132.2%
Linen fabric (mill sq. meters)	788.8 (77.4%)	No data
Footwear (mill pairs)	819.1 (18.4%)	+141.2%
Refrigerators (mill)	6.2 (11.5%)	+167.3%

In the pipeline *(above)* The source of Siberian oil is very distant from its users, so vast networks of pipes have been built to transport it. Sub-zero arctic conditions and the remoteness of the pipelines make routine maintenance extremely difficult.

RESOURCES TO FEED RAPID GROWTH

The story of how resources have been exploited in the region is dominated for most of the 20th century by the collective aims of the Communist Party of the Soviet Union (CPSU). Through a series of five-year plans the Party channeled labor, finance and resources into building up state-controlled heavy industry, especially the sector manufacturing machinery to equip new industrial plants, farming and construction. Ambitious targets of production set by the CPSU in 1922 forced a dramatic growth in the search for and exploitation of energy supplies. Huge oil and gas discoveries meant that between the 1920s and the 1980s the region's energy output soared eighty-fold. Hydroelectric and nuclear power also added to the region's ample energy resources.

Almost one-quarter of the world's reserves of coal are found in Northern Eurasia. The largest coal-producing area has always been the Donets basin (Donbas) in the Ukraine, which produces 190 million tonnes per year. Recently, however, extraction has become more difficult and expensive as the remaining exploitable seams run very deep. The second most important area of coal production is the Kuznetsk basin (Kuzbas), which is in Siberia, Russia's inhospitable central zone that runs from the Ural mountains toward the Pacific Ocean. Here the seams are thick, continuous and relatively near the surface. Fields at Karaganda in Kazakhstan and Pechora in the far north of Russia also contribute significantly to the total output. In addition, there are two fields in the Tunguksa and Lena basins in Siberia that remain unexploited because of the brutally cold climate. Coal reserves in the region are so vast that finding a way to exploit these fields is not yet essential.

The region is the world's second largest oil producer after the Middle East. Almost 185 million tonnes of crude oil are exported annually. Before the 1930s practically all the oil produced in the region came from the Baku fields in Azerbaijan. Since then reserves have been exploited at four large fields in the Ural–Volga area of Russia and at the Tyumen complex in western Siberia (which now produces more than 50 percent of the region's oil). Natural gas is proving just as important as oil, especially as an export.

Hydroelectric power was first developed in 1928 on the Don and Volga rivers in the western part of the country. The Dnieper dam project began production in 1932, and since 1955 gigantic stations have been constructed in south-central and eastern Siberia, at Krasnoyarsk on the Yenisey river and at Bratsk and Ust-Ilimsk on the Angara river.

Nuclear power produces about 10 percent of all electricity generated in the region, although almost all the reactors are the standard slow reactor type, overtaken by new technology. Most of the stations are dispersed in the energy-deficient west from the Kola Peninsula in northwestern Russia to Armenia in the south, and are run on uranium supplies from Krivoi Rog in the Ukraine and Narva in Estonia. Uranium is also found in Kazakhstan, eastern Siberia and the Tien Shan mountains near the Chinese border.

A treasure trove of minerals

At Norilsk in the Arctic Circle, there are huge reserves of platinum, nickel, gold, copper, cobalt and silver. Lead and zinc reserves are also vast in Kazakhstan at Leninogorsk and Ordzhonikidze in the northern Caucasus Mountains. The world's largest deposit of iron ore is in Russia at Kursk Oblast in the central plateau, and this together with six other large reserves, including those at Krivoi Rog and Karelia-Kola, yield over 236 million tonnes a year. Nonmetallic minerals include apatite concentrate (used to make fertilizers) mined from the reserves in the Kola Peninsula bordering Finland in northwestern Russia. Asbestos and potash are also mined in large quantities in the central Urals and eastern Siberia. After Zaire, the region is the second largest diamond producer with most gemstones coming from mines in eastern Siberia.

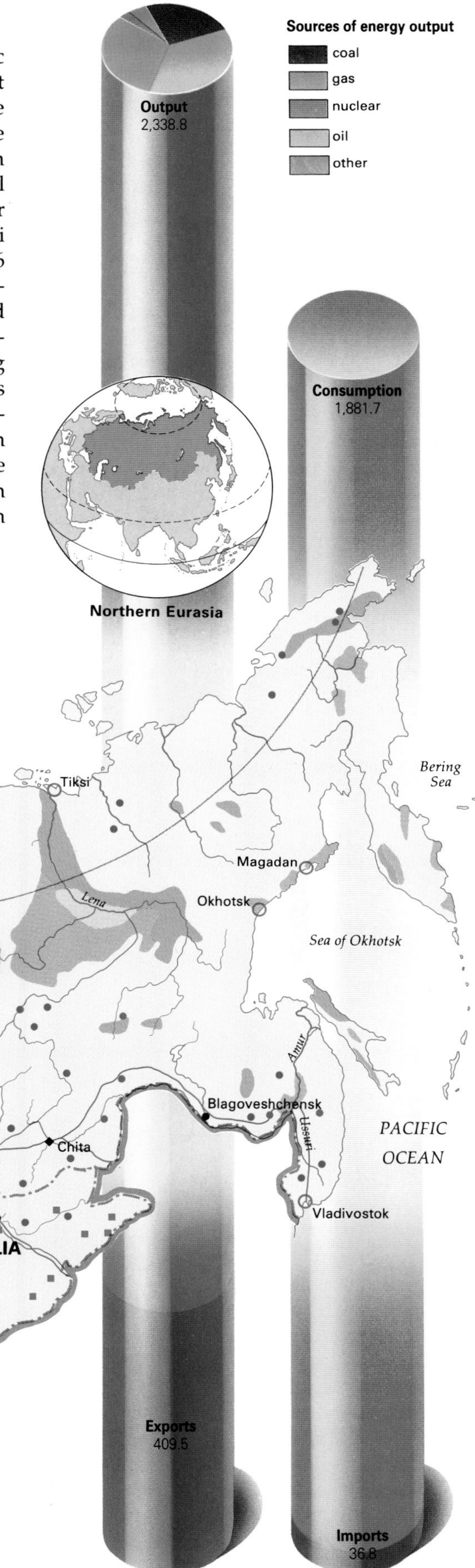

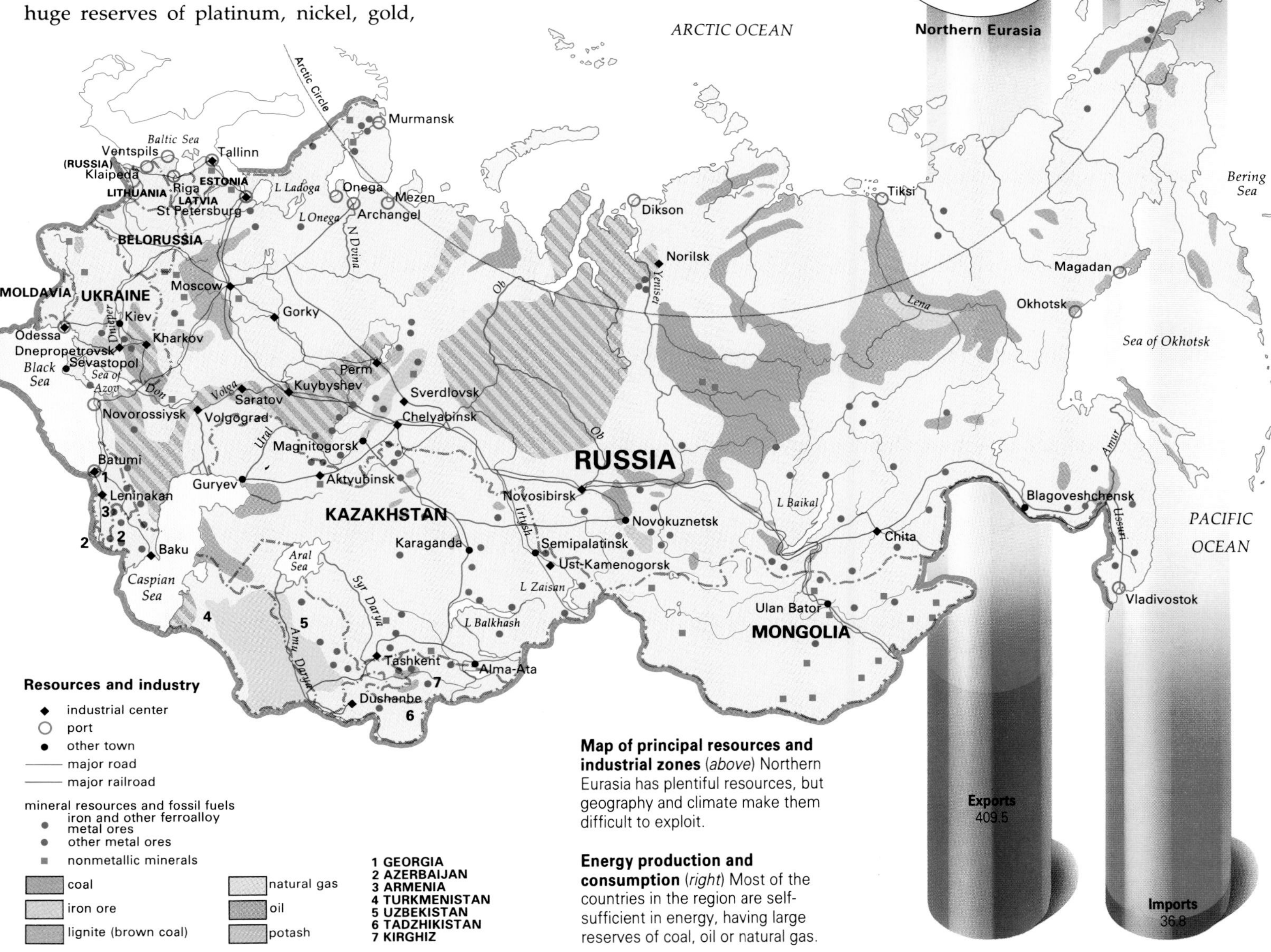

Map of principal resources and industrial zones *(above)* Northern Eurasia has plentiful resources, but geography and climate make them difficult to exploit.

Energy production and consumption *(right)* Most of the countries in the region are self-sufficient in energy, having large reserves of coal, oil or natural gas.

ACHIEVING INDUSTRIAL SELF-SUFFICIENCY

It took the communist revolution of 1917 to quicken the pace of industrial growth in the region, which prior to this had a predominantly rural society. Once the controlling CPSU had decided that the state must be self-sufficient in manufactured goods, all natural resources were declared to be state-owned and then systematically exploited to meet the aims of the new political creed.

The overriding economic aim of early communist leaders under Vladimir Ilich Lenin (1870–1924) was to achieve industrial self-sufficiency. Without it they perceived that they would never maintain independence from the capitalist West. Under Joseph Stalin (1879–1953), large-scale heavy industries such as steel, metal processing and machinery manufacture took priority over the production of consumer goods such as clothing or household appliances. In fact, in the period from 1920 to 1980, engineering, industrial construction and heavy industry grew nine times faster than the consumer industries. By any standards the transformation of the country was extraordinary: during the same 60-year period over 40,000 new mines, power stations and manufacturing plants were opened.

Spreading the message

Another influential political aim was to spread the benefits of communism across the whole region, by giving each of the many distinctive areas its own industrial heartland and urban center. It was a herculean task and was achieved largely through military force and strong political conviction. Industrial processing plants were deliberately situated near raw materials, no matter how inhospitable the climate. Railroads and roads were built to open access to resource-rich areas such as Siberia, and to make it feasible to develop manufacturing complexes close to where resources were mined and processed, and distribute the goods from there. The country's largest tractor factory, for example, was constructed at Chelyabinsk in the Urals, and flourished as an industrial center because of its proximity to the Trans-Siberian Railroad.

After World War II the Soviet Union embarked on a period of building up its defense industries: weapons, tanks and aircraft. At the same time it developed a policy extending its political influence and the Soviet model of industrialization became widespread. Latvia, Lithuania, Estonia and Moldavia were annexed in 1945, while Mongolia, eastern Europe and parts of east Asia and Central America began to develop centrally controlled heavy industry under Soviet guidance. China was also heavily influenced until the mid 1950s. During its expansionist period the Soviet Union initiated a program of space research. All these ventures required even greater exploitation of the region's natural resources and energy.

Developing the outlying regions Sumgait in Azerbaijan, north of Baku, had no tradition of steelmaking until Soviet economists decided to locate the industry there. Creating industrial centers from scratch was a hallmark of central industrial planning.

The central importance of Siberia

Before 1913 industry was concentrated in four main areas, three of them – Moscow, St Petersburg and the fertile triangle of the Russian Republic – were in the west of the country. The other main industrial area was around the Baku oilfields of Azerbaijan. The rapid industrialization that began with the implementation of the first Five-year Plan (1928–32) changed this pattern in two significant ways. First, Siberia was developed enormously because of its substantial natural resources, better communications, transportation networks, and an increased labor force. In addition, new industrial clusters or "nodes" developed in far-flung regions where energy, forest resources and metals

THE SPACE PROGRAM

Until the mid 1980s the Soviet space program was the flagship of technological and scientific prowess in the region. This meant that the program benefited from the best scientists and resources that the state could provide, until economic accountability became an issue in this and other Soviet research programs. Although initial research started in the 1920s, real progress began after World War II when the Soviet Union acquired German expertise in rocket propulsion. However, it was the work of the brilliant rocket engineer, Sergey Korolyov, that gave the space program the technological capability to lead the world.

In October 1957 the launch of Sputnik 1 was a pioneering achievement, as was the first manned flight with Yury Gagarin in April 1961. The Soviet moon program culminated with Luna 17, which put a space vehicle, Lunakhod, on to the moon's surface in November 1970. Venera probes were sent to Mars and Venus in the 1980s, but thereafter scientists decided to concentrate on launching large Salyut space stations into orbit. Serviced by Soyuz spacecraft, these manned stations enable scientists to carry out research over longer periods, monitor the weather, supply communications facilities, and collect data for geological prospecting.

Most of the highly specialized equipment for spacecraft and space stations is manufactured in Russia, particularly in Moscow. Launches are carried out at the Baikonur cosmodrome in southern Kazakhstan.

Generating business Hydroelectricity projects boost the construction industry as well as providing energy for additional growth. The Ust-Ilimsk project in central Russia has been under construction since 1974, and will bring a fresh impetus to industry along the river.

attracted massive political and financial commitment until the twelfth Five-year Plan (1986–90), which put many of the projects that were designed for these regions into abeyance.

Trends in manufacturing have changed noticeably since early industrialization. A good example of this can be seen in the industrial area around Moscow, which before 1940 had a concentration of metal processing, machine manufacturing and vehicle industries (notably the giant ZIL truck, bus and automobile plant). Since 1950, however, the area has specialized in skill-related industries such as research-intensive electronics, aerospace and nuclear technology, though in several cases such industry has been situated in new satellite towns and cities some distance from Moscow.

During the 1970s the Soviet Union accepted some technical help from the West and by 1987, under the leadership of Mikhail Gorbachev, joint ventures with Western firms were officially sanctioned. Between 1987 and the dissolution of the Soviet Union in 1991 over a thousand such enterprises were officially registered and some three hundred began operating. How these ventures would develop during political reorganization remained an open question.

Other longterm projects, initiated at state level by the Soviet Union, have had considerable effect on the distribution of industry in the region. A case in point is the Baikal-Amur Magistral (BAM) railroad project, which was opened in 1985 after 20 years of construction work. Running to the north of the existing Trans-Siberian railroad, it is a principal route for exporting freight from eastern ports, though permafrost conditions still limit use of the track. This project has consolidated the importance of Komsomolsk-na-Amure, which was built from scratch in 1932 and has become one of the major areas in the Far East for the production of steel, oil-refining and pulp and paper mills. However, despite government incentives to persuade workers to migrate east, the bulk of industry still remains in western and central Russia, where the labor force is well established and the industries wide ranging.

PEOPLE AS INDUSTRIAL FODDER

The states of the former Soviet Union together with Mongolia house a huge range of nationalities, and after 70 years of communism newly emergent peoples are struggling for the right to express their regional identities. Under the leadership of Lenin and Stalin, people were often moved against their will from a local agricultural way of life to another area where they formed part of the workforce establishing an industrial base for the new communist state. By forcing industrial progress on the people Stalin increased production dramatically, opening 6,000 new enterprises during the first 10 years of his regime; but the human cost is not recorded. During the 1930s, for example, development of a large iron and steel complex transformed a Russian steppe village into the city of Magnitogorsk, making it a major producer of military equipment. The local population, along with imported labor, was drafted to work in the booming city.

Working for the state
In order to meet industrial goals, the government employed forced labor during the 1930s and afterward, especially for geographically remote projects such as the White Sea Canal, which cuts through Onega near the Arctic Circle. It is estimated that between four and five million people were forcibly put into labor camps and made to build railroads, canals and mines in the years up to 1939. During World War II when most men were conscripted, women were put to work in industry.

More than 10 million people were evacuated eastward to keep Soviet industry running, but destruction of industrial units by the Germans was massive. More than 82,000 enterprises were destroyed, 55 percent of steelmaking capacity and 60 percent of coal output were lost. After the war people were exhorted to work even harder, but cumbersome centrally planned policies were failing to achieve results. In the 1960s Siberia was singled out as a priority area because of its abundant energy resources, and there were government incentives for people to live and work there.

The continued dominance of manufacturing in the machinery and arms sectors led to chronic shortages of consumer goods, and discontent among the people flared up in demonstrations that were repeatedly quashed by the Red Army. Although Nikita Krushchev (1894–1971), leader of the Soviet Union 1955-64, expanded production of consumer goods to some extent, his priorities were to develop the oil and gas-based industries, plastics and fertilizer manufacture. His attempts to integrate industry and agriculture had only limited success, leaving many rural areas with plentiful labor but few industries. Although government-run scientific studies called for full use of surplus labor resources in the 1960s and 1970s, local Party bosses lost sight of this aim in an effort to compete with each other in diverting industry to their areas. This led to "tugs of war" between those people who supported regional specialization and those who were in favor of regional self-sufficiency.

TRADITIONAL CARPETMAKING

Northern Eurasia is one of the leading producers of handwoven rugs and carpets. There is a great tradition of craftsmanship that dates back to at least the 3rd or 4th century BC, a fact attested to by the discovery of what is thought to be the world's oldest known woolen carpet unearthed in a frozen Scythian tomb in southern Siberia.

The best known and most important regional carpets are the oriental style rugs woven in the eastern Caucasus Mountains and central Asiatic districts. Historically these rugs have always possessed a distinctive character – never as fine or densely knotted as Persian or Turkish carpets – but rather more coarsely woven with strong colors and innovative designs. Even now, many of the carpets produced in the region are woven on hand-held looms, usually by women and children who work either from home or in cooperative workshops and studios.

No one area has an absolutely pure regional pattern; the finished product is usually the result of a composite influence of different regional designs and colors. For instance, carpets woven in Armenia and Azerbaijan often display Persian, Anatolian or even Chinese influences, while the well-known Dagestan rugs display designs from Baku, Chichi and Kuba.

In central Asian districts carpets renowned for their durability and unique patterns are woven by various nomadic tribes, such as the Ersari, Salor, Tekke and Yomut. These rugs, sold in centers such as Ashkabad in Turkmenistan, Samarkand in Uzbekistan and Khodzhent in Tadzhikistan, owe much to the Islamic culture of the old silk road, which in medieval times was a trade route for silk from China.

The new silk road Tadzhikistan and Uzbekistan are two of the world's largest producers of cotton and silk. After 1990, a traditional Islamic weaving culture producing high-quality carpets and rugs reemerged in these independent states.

Traditional carpetmaking was preserved as a craft by the communist state to earn hard currency from exports and to provide jobs and cultural continuity among ethnic minorities. The combination of Islamic revival and political independence in the southern republics is likely to stimulate even further development of the industry.

Fighting for survival *(above)* The defense industry in the former Soviet Union received approximately 65 percent of the region's annual budget. Military factories, unlike civilian ones, were accorded the best machinery and strictest quality control. Huge amounts of money were allocated for the development and perfection of armaments such as the MiG fighter, once the most effective aircraft in the Soviet arsenal. With the breakup of the Soviet Union defense funding and the former military-industrial complex have been severly affected. MiG factories such as this one must now seek foreign customers or otherwise face closure.

Trucks made by women *(left)* Gender barriers in most occupations have been eliminated in the region, due more to labor shortages than ideology. Under Soviet rule, state-subsidized childcare provided by employers made it the norm for mothers of small children to continue working.

The more liberal policies of Mikhail Gorbachev, who took up leadership of the Soviet Union in 1985, sowed the seeds of reform and change. He tried to move labor from heavy industry and defense manufacturing to the construction of better social services and improved training and education. However, nationalist uprisings led to unresolved strife over the control and use of natural resources, and combined with longstanding discontent to cause political upheaval. With the breakup of the Soviet Union, the question of which national groups have the right to exploit the vast mineral, manufacturing and labor resources of the region is even more crucial and potentially dangerous.

Women in the workforce

Under the Soviet regime women formed an essential part of the labor force, even in heavy manual trades such as building and construction work. The rapid entry of women into the urban industrial sector was accompanied by welfare programs superior to those in the West. Workplace creches were provided, and even at the end of the 1980s were still more plentiful than in most Western countries. Maternity leave with the option of returning to a secure job was introduced as early as 1918, and in the factories women were allowed to take half-hour breaks every three hours to feed their babies.

Such measures were aimed at maintaining maximum productivity from a large sector of the potential workforce. Following the two world wars, women comprised 56 percent of the region's population, and they now dominate the labor markets in light industry and education. There are also far more women economists, engineers and doctors than in the West. However, their professional status and salaries tend to be much lower.

The industrial Volga valley

The Volga river in Russia is arguably the most important waterway in the region, and has become the focal point of a giant industrial development. In total, the Volga measures 3,500 km (2,200 mi) and flows through an area containing some 40 million people. The industries along its banks produce 18 percent of the region's total industrial output – as much as the entire Ukraine. Two major advantages have contributed to this prominence: the river is a cheap and convenient way to transport heavy goods between the north and south, connecting most of the industrial centers in the area, and the force of the river can be harnessed to generate hydroelectric power.

When Soviet-planned industrialization began in 1928 experts considered the Volga valley to be lacking in sufficient energy resources. At first it looked as if this prediction was accurate. Nonrenewable reserves of oil and gas were rapidly depleted between 1937 and 1975, but hydroelectric power has increased the energy potential of the area far beyond any early expectations. The river has been transformed into a string of lakes with eight HEP stations providing 10,000 megawatts of generating capacity.

High-energy manufacturing

Despite these initial doubts, rapid manufacturing growth began in the Volga in 1933 and has been sustained ever since. The area has acted like a magnet, drawing investment and labor toward central Russia. During World War II as the Nazis overran the Ukraine, most of the major machinery, vehicle, armaments, aircraft and chemical factories were moved from the Ukraine to the Volga region for strategic reasons. As the industries expanded, new satellite towns were built up around Gorky, Yaroslavl, Kuybyshev and Volgograd (formerly Stalingrad). After 1945 oil refineries were set up to process local oil, and petrochemical plants were established nearby, making artificial fibers, tires, fertilizers and plastics.

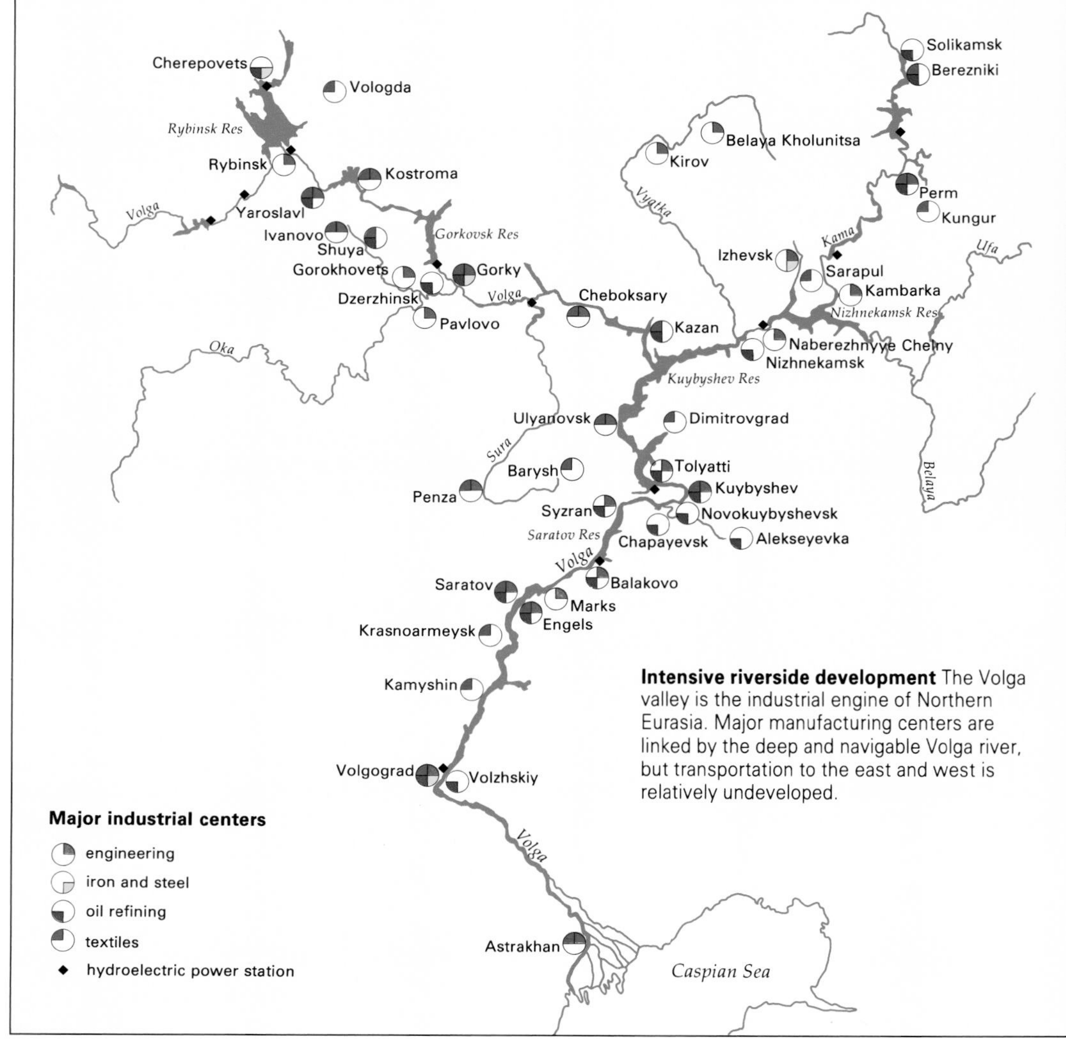

Intensive riverside development The Volga valley is the industrial engine of Northern Eurasia. Major manufacturing centers are linked by the deep and navigable Volga river, but transportation to the east and west is relatively undeveloped.

With the development of hydroelectric power came energy-intensive industries including smelting and electroplating. The largescale migration of families from the countryside created a growing male workforce for local heavy industry, and additional pools of female labor for work in food processing, textiles, clothing and shoe manufacture, the fur trade and a range of other light industries.

In addition, the Volga's geographic location between Europe and the western republics on one side (the potential market), and central Asia, the Urals and Siberia on the other (rich in energy and

River of steel This iron and steel mill is typical of the linked concentration of heavy industry in the region. Supplies and raw materials arrive by water from plants nearby and steel bars are shipped along the river for further processing.

resources) has encouraged the development of engineering assembly plants, especially of transportation equipment and, since 1965, motor vehicles.

East–west transportation problems

Although the Volga region has much in its favor as a flourishing industrial area, it is hampered by an inadequate transportation system with which to distribute its products to markets. The Soviet developers had overlooked this, regarding transportation, like all service industries, as nonproductive and low priority. As a result of neglecting distribution services, a good deal of labor, food, materials and energy have been squandered because rolling stock or haulage systems are inadequate, slow, unreliable and out of date. The situation is not helped by the huge distances between major cities in the region, or by the hostile physical environment of mountains, permafrost, marshes and deserts.

Although the mighty Volga river carries two-thirds of inland waterway traffic and is a significant link for boats from the connector canals between the Black and Caspian Seas in the south and the Baltic, White and Barents Seas in the north, it has always been a barrier to the east-west flow of commodities. The river is crossed by seven railroads and many pipelines, but few serve the river traffic and vice versa. Most Volga industry uses rail, pipeline or powerline transportation and though there is some major growth in the production of motor vehicles in the area, the condition of surrounding roads is poor and their network thin.

Refining history

The modern international oil industry can trace its roots back to the oil fields of Azerbaijan. Oil wells in the Baku region, between the Caucasus Mountains and the Caspian Sea, were supplying fuel to Europe and beyond as early as the middle of the 19th century. At the beginning of the 20th century, Azerbaijan was the world's leading producer of oil, accounting for 95 percent of Russian production and over 50 percent of the world total. In 1901 the world's first modern refinery was built near Baku.

The combination of drilling and refining industries in Azerbaijan has helped the area to develop other aspects of petroleum-related manufacturing such as chemical, fertilizer, rubber and plastic production. Baku and the surrounding area also manufacture equipment for the oil and gas industries.

These industries survive although Azerbaijan contributed less than 10 percent of total oil output in the former Soviet Union in 1990. Baku was surpassed as the leading Soviet oil center by the Volga-Urals oilfield in 1950. This field in turn relinquished leadership to the west Siberian oilfield in 1978 that now yields 70 percent of production throughout the region. In 1991, however, British Petroleum gained the right to drill for oil in the Caspian Sea and a renewed period of exploration, drawing on North Sea experience, may help to revitalize the industry in Azerbaijan.

A dense forest of oil derricks stretch across the horizon, dominating the environment around Baku and throughout most of eastern Azerbaijan.

INDUSTRIALIZATION THROUGH OIL

OIL BUT NO WATER · PROMOTING DIVERSIFICATION · RETRAINING THE WORKFORCE

Oil and natural gas are the dominant resources of the Middle East. They provide raw materials for other industries, such as plastics and chemicals, and the revenue from exporting them supplies the capital investment needed for wider industrial development. The mining of other mineral resources is also significant – especially in those countries that are not endowed with oil. Any program of industrialization in the region has to overcome fundamental difficulties, including civil strife, and Israel and Turkey are the only countries so far to have developed broad-based manufacturing industries. The industrial experience of the workforce in these newly wealthy countries is limited, and both technical expertise and labor have to be imported from other developed and developing nations.

COUNTRIES IN THE REGION

Afghanistan, Bahrain, Iran, Iraq, Israel, Jordan, Kuwait, Lebanon, Oman, Qatar, Saudi Arabia, Syria, Turkey, United Arab Emirates, Yemen

INDUSTRIAL OUTPUT (US $ billion)

Total	Mining	Manufacturing	Average annual change since 1960
237.3	83.4	55.6	+4.9%

INDUSTRIAL WORKERS (millions)
(figures in brackets are percentages of total labor force)

Total	Mining	Manufacturing	Construction
11.77	1.33 (2.5%)	5.55 (10.6%)	4.9 (9.3%)

MAJOR PRODUCTS (figures in brackets are percentages of world production)

Energy and minerals	Output	Change since 1960
Oil (mill barrels)	5966.6 (26.3%)	+288.5%
Natural gas (billion cu. meters)	102.2 (5.3%)	+604%
Marble (mill cu. meters)	3.9 (40%)	No data
Magnesite (mill tonnes)	3.4 (18.8%)	No data
Borate (mill tonnes)	2.04 (52.8%)	No data
Manufactures		
Wool yarn (1,000 tonnes)	33.9 (10.3%)	-38.4%
Cement (mill tonnes)	66.5 (6.1%)	+1724%
Steel (mill tonnes)	9.8 (1.3%)	+2700%
Nitrogenous fertilizer (mill tonnes)	5.1 (5.5%)	N/A
Polyethylene (1,000 tonnes)	357 (1.6%)	N/A
Jet fuels (mill tonnes)	7.4 (5.2%)	N/A
Motor gasolene (mill tonnes)	26.0 (3.6%)	N/A
Liquefied petroleum gas (mill tonnes)	20.1 (13.1%)	N/A

N/A means production had not begun in 1960

Refining in Abu Dhabi *(above)*, the wealthiest member of the United Arab Emirates. Since oil was discovered in 1958 the government has taken a strong interest in the industry, but large concessions are also owned by private and foreign-owned companies.

OIL BUT NO WATER

The Middle East has by far the world's largest reserves of crude oil – in 1989 it was estimated that the region had 65.2 percent of the world's total stock. Saudi Arabia is the largest producer in the region and possesses over one-quarter of the world's oil. Natural gas reserves, of which Iran has by far the greatest share, are also substantial and the region is the world's second largest producer – after the states of the former Soviet Union – with 30.7 percent of global reserves.

Although the region has an abundance of fuel resources, there is a widespread scarcity of water – a resource that is required not only for agriculture, but also for industry and the needs of growing urban populations. Oil-rich countries such as Saudi Arabia and Kuwait are able to afford costly distillation or desalination plants to purify seawater. The Tigris and Euphrates rivers provide Iraq with some hydroelectric energy, though this is limited by seasonal variations in the flow. The fact that the ground is generally very level is also a problem, since it takes fast-running water falling from a height to run the generators. Syria, Iran and Turkey have also built some hydroelectric stations, and Turkey's plans to dam the Euphrates and control the flow of water have led it into dispute with Iraq.

Apart from oil and gas, the Middle East is relatively poorly endowed with natural resources, though there are a few notable exceptions. Both Iran and Turkey have broad-based mining industries producing coal, iron ore, chromite and copper, as well as other metallic and nonmetallic minerals. There is a large phosphate belt that covers parts of Jordan, Syria, Iraq and Saudi Arabia, and all these countries except Saudi Arabia are phosphate producers. Israel and Jordan both produce potash from the southern end of the Dead Sea. Many parts of the region are believed

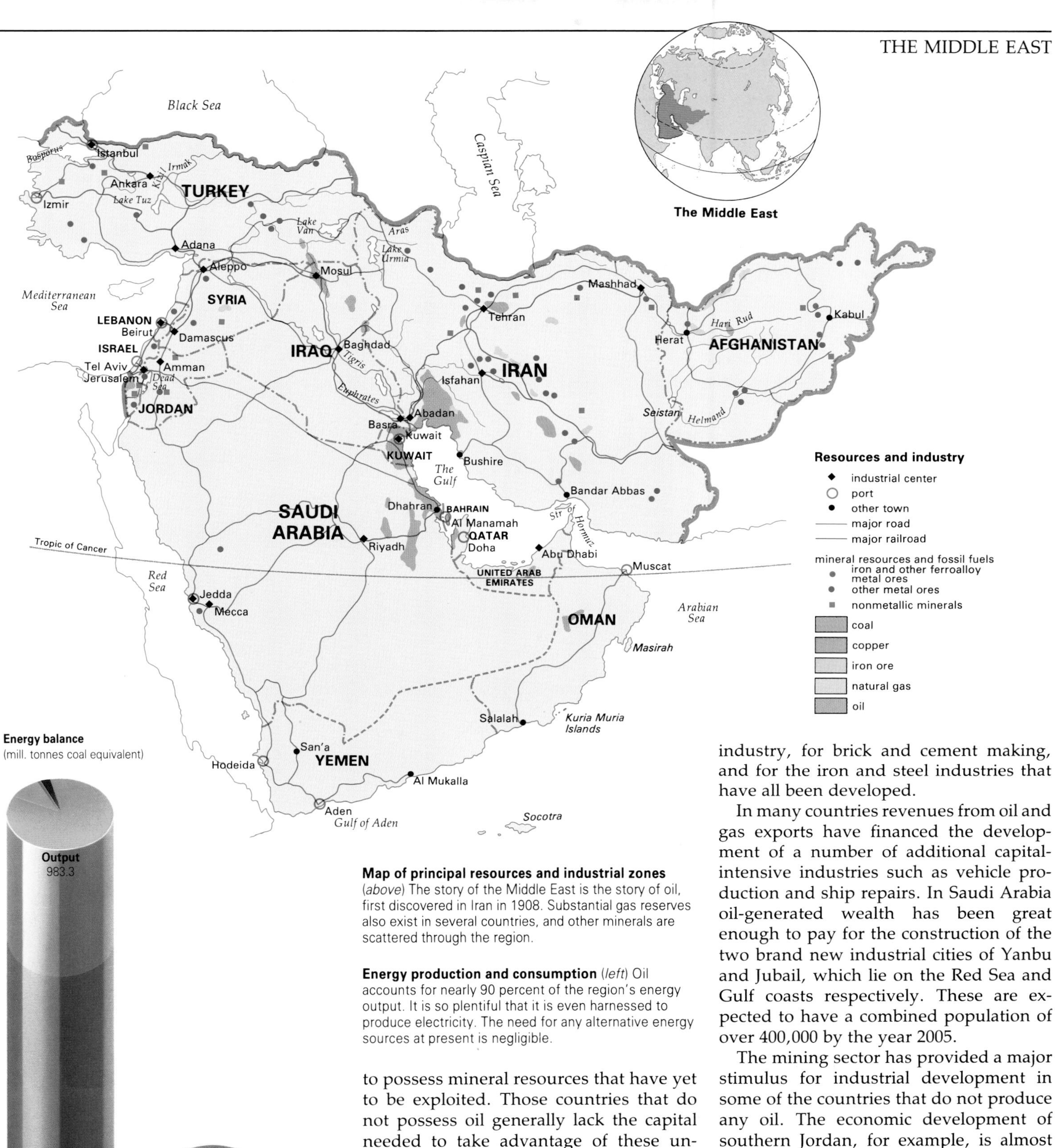

Map of principal resources and industrial zones (*above*) The story of the Middle East is the story of oil, first discovered in Iran in 1908. Substantial gas reserves also exist in several countries, and other minerals are scattered through the region.

Energy production and consumption (*left*) Oil accounts for nearly 90 percent of the region's energy output. It is so plentiful that it is even harnessed to produce electricity. The need for any alternative energy sources at present is negligible.

to possess mineral resources that have yet to be exploited. Those countries that do not possess oil generally lack the capital needed to take advantage of these untapped assets.

Black gold

Oil and gas have played a crucially important secondary role in developing significant sectors in each of the region's oil-producing countries. Large factory complexes producing petrochemicals, chemicals, fertilizers and plastics, all byproducts of the oil industry, have grown up around the main processing plants. Oil and gas also provide cheap fuel for the energy-intensive aluminum-processing industry, for brick and cement making, and for the iron and steel industries that have all been developed.

In many countries revenues from oil and gas exports have financed the development of a number of additional capital-intensive industries such as vehicle production and ship repairs. In Saudi Arabia oil-generated wealth has been great enough to pay for the construction of the two brand new industrial cities of Yanbu and Jubail, which lie on the Red Sea and Gulf coasts respectively. These are expected to have a combined population of over 400,000 by the year 2005.

The mining sector has provided a major stimulus for industrial development in some of the countries that do not produce any oil. The economic development of southern Jordan, for example, is almost entirely based on its phosphates and potash reserves. The country produces cement and other building materials, and the Jordan Fertilizer Industries Company, based in the country's only seaport, Aqaba, exports phosphatic fertilizers. In addition, the Arab Potash Company's state-of-the-art plant literally "harvests" potash from the Dead Sea. Both Iran and Turkey have well-developed metallurgical industries, and domestically produced metals provide some of the raw materials for a variety of other industries.

PROMOTING DIVERSIFICATION

The industrialization of the Middle East during the colonial period (late 19th and early 20th century) was based on food processing and extracting mineral resources. Only in Turkey, as a result of the policies introduced by Kemal Atatürk (1881–1938) after World War I, was a more balanced program set in place to create a diversified industrial and manufacturing base. Iran attempted a similar pattern of development during the dynasty of the Pahlavi Shahs (1921–79) but with little noticeable success.

In more recent years the immense wealth of countries within the region has put them in a much more powerful position and enabled them to make provisions for an industrial economy after the oil runs out.

The price of independence

Among the major oil producers, the first phase of industrial development was based on the exploration, extraction and processing of crude oil. After the price of oil soared in 1973, the Gulf oil producers sought to capitalize on their increased revenues by developing more refineries and export-oriented industries such as petrochemicals and plastics. Their plan was to generate still more revenue from their raw materials by processing them at source. Unfortunately, many of these facilities came into operation in the mid 1980s just as there was both a general economic recession and massive overcapacity in the petrochemicals industry across the world. This led to growing tensions between the Gulf producers and their intended markets (and competitors) in the industrialized world.

The 1970s and 80s also saw the oil producers diversifying into heavy industries, including ship repair, using locally produced aluminum, iron and steel. Western protectionism hit these enterprises hard in their critical early years and their future in the 1990s is still an open question. Other problems for these fledgling Gulf industries include the small domestic market (most goods have to be made for export), the absence of a local skilled workforce, and the lack of the necessary economies of scale. All these factors push up production costs so that the products are significantly more expensive than imports.

The aluminum industry in Bahrain *(above)* With its relatively low reserves of oil, Bahrain had more incentive than its neighbors to develop other resources. The aluminum-smelting industry is one of the largest in the Gulf not related to oil.

A traditional craft continues *(right)* In both Turkey and Iran – ancient Persia – carpet weaving is an important rural industry and contributes to national exports. Designs and techniques are often passed down the generations from father to son.

The Middle Eastern countries that lack oil resources have had to follow a different route to industrialization. Jordan, Lebanon and particularly Yemen relied heavily in the early days of the oil boom on refining crude petroleum produced by their neighbors. Although this remains an important activity, much more refining is done now by the oil producers themselves, and the other states are forced to turn to processing their own limited natural resources and to setting up new manufacturing ventures. Jordan is exploiting its phosphates to produce chemicals and fertilizers, while Syria has a more diverse range of manufacturing including textiles, cement and food.

Industry and political upheaval

In 1979 a popular revolution in Iran toppled Shah Muhammad Reza Pahlevi (1919–80), and put the Islamic leader Ayatollah Khomeini (1901–-89) in his place. In the wave of Islamic fundamentalism that followed the revolution, the country sought to reduce its dependence on trade with non-Muslim countries. In the 1980s and 1990s this meant promoting more heavy industry, especially in the area of arms and defense where Iran had formerly been dependent on the West, and developing more sophisticated petrochemicals industries.

The Iran–Iraq war of 1981–88 set back this program of self-reliance, and for some time actually increased both nations' dependence on oil revenues from their principal Western markets and allies. Over the following few years Iraq's reliance on oil revenues was a factor leading to its dispute with Kuwait over prices, and ultimately to its invasion of that country in August 1990. Kuwait's oilfields and refineries were severely damaged, and Iraq's subsequent defeat in the ensuing war has left much of its manufacturing plant and service industries in ruins.

Manufacturing in Turkey and Israel

Turkey's large population and its wealth of mineral resources have made it the region's industrial giant. It is almost the only Middle Eastern state to have successfully developed a full range of light and high-tech industries, heavy industries and services. In the 1980s and early 1990s manufacturing has been Turkey's fastest growing economic sector and accounts for over three-quarters of total exports. Textiles and leather goods are the most profitable, but machinery, metalwork, mining and quarrying all contribute to continuing development.

Industrial expansion in Israel was boosted in the second half of the 20th century by an influx of highly skilled and motivated Jewish immigrants after the country achieved independence in 1948. They formed the nucleus of a thriving workforce that brought about a radical change in Israel's manufacturing and food-processing industries.

With only a small domestic market, Israel's industrial production is largely for the export market. Food processing contributes the largest share of manufacturing output, followed by textiles, chemicals and metals. Heavy industry, especially the high-technology armaments sector, has also expanded rapidly. These defense-related ventures have a spinoff effect on nonmilitary areas of the electronics industry, stimulating new areas of development and growth. In common with Turkey, Israel has developed a significant tourist industry, which, in turn, is creating expansion in transportation, construction, catering and other services.

Born out of necessity, Israel's per capita spending on defense is among the highest of any country in the world. Here work is being completed on the Lavi prototype for a multirole fighter aircraft. This was later abandoned because of the high cost of development.

TEXTILES IN TURKEY

The textile industry is one of the flagships of Turkey's economy. Textiles form the largest element of the manufacturing sector, contributing approximately 20 percent of output and employing one-third of workers.

The textile industry also accounts for a significant part of Turkey's exports – approximately one-third of the total in the early 1990s. As well as both synthetic and natural fibers, Turkey also exports higher value items such as clothing, leather goods and carpets.

Turkey's export trade with Europe has been threatened by disputes with the European Community (EC). However, the creation of the European single market in 1992 will put Turkey in a potentially powerful position to expand its manufacturing base again. Strategically situated on the doorstep of the EC, and an excellent source of cheap labor, Turkish textile firms are an attractive investment proposition for foreign companies.

RETRAINING THE WORKFORCE

Imported labor (*above*), both of technicians and of unskilled workers, accounts for a high proportion of the workforce throughout the region. In Qatar 90 percent of all workers are expatriates. This Pepsi plant employing a high percentage of Pakistanis is in Saudi Arabia.

Although the oil-producing countries of the region – notably the Gulf states – are extremely wealthy, the general workforce has little industrial experience. Often, too, wealthy oil-rich states tend to be very sparsely populated, so one obvious obstacle to widespread industrialization is the very small amount of skilled manpower available to run it. The combination of the lack of an industrial working class and little or no local unemployment means that the Gulf countries have had to employ foreigners to build and run their new industrial developments. In Saudi Arabia, for instance, foreign workers total 20 to 25 percent of the population.

Senior managers are mainly Gulf nationals who are usually – not always – "shadowed" by Western expatriates who actually run the plants. Most specialist technical workers are Americans or Europeans. By contrast, the lower paid, more arduous jobs are done by immigrants from other countries, particularly the Indian subcontinent and Southeast Asia.

Running on imported labor

These imported workers provide a cheap and reliable workforce. As their work permits are issued specifically for a single company, they are virtually captive. If they were to create trouble or wish to transfer to another company, they would be deported. Despite the hardships and rough treatment that they often encounter, they have a powerful incentive to work in the Gulf. Salaries may be as much as ten times what they could earn at home. Before the Iraqi invasion, Kuwait employed large numbers of foreign laborers. However, in the aftermath of the fighting, the attitude of these workers toward the host country underwent a change – many felt that the loyalty they had shown through their years of service was not repaid in a time a crisis.

The lack of industrial training, the social and religious customs of the region, as well as a host of other factors, make it very difficult to recruit Gulf nationals to poorly paid, semiskilled industrial jobs. It is therefore quite common to find factories where Gulf nationals are employed as relatively well paid guards and drivers, even if they are not actually required. Despite the waste and inefficiency of such a policy, it is seen by the Gulf states as a relatively small price to pay in order to satisfy the local community and government while at the same time getting the job done by an imported workforce.

The demographic trends show that this state of affairs is likely to change over time. In Saudi Arabia, for example, the high birth rate means that over 45 percent of the population is under 15 years old, and nearly three-quarters of them live in cities. As the new generation grows into a workforce it should be able, through education and training, to take on more of the responsibility for running the state's new industrial economy.

State and private sectors

Until relatively recently most Middle Eastern industry was under state ownership and control – either because the state was the richest source of investment, or due to an enthusiasm for nationalization on the part of newly independent governments. This extension of public ownership often led to poor productivity, and during the 1980s many countries allowed much of their industry to be privatized, often with the cooperation of companies from abroad. As a result, the widespread hostility to foreign, and particularly Western, investment and influence died down for a time, and by the early 1990s there was a much more varied pattern of ownership in the region.

This opening up of manufacturing industry to the domestic or foreign private sector was mainly prompted by financial constraints on the host governments, or by their sheer inexpertise in technical

A woman worker (*below*) checks batches of saline solution – a branch of the pharmaceuticals industry that makes use of one of Jordan's principal natural deposits, salt. As more and more women are educated to university level, they are able to find jobs as technicians.

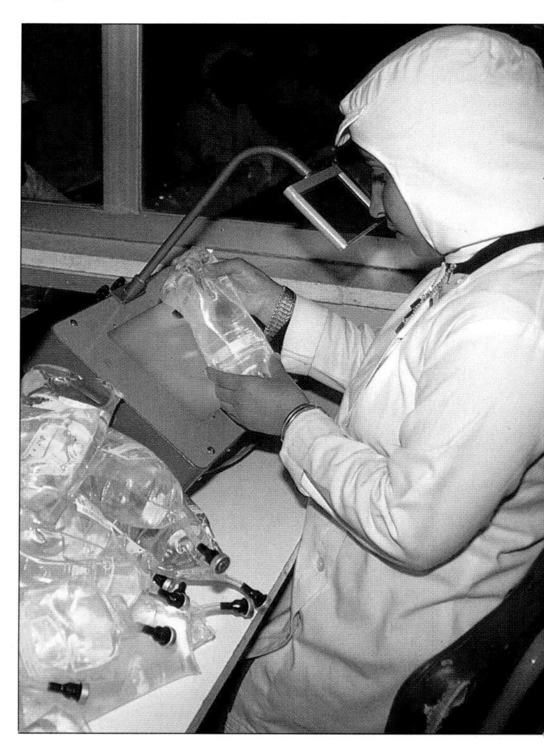

A giant phosphate plant *(above)* at Aqaba on the Red Sea. Phosphates – which are used in fertilizers – provide a major source of export revenue, particularly in those countries (Jordan and Syria) that do not share in the Middle East's substantial oil deposits. In a plant such as this the phosphates are processed on site, thereby adding value to the export.

matters. Even in the most liberal economies, however, the state still tends to play a major and direct role in the capital-intensive heavy industries as these need a greater level of investment than the private sector can usually provide.

Turkey and Saudi Arabia tend to be an exception to this rule, and a number of private manufacturers there have grown into large and successful companies. In general, though, most private-sector industrial companies in the region are relatively small. Typically they produce food and drinks, furniture and other household goods using imported rather than local raw materials. Many are still in their infancy, and are little more than assembly units.

DO TRADE UNIONS HAVE A PLACE?

With the exception of Israel, and to a lesser extent Turkey, trade unions in the Middle East tend to be very weak. Partly this is because in most countries industry is still a relatively minor employer; partly it is because the idea is alien to the local culture. Some countries, such as Iraq and Syria, have passive official unions that are linked to the ruling party. Independent trade unions, however, are often seen as a potential threat to the government's national aims.

Israel is unique in the region for having an industrial sector controlled predominantly by trade unions. The food-processing sector has been administered for some years now by the kibbutzim agricultural cooperatives. The largest labor organization, however, is the Histadrut trade union federation, founded by David Ben-Gurion (1886–1973) at the end of World War I. Ben-Gurion, who was later the architect of Israel's independence and its first prime minister, formed the organization as a nucleus to organize Jewish workers moving to their new national home in Palestine.

Histadrut is now immensely powerful, owning about 15 percent of the nation's industrial enterprise including the clothing, shoe, plastic and rubber manufacturers. It also controls most of Israel's largescale industrial development including the construction sector. This is done through its Hevrat Ha'Ovdim ("workers' society"), a central holding company, with 170 constituent companies. It negotiates collective bargaining agreements and salary increases, and administers employee benefit programs including pension funds. In addition, it supports salary incentive schemes to promote higher productivity and to improve the performance of the industries under its direct influence.

The story of oil

Oil is a naturally occurring compound found beneath the earth's surface. Its origins go back more than 500 million years to the rotting remains of aquatic animals buried in layers of sand and mud. These layers gradually became sedimentary rock. The long process of decomposition eventually produced oil in liquid, solid or gaseous form. It seeped up toward the earth's surface but became trapped in layers of porous rock, where it formed pools. A collection of pools in one area make up an oil field. Many fields clustered in a single geographical location are called an oil province.

Before drilling was invented, oil often seeped to the surface and formed tar pits or oil lakes. Archaeological evidence in the Middle East indicates that it was used thousands of years ago in shipbuilding and in the construction of roads. However, there was no great demand for it until the automobile was invented. Much of the precious fuel was wasted, though by the mid 19th century it was being used to make kerosene (paraffin). The first automobile fuel was a waste product composed of crude oils too light for kerosene.

Oil now supplies 39 percent of the world's energy. There are more than half a million wells in 100 countries, on every continent except Antarctica. With 80 percent of the world's land-based reserves already tapped, offshore reserves are becoming increasingly important. Crude oil from the well needs to be refined before it can be used as fuel or in other products, including plastics and synthetic rubber. Refining also removes related minerals such as sulfur that increase atmospheric pollution.

Vast reserves in the Middle East

The Middle East's commercial oil industry began in Iran in 1908 with the discovery of the Masjid-al-Sulaiman ("Temple of Solomon") well by a British consortium. The discovery of oil in Bahrain in 1932 accelerated exploration, on the other side of the Persian Gulf, and in 1938 oil was discovered in Saudi Arabia by the Arabian American Oil Company (Aramco), an American joint venture. Intensive exploration subsequently led to the discovery of Ghawar, the world's largest field, in 1948.

During the 1930s and 1940s vast amounts of oil were discovered in the region, and the low costs of extracting it yielded high profits for the companies involved – mostly British and American. By the late 1940s, many Middle East governments began to feel that the oil companies had achieved something akin to colonial authority over the host countries, leaving them with no managerial control over production. Finally, in 1960, the Gulf states were among the founder nations of the Organization of Petroleum Exporting Countries (OPEC), set up to coordinate the petroleum policies of third-world oil-producing nations.

Saudi Arabia is the world's largest oil producer. Its reserves dwarf those of Kuwait, Bahrain, South Yemen and other smaller Middle East countries where refining and related industries are more important. A large refinery can have a capacity of as much as 200,000 barrels a day. Oil from different regions has distinct and measurable characteristics. Iranian oil is a medium-weight oil yielding about 19 percent gasoline and 50 percent fuel oils; it is similar to other Gulf oils but very unlike Bachaquero Heavy crude from Venezuela, which yields less than 10 percent gasoline and about 70 percent fuel oils.

Onshore exploration drill When a geological survey suggests that oil is present in the rock strata below, an exploratory well is drilled by a portable rig.

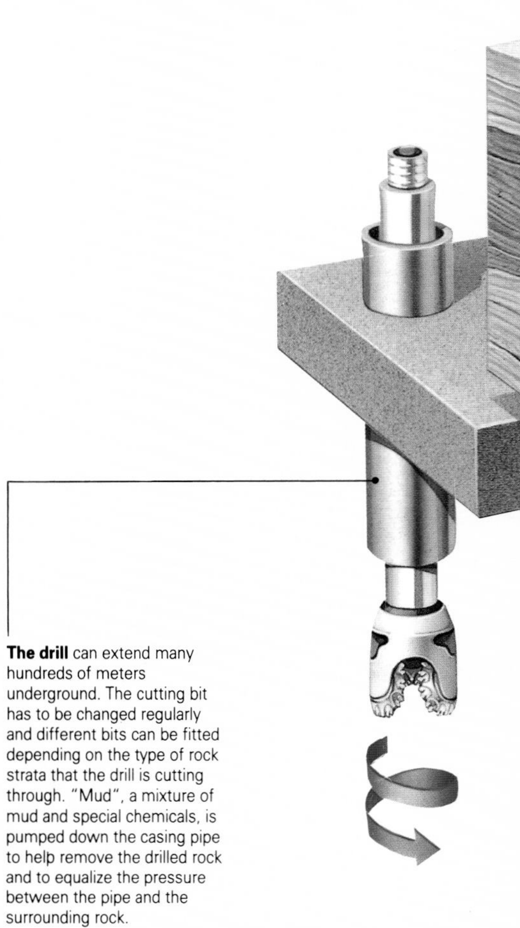

The drill can extend many hundreds of meters underground. The cutting bit has to be changed regularly and different bits can be fitted depending on the type of rock strata that the drill is cutting through. "Mud", a mixture of mud and special chemicals, is pumped down the casing pipe to help remove the drilled rock and to equalize the pressure between the pipe and the surrounding rock.

Wealth from beneath the desert An oil drilling station stands amidst the endless dunes of Saudi Arabia's Rub al-Khali, the largest sand desert in the world. Its oil reserves far outstrip those of the other oil-rich countries of the Middle East.

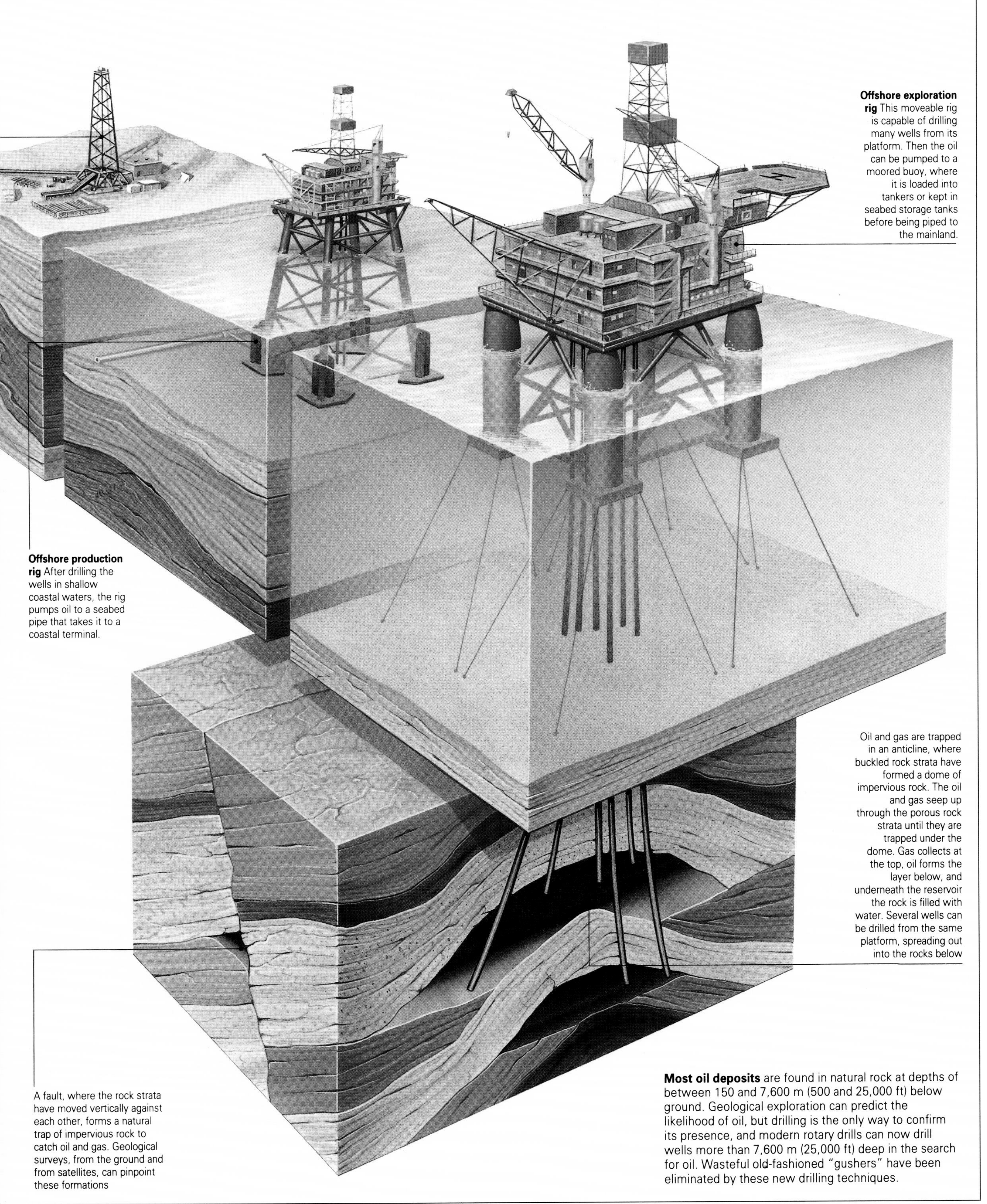

Most oil deposits are found in natural rock at depths of between 150 and 7,600 m (500 and 25,000 ft) below ground. Geological exploration can predict the likelihood of oil, but drilling is the only way to confirm its presence, and modern rotary drills can now drill wells more than 7,600 m (25,000 ft) deep in the search for oil. Wasteful old-fashioned "gushers" have been eliminated by these new drilling techniques.

RICHES BENEATH THE DESERT

FUEL FOR NEW GROWTH · FROM NOMADISM TO MANUFACTURING · THE RISE OF PRIVATIZATION

Northern Africa possesses substantial natural resources, many of which have yet to be exploited. The most abundant are crude oil, natural gas, phosphates and iron ore. Other metal deposits are plentiful, but the desert conditions make exploitation difficult. In such an arid land water is also highly valued: without it many heavy and light industries would be unable to function. The revenue from exports of oil and gas is being used to increase industrialization in the region, particularly in Algeria, Morocco, Tunisia, Libya and Egypt. In its early stages industry is largely confined to mineral extraction, petroleum refining, iron and steel, and the manufacture of textiles, building materials, food processing, vehicles and footwear. Tourism, and with it the construction industry, is expanding in Morocco, Tunisia and Egypt.

COUNTRIES IN THE REGION

Algeria, Chad, Djibouti, Egypt, Ethiopia, Libya, Mali, Mauritania, Morocco, Niger, Somalia, Sudan, Tunisia

INDUSTRIAL OUTPUT (US $ billion)

Total	Mining	Manufacturing	Average annual change since 1960
59.8	19.4	30.1	2.6%

INDUSTRIAL WORKERS (millions)
(figures in brackets are percentages of total labor force)

Total	Mining	Manufacturing	Construction
7.34	0.62 (1.0%)	4.32 (7.1%)	2.4 (3.9%)

MAJOR PRODUCTS (figures in brackets are percentages of world production)

Energy and minerals	Output	Change since 1960
Oil (mill barrels)	1133.2 (5.0%)	+1193%
Natural gas (billion cu. meters)	50.0 (2.9%)	+333.3%
Iron Ore (mill tonnes)	9.5 (1.7%)	+42%
Natural phosphate (mill tonnes)	10.4 (19.6%)	+211%
Manufactures		
Cotton yarn (1,000 tonnes)	329 (2.1%)	+86%
Cotton woven fabrics (mill meters)	907 (1.8%)	+134%
Silk fabrics (mill sq. meters)	24.5 (1.0%)	No data
Manufactured tobacco (1,000 tonnes)	46.6 (20.7%)	No data
Footwear (mill pairs)	151.2 (3.4%)	+47%
Superphosphate fertilizer (mill tonnes)	1.5 (7.0%)	N/A
Liquefied petroleum gas (mill tonnes)	6.8 (4.4%)	N/A
Cement (million tonnes)	27.2 (0.25%)	+938%

N/A means production had not begun in 1960

FUEL FOR NEW GROWTH

Northern Africa possesses about 3.8 percent of the world's total known reserves of crude oil (a small amount compared with 65.2 percent in the neighboring Middle East). The largest reserves are in Libya, followed by Algeria, Egypt and Tunisia. Oil production in 1989 totalled over 150 million tonnes.

Reserves of natural gas are estimated at about 4,200 billion cubic meters (also about 3.8 percent of the world's total). Algeria possesses the largest reserves, followed by Libya and Egypt. In 1989 Algeria was the world's fourth largest producer after the Soviet Union, the United States and Canada. Besides drilling for oil and gas from known reserves, international companies have also undertaken exploration work in most of the other countries in the region. In southern Sudan, the California-based multinational Chevron and the French petroleum company Total discovered reserves amounting to over 250 million barrels of oil, with the possibility of considerably more. However, most exploration work has had to be suspended because of the civil wars in Sudan, Ethiopia, Somalia and Chad.

Considerable amounts of hydroelectric power (HEP) are produced at the Aswan High Dam in Egypt, providing electricity for the country's heavy and light industries. Production has been constrained, however, by demands for water for irrigation. There are also huge reserves of subterranean fossil water below the Sahara that are now being exploited by Libya for industry and irrigation. In Djibouti – a tiny country starved of natural resources – two geothermal wells have been drilled to provide power.

Phosphates are produced and exported by Morocco, Tunisia, Algeria and Egypt. There are also reasonably large reserves of iron ore in the region, but so far only Mauritania, Algeria and Egypt have begun to exploit them. Mauritania's annual iron-ore production fell as a result of the weak world market from 11 million tonnes in the mid 1970s to about 9 million tonnes in the early 1990s. In northeastern Algeria approximately 4 million tonnes are extracted each year.

Egypt extracts high-grade iron ore at the Bahariya Oasis in the Western Desert. This is then transported 350 km (220 mi)

Energy balance (mill. tonnes coal equivalent)

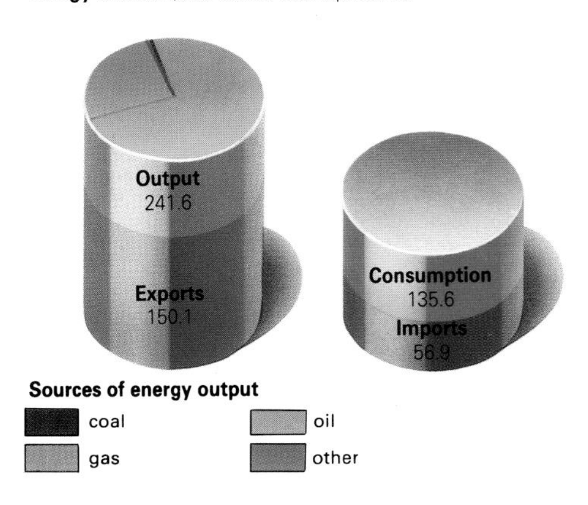

Sources of energy output

coal
oil
gas
other

Energy production and consumption (*above*) There are large reserves of oil and natural gas in the region and underdevelopment means production outstrips consumption. Oil output is nearly three times greater than gas.

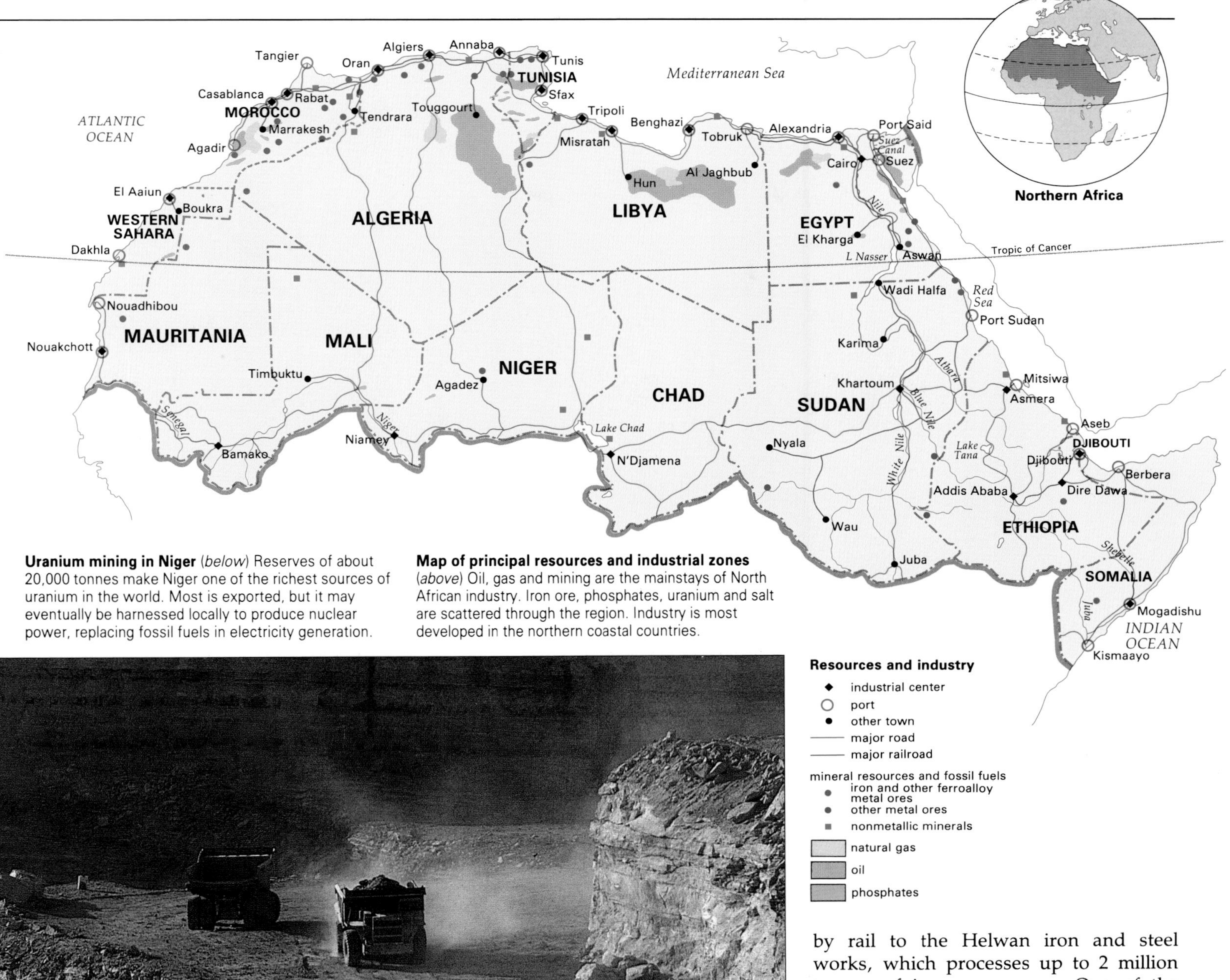

Uranium mining in Niger (*below*) Reserves of about 20,000 tonnes make Niger one of the richest sources of uranium in the world. Most is exported, but it may eventually be harnessed locally to produce nuclear power, replacing fossil fuels in electricity generation.

Map of principal resources and industrial zones (*above*) Oil, gas and mining are the mainstays of North African industry. Iron ore, phosphates, uranium and salt are scattered through the region. Industry is most developed in the northern coastal countries.

by rail to the Helwan iron and steel works, which processes up to 2 million tonnes of iron ore a year. One of the world's largest iron-ore deposits is the estimated 1,815 million tonnes at Wadi Ash Shati in Libya's Sabha district. However, the combination of the weak world market for steel and the major problems involved in transporting the ore all the way to the Misuratah iron and steel complex on the coast currently make its development uneconomic.

Mineral potential

Northern Africa also has significant deposits of uranium, lead, zinc, antimony, cobalt, manganese, copper and salt. However, comparatively few of these have been developed commercially: deposits vary in quality and are often in difficult locations, either physically or politically, such as the uranium deposits on the disputed Libya–Chad border. Successful uranium production takes place in Niger – in 1986 it accounted for 80 percent of total exports – and there is small-scale gold and chromite mining in Sudan.

Old meets new in Egypt *(left)* The construction industry has boomed since World War II as the country's major cities have expanded rapidly to accommodate a growing population. This apartment development in an affluent residential district incorporates traditional features of design – flat roofs, high ceilings and windows and shutters to keep out strong sunlight. Increasing tourism in Egypt has also stimulated the demand for property and, for the moment, demand outstrips the industry's power to deliver. The construction industry has faced some major challenges in recent years with the rate of new building lagging well behind urban population growth.

Industrial transformation in Mauritania *(right)* The port of Nouadhibou is rich in hematite iron ore and has developed a thriving extraction industry, linked to other mining centers in the country's interior by hundreds of miles of railroad. Of all minerals used in modern industry, hematite is probably the most important. Its name, derived from the Greek word for blood, refers to its deep red color which stains the earth wherever it is found. In addition to being the principal source of iron, powdered hematite is used as a polishing agent (jeweler's rouge) and as a pigment in paint.

FROM NOMADISM TO MANUFACTURING

Until the 19th century Northern Africa was a land dominated by desert with very little manufacturing industry. The majority of the population were nomadic or lived in rural communities where goods were made in small family workshops by hand. Today, traditional manufacturing skills are still practiced in some rural areas: coarse wool is woven into cloth, while the countless tanneries process the hides of cattle, sheep and goats, and skilled craftsmen make leather shoes, saddles and items for the tourist trade.

Change through colonialism

The massive influx of European settlers to Algeria, Morocco, Tunisia and Egypt during British and French colonial rule in the 19th century began the long process of change. Colonization created a large domestic market for cheap products, particularly food and textiles, as well as a foreign market in Britain and France for exported raw materials. At first, raw materials such as cotton were exported to Britain for processing, and finished cloth was imported; but this pattern began to change when the colonizing powers introduced cotton processing (unsuccessfully) in an effort to supply their armies with cheap cloth. The colonizers not only intensified exploitation of the region's natural resources, but also introduced new and improved technology and transportation systems, paving the way for 20th-century industrialization.

By 1962 Algeria, Morocco and Tunisia had all gained independence from France and they began to reform their industrial sectors. Foreign companies producing vehicles and electrical goods (particularly televisions, washing machines and radios) were encouraged to set up plants employing the local workforce. Algeria, Egypt and Tunisia then diversified into heavy industry. Until 1979 Algeria concentrated on producing iron and steel, petrochemicals (polyethylene, PVC, caustic soda and chlorine), and processing its large reserves of natural gas, some of which is used in the production of nitrogenous fertilizers.

Since then, the Algerian government has embarked on a radical reform program. In an effort to improve living standards it has invested heavily in the production of consumer goods and the development of agriculture. Polyester textiles, leather, wine, tobacco, cork, paper, electrical goods and building materials are all produced locally by manufacturing and food-processing industries. Algeria is also one of the few countries in Africa to manufacture vehicles for both domestic and commercial use. However, its first priority is to increase natural gas exports to Europe by expanding the capacity of the Transmed undersea pipeline, which runs to Italy.

Industry in Egypt is primarily directed toward the home market, and is dominated by food processing, textiles (mainly of cotton), iron and steel, aluminum

chemicals, tools, furniture and glass. Foreign-owned companies employing the local workforce are well established, while private-owned companies are playing an increasingly important role in Egyptian industry. However, the major industries – oil, textiles, iron and steel – were all nationalized by the government under President Gamal Nasser (1918–70). He also nationalized the previously Anglo-French-run Suez Canal, central to importing and exporting in the region, so that it was controlled by the state.

Morocco, by contrast, has always sought to promote industries aimed at the export market, primarily food processing (including fruit canning) and textiles. With relatively few capital resources, a small domestic market and a lack of technical expertise, heavy industry is restricted to fertilizer production from phosphates, which is subsidized by the government. In general, government policy has been to reduce its own role in industry, and to encourage private and foreign investment in, for example, automobile assembly plants and soap and cement factories. Tunisia's liberal economic policy has also succeeded in attracting both Arab and Western investment in companies producing goods for the European market.

The growing importance of oil

The development of industry in Libya has followed a similar pattern to that of the oil-rich countries of the Middle East. Oil was discovered in the late 1950s, but exploitation was at first considerably hampered by the desert climate and terrain. Until 1971 food processing, particularly drying and preserving fruit, still accounted for over half of Libya's industrial revenue. However, as petroleum refining rapidly increased, the revenue derived from exports soared, and was used to build extremely expensive aluminum and iron and steel plants. As in the Middle East, these rely on imported raw materials.

Sudan and Ethiopia both have the potential to develop an industrial sector, but drought and long-running civil wars continue to consume the money needed to exploit these countries' resources. In Sudan the small manufacturing sector produces cotton, building materials, petroleum products and sugar. Ethiopia produces coffee, with the help of foreign aid, as well as footwear and tires.

POWER FROM THE DESERT

Egypt was thought to have no natural gas until as late as the 1970s, when discoveries were made in the Western Desert, the Red Sea and off the Mediterranean coast near Alexandria. As a resource, gas had always been little exploited in the region. Unlike oil it is difficult to export and requires very large investment for little return. However, many of the North African countries now use gas for fuel as a substitute for oil, particularly in power stations. This enables them to export their crude oil, which brings in much needed revenue.

The extensive gas fields that were discovered in Egypt proved to be an invaluable power source for the country in the late 1980s, when hydroelectric power could no longer meet the huge demand, the urban population was growing rapidly, and the country's balance of payments were becoming a serious problem. The total gas reserves in the country are large enough to supply a gas grid system in Cairo, and Egypt now has plans to expand its gas-fired power plants. In the early 1990s a 300 km (186 mi) gas pipeline was being laid from the Abu Gharadeq gas fields in the Western Desert all the way to Cairo. Were it not for this unexpected discovery, expensive alternative energy resources would have been the only solution to the country's fuel shortages.

THE RISE OF PRIVATIZATION

Since gaining independence from France, the governments of Algeria and Tunisia have played a significant part in transforming their industries from smallscale enterprises to major revenue earners. Initially the departure of the French delayed the process of industrialization, leaving a dramatically reduced skilled workforce, a smaller market for consumer goods and a reduction in capital for investment. Foreign firms were reluctant to invest in industries that might be nationalized by the region's predominantly socialist government.

During the 1980s, however, most of the countries of Northern Africa embarked on privatization programs. Morocco and Tunisia were the first to develop a thriving private sector, at a time when most of the other countries in the region were dominated by the public sector. In Algeria in the early 1980s, for example, the government used the revenue from oil to set up manufacturing projects and subsidize food and energy, while the private sector dwindled. Entrepreneurs used their influence to gain control of large government projects rather than private ones.

This pattern changed in the late 1980s when the Algerian government pushed measures through parliament that radically altered the way that firms operated. Although they concentrated on government-run enterprises, they also included a private investment law allowing private companies to undertake joint ventures with foreign firms. The main aim of the reforms was to give the public-sector firms a far greater degree of autonomy. Since 1989 Algeria has seen the emergence of new state-owned but autonomous companies known as Public Economic Enterprises.

Reaction from the workforce

Numerous labor disputes occurred during this period of change, prompted by a deep suspicion among public-sector employees of the get-rich-quick philosophy of the private sector. In the new market economy much will depend on how free a hand the private companies will have and whether foreign investors believe they have a secure future in Algeria.

Libya and Ethiopia have also begun to reduce government control of industry in recent years. In Libya the government – for ideological reasons – closed down the private retail sector during the 1980s, replacing it with a network of large hypermarkets. However, these were unsuccessful. Libyan men were unwilling to give up buying and selling in the local markets. To prevent a complete breakdown of retail distribution the government was forced to reopen the private markets and small shops, abolish its own trading organizations, and promise to privatize the small- and medium-sized public-sector companies.

The Egyptian government's interventionist role in industry goes back more than a century: Muhammed Ali (1805–48) – the founder of the modern Egyptian

Dyeing textiles in Morocco *(above)* Traditional dyeing methods using dried insects to produce this deep red are characteristic of the region and probably spread west from India.

Somalian mat weaver *(right)* The long grasses are woven – usually by women – into mats and baskets that have long been used locally for food and storage. Now they find their way to western shops.

Metalsmiths, Cairo *(above)* Craftsmen and small businesses have always flourished in cities such as Cairo and Alexandria. Traditionally, each craft located in a specific area of the city. Skilled city workers like these have a higher standard of living than those in rural areas.

state, imposed rigid state control on local manufacturing. More than a century later, in 1956, President Nasser nationalized not only the Suez canal, but also much of the industrial sector.

His successor, President Sadat (1918–81), tried to reverse this with his 1974 "infitah" (open door) law, which supported private enterprise and enabled the owners of many private companies to become immensely wealthy. A new class of some 21,000 millionaires was created, known as the "Mafia fat-cats". President Hosni Mubarak has since tried to take a middle line by continuing to privatize some industries, while keeping government control on others.

Expansion through decentralization

Morocco's 1988–92 development plan aimed to achieve the privatization of some 300 major and profitable companies formerly under government control. A new Ministry of Privatization was created in 1989 for the purpose, and efforts have been made to decentralize industry from around Casablanca and the coastal strip running up to Rabat and Kenitra. Special tax incentives are offered to industries that relocate outside this area.

The decade between 1980 and 1990 was United Nations Industrial Development Decade for Africa, during which African multinational industrial corporations were encouraged and the foundations were laid for the region's self-sustained industrial development. Unfortunately, between 1983 and 1985 drought and famine, particularly in Ethiopia and Sudan, meant that finance from abroad had to be used on food rather than investment in industry.

Despite all these changes, however, the most important industries in Northern Africa – oil, gas, petrochemicals, phosphates, iron ore and textiles – are all government run, and there are few signs that this will change. It is these industries that earn the vast majority of the hard currency that enables the governments to finance their imports, repay their debts and function as nation states.

SMALL INDUSTRIES AND HANDICRAFTS IN SUDAN

In the Sudanese economy, the rural-based smallscale and handicraft industries account for 95 percent of all enterprises and 27 percent of employees. Traditional industries – blacksmithing, woodwork, tannery and leatherwork, plaiting palm leaves and straw, artistic handicrafts, carpetmaking and pottery – rely totally on local resources. The modern small industries, which are mainly urban based, include carpentry, electrical installation, automobile and machinery repair, metal workshops and the clothing industry.

Traditionally many jobs were practiced only by certain social groups, but this is now changing. In Darfur, local people are taking up blacksmithing, once a trade that was looked down on by the Sudanese. Men are now taking up weaving mats, baskets and food covers, traditionally considered to be women's work.

Most products are made to order and are sold by the craftspeople directly from their workshops for local consumption, generally to low-income families. With the exception of tanneries and some forges, few rural industries produce goods for other larger industries. The level of technology is usually very low, and the workshops are labor intensive, using little capital.

Phosphates and fertilizers

Phosphate – a major component in the production of fertilizers – is by far the most important mineral resource in Northern Africa after oil and gas. In the late 1980s Morocco, Tunisia, Algeria and Egypt produced some 30 million tonnes of phosphate (20 percent of the world total). Exports from these countries amounted to as much as 15 million tonnes a year.

Phosphate rock is found in sedimentary deposits that were originally laid down on the ocean floor. The largest deposits in the region are in Morocco and the disputed Western Sahara (currently occupied by Morocco), which contain 50 to 75 percent of the world's known phosphate reserves. Morocco also boasts the largest phosphate mines in the world, at Khouribga and Youssoufia in the west of the country. There are also sizable reserves at Bou Craa in Western Sahara. Phosphate in Morocco is mined and marketed by the government-owned Office Cherifien des Phosphates (OCP). In 1988 it produced more than 25 million tonnes of phosphate rock concentrates, just under half of which were exported to western and Eastern Europe and to central and South America.

There are smaller reserves along the Algeria–Tunisia border, at Jebel Onk and the Gafsa basin respectively. In Tunisia the phosphate is very low grade, and it has proved more economic to process it at home than to export the raw material. In 1988, for example, 6 million tonnes of phosphate rock were extracted, but only 1 million tonnes were exported, the rest being used to supply the domestic fertilizer industry. In the same year Algeria's production rose by 24 percent to 1.3 million tonnes; 877,000 tonnes was exported mostly to Europe.

There are further phosphate reserves in Egypt's Western Desert and along the Nile Valley and Red Sea coast, but most is relatively low grade and has not yet been fully exploited. Much of it is used in the manufacture of fertilizers, and only about a quarter is exported.

From soft drinks to flameproofing

In the chemical production of fertilizers, phosphorous – an essential nutrient required by plants for good root growth – is obtained from phosphate rock by the application of huge quantities of sulfuric acid. This produces phosphoric acid and superphosphate, a substance that is soluble in water and contains about 20 percent phosphorous. A more concentrated fertilizer, triple superphosphate, is obtained when phosphoric acid is applied to phosphate rock. Triple superphosphate contains about 50 percent phosphorous. Phosphoric acid is also used in the manufacture of soft drinks, while phosphates are used for diverse products ranging from baking powder and flameproofing to detergents.

Exporting phosphates from Safi, Morocco *(above)*, one of six ports built to support the giant phosphate industry, whose annual production topped 25 million tonnes in the 1980s making the country the world's leading exporter. About 30 percent of the cargo is handled by the national shipping company, Comanav.

Tunisian mine at Metlaoui *(left)* Tunisia has much smaller phosphate resources than Morocco and most of it is low-grade making it unsuitable for export. The majority is used in local fertilizer production, where it supports a significant industry. Most refining and processing is concentrated in the northern part of the country.

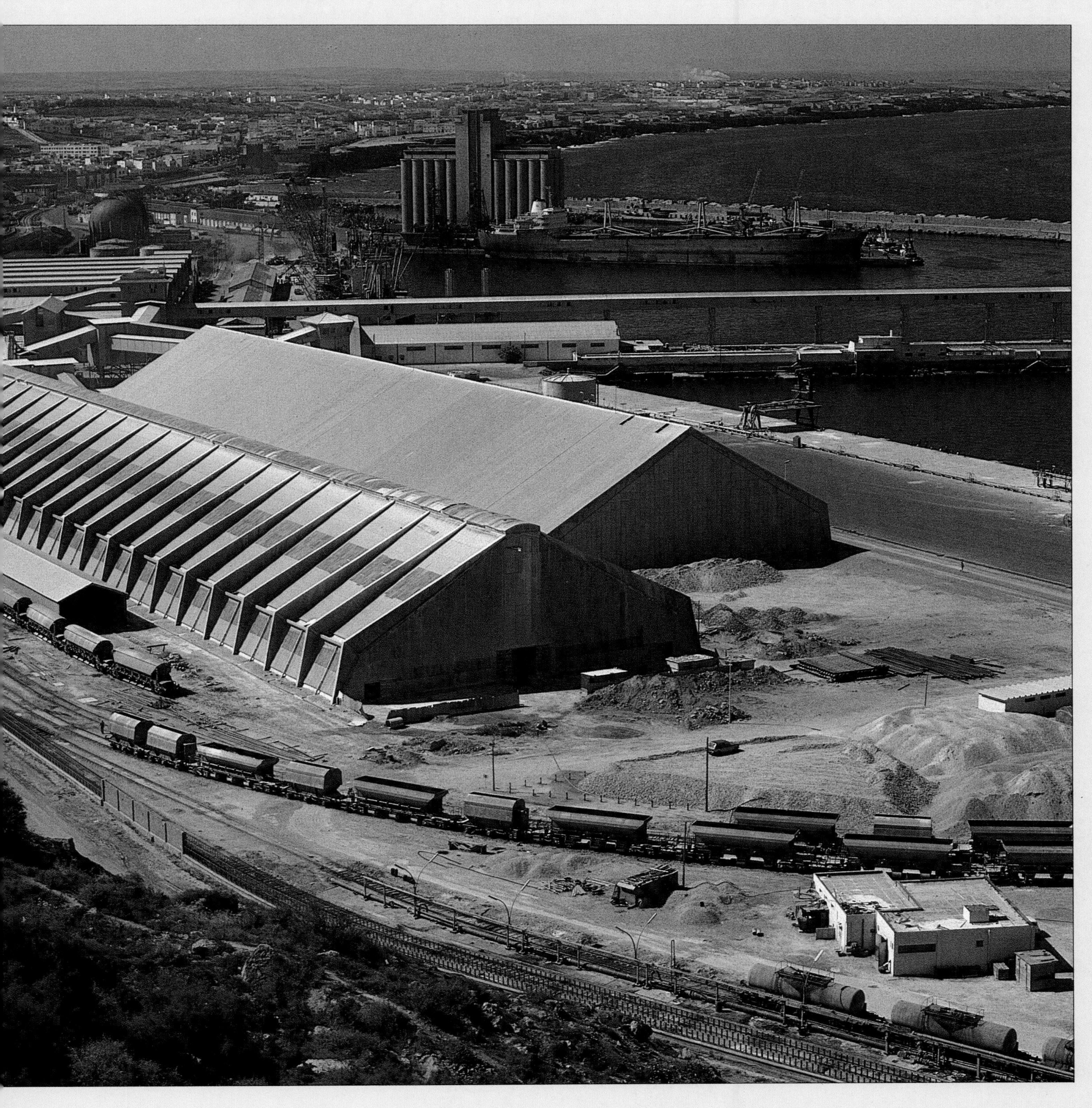

The phosphate industry has had a dramatic effect on the producing countries, not just in terms of exports. In the relatively underdeveloped rural areas where the mines are located, the industry provides a major source of employment, and in both Morocco and Tunisia it dominates road and railroad freight, as well as international shipping.

All the producing countries have made efforts to develop their phosphate-based industries. Morocco, for example, had an optimistic ambition to become the largest producer of phosphoric acid and fertilizers in the world. OCP has four acid plants on the Atlantic coast at Safi and Jorf Lasfar, which are now being expanded and integrated to produce the world's largest solid fertilizer plant. In 1988 Morocco produced 4.1 million tonnes of phosphoric acid.

The Tunisian phosphate industries are dominated by the government-owned Gabes Chemical Group; exports are worth over three times as much as raw phosphate exports. By comparison the Algerian domestic fertilizer industry is still relatively small, though it is being expanded. In Egypt, which uses more fertilizer per hectare than any other country in Northern Africa, the phosphate-based industries mainly supply the extensive domestic market.

RESISTING FOREIGN DOMINATION

TAPPING SOURCES OF ENERGY · OVERCOMING A COLONIAL HERITAGE · A NEW GENERATION OF INDUSTRIALISTS

Central Africa's main resources include oil, coal, precious metals, bauxite, iron ore and barium (a toxic silver-white metal). Although some of these are plentiful, few countries in the region have been able to exploit them efficiently. Nigeria, Congo and Gabon are among the largest oil-producing countries. Fishing, agriculture and crafts flourish in some areas but have been superseded elsewhere by foreign-owned manufacturing plants that employ local labor but use few local resources. Development has been hampered by a limited workforce (parts of the region are very sparsely populated) and by some conflict of interest between governments and foreign companies in deciding how to develop resources. However, the region's vast areas of unspoilt land are attracting more and more tourists.

COUNTRIES IN THE REGION

Benin, Burkina, Burundi, Cameroon, Cape Verde, Central African Republic, Congo, Equatorial Guinea, Gabon, Gambia, Ghana, Guinea, Guinea-Bissau, Ivory Coast, Kenya, Liberia, Nigeria, Rwanda, São Tomé and Príncipe, Senegal, Seychelles, Sierra Leone, Tanzania, Togo, Uganda, Zaire

INDUSTRIAL OUTPUT (US $ billion)

Total	Mining	Manufacturing	Average annual change since 1960
22.8	10.6	12.2	+2.3%

INDUSTRIAL WORKERS (millions)

Total (mining, manufacturing, & utilities)	% of labor force
6.6	7.1%

MAJOR PRODUCTS (figures in brackets are percentages of total world production)

Energy and minerals	Output	Change since 1960
Oil (mill barrels)	700	+70%
Bauxite (mill tonnes)	18.0 (20%)	+350%
Copper (mill tonnes)	0.5 (6%)	+8%
Diamonds (mill carats)	25.9 (27%)	+28%
Manufactures		
Textiles (mill sq. meters)	677.3 (1.0%)	+123%
Palm oil (mill tonnes)	1.5 (8.1%)	No data
Tobacco products (billion units)	9.7 (2.1%)	+79%

TAPPING SOURCES OF ENERGY

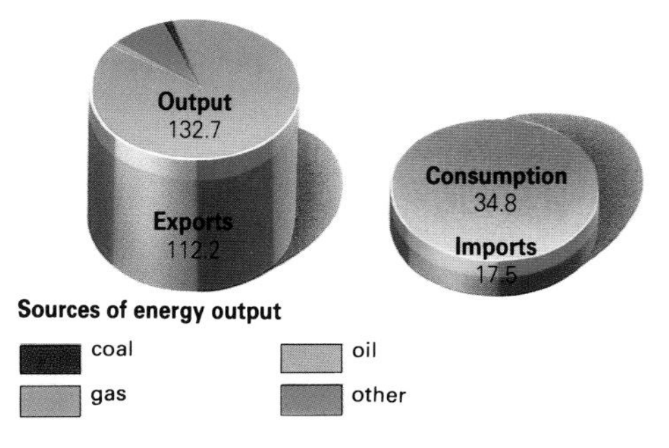

Energy production and consumption The lack of major industries and scarcity of power distribution networks in the area means that most of the region's energy resources are exported. The area's extensive river system is boosting hydroelectricity.

Oil, found throughout the region, is increasingly important as a fuel for energy. When it can be exploited effectively, it is also a valuable export. Nigeria, where the first commercial discoveries were made in the late 1950s near the Niger river, has enough oil to support output at the current rate until about 2020. Valuable reserves of natural gas, the largest in the region, are also found here. Although Gabon only began to exploit its oil in 1957, it was the largest oil producer in sub-Saharan Africa by 1985, and new reserves are still being discovered. Congo exhausted its crude oil in the early 1970s, but opened new fields during the 1980s after fresh discoveries were made at Emeraude and Loango, north of the mouth of the Congo river.

Other smaller deposits are scattered around the region. Ghana, with offshore reserves on the continental shelf, began producing in 1978. Oil has also been drilled off the Seychelles and is present in Senegal, but is uneconomic to extract using current methods. Tanzania has recently discovered offshore oil and natural gas reserves in the Songo Songo Island area south of Dar es Salaam.

Although petroleum is the dominant

energy source, generating approximately 90 percent of power in Central Africa, there is considerable potential for renewable sources of energy to be exploited as well. Many of the rivers in the region could be harnessed to generate hydroelectric power, and there is considerable potential for making use of solar power.

Mining coal and iron ore

Nigeria has some of the most plentiful mineral resources in Central Africa, including tin, iron ore and coal. However, coal mining is still underdeveloped because there are so few major users in the region. High-grade iron ore has been discovered in eastern Senegal, and mining began in 1989. One of the largest deposits in the world exists at Mekambo, in northeastern Gabon, mined by a joint European and American consortium. In Guinea, mines that had been worked in 1953 were abandoned 20 years later, and much richer deposits at Mount Nimba in Ivory Coast are now being developed by a multinational consortium. Iron ore was Sierra Leone's most important mineral export until 1975, and has been the main extraction industry in Liberia since 1961. Unexploited reserves are in Benin, Uganda and Cameroon.

River of oil (*left*) Nigeria is one of Africa's leading suppliers of oil. The main oil-producing areas are Oloibiri and Afram in Rivers state. Resources here are particularly valued by the industry because of their low sulfur content. With familiar oilfields being explored to exhaustion, Nigeria can only expect present rates of production to last until 2020, though improved technology is being subsidized by the government to reach previously inaccessible reserves and so extend the profitability of the industry.

Lands of diamonds and gold

Diamonds are a valuable resource mined throughout the region. Although most are suitable for industrial use rather than for jewelry, smuggling is still a problem. This is also the case with the gold-mining industry in Tanzania, and in the Congo where gold is mined in small quantities. In Guinea mining is forbidden by individuals because of the risk of illegal profiteering from the metal. Gold is Ghana's principal mineral export and is mined at four sites, three of them government subsidized and the fourth, Obuasi in the central south, operated by a subsidiary of the British multinational Lonrho.

The region has good supplies of other minerals including bauxite, phosphates, limestone for cement manufacture, manganese, copper, lead, nickel, salt, zinc and tin. Burundi has recently been discovered to have the world's richest deposits of vanadium, used to make steel alloys, and quantities of uranium, which are still being investigated.

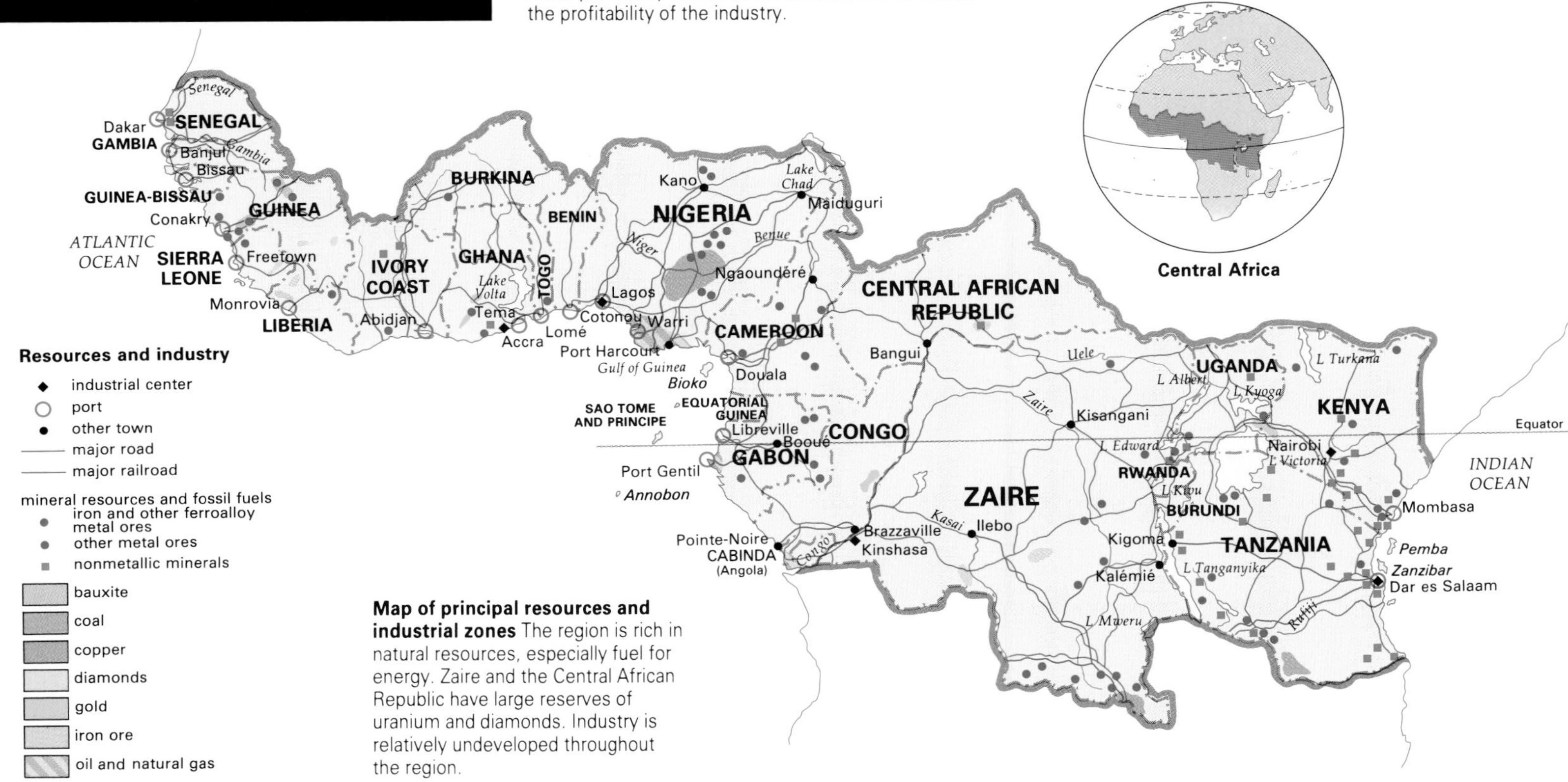

Map of principal resources and industrial zones The region is rich in natural resources, especially fuel for energy. Zaire and the Central African Republic have large reserves of uranium and diamonds. Industry is relatively undeveloped throughout the region.

OVERCOMING A COLONIAL HERITAGE

During the years of colonial rule, the region's natural resources were exploited in bulk by the colonizing countries to support industries at home. Since independence in the 1960s the countries of Central Africa have found it difficult to overcome this legacy and develop industries that use the region's resources to maximum commercial advantage. Extraction industries still export raw materials to manufacturing industries in other parts of the world, rather than adding value to the resources by using them to manufacture goods within the region. Foreign-owned companies have set up manufacturing plants that make use of local labor but import most of their own materials, so that the host country usually benefits very little.

Most of the largescale manufacturers in the region fall into this second category, known as "import substitution industries". They are attracted by a cheap local workforce, but profits revert to the company's headquarters, and very little is reinvested in the region. Almost every country in Central Africa has some kind of import substitution industry using mainly imported materials to produce beer, cigarettes, soft drinks, textiles and foodstuffs. The machinery too is mostly imported and the work is supervised by foreign managers.

More recently, parts of the region rich in natural minerals have developed a locally controlled manufacturing sector that processes these resources to produce consumer goods for their domestic markets. Zaire, a major exporter of minerals, also produces food, beverages, tobacco, textiles, footwear and chemicals for its own consumption. Locally controlled industry is not so easily developed across the whole region. While some countries are trying to limit imports to stimulate the use of domestic resources, others such as Gambia are so poorly endowed that they must import items as basic as matches and chickens.

Food-processing industries

The leaders of many newly independent Central African countries saw industry as the key to progress and ignored agriculture, which declined in output just as the oil crisis of the 1970s hit the oil-exporting countries. To compensate, Kenya, Tanza-

Enriching industry *(above)* Uranium, extensively mined in the Central African Republic and Gabon, used to be exported in its raw state. New enrichment facilities such as this one in Gabon add value to the resource, and increase export value.

Pumping hot air *(below)* While many of the countries in the area are self-sufficient in fossil fuels, others have to rely on alternative sources of energy. Here government and industry officials inspect a wind pump for Kenya's wind-powered farms.

TOURISM IN THE SEYCHELLES

Natural beauty is the most important resource of the 115 islands that make up the Republic of the Seychelles. Lying 1,600 km (1,000 mi) off the coast of Kenya, their beaches, lagoons and tropical climate draw tourists from all over the world who contribute over 20 percent of the republic's income.

A popular holiday destination in the 1970s, the Seychelles experienced a tourist slump in 1980–82 after a period of political turbulence. In 1981 a new tourist board began to market the islands as a holiday resort for visitors from all over the world. Prices were reduced to attract more people, and in 1982 new legislation allowed hotels to run casinos. Tourism is again buoyant, helped by improved air links with many countries including Singapore and Italy. The islands also attract many visitors from South Africa.

The Seychelles have few industrial alternatives to tourism. The islands' small population (below 70,000), its lack of easily accessible raw materials and its overvalued currency have not encouraged foreign investment, and the potential for development of other resources is minimal.

Stretching the rubber industry Originally, rubber plantations provided the region with an important cash crop. Recently, however, rubber industries such as tire manufacturing use increasing amounts of synthetic rubber to meet rising demand and lower costs.

nia, Ivory Coast, Ghana and Nigeria capitalized on the "beverage boom" of the 1980s, during which they earned considerable revenues from coffee, tea and cocoa. They used this income to invest in capital-intensive industry and to build up their local services. Later, beverage prices fell, leaving many countries with half-finished industrial developments and unpaid loans. Cocoa for export is still the economic mainstay of the islands of São Tomé and Príncipe, which were major world producers in the early 1900s. Before independence their output was about 10,000 tonnes a year. Production fell as poor soil fertility, outdated techniques, a shortage of labor supply and a slump in world prices took their toll. Exports have had to be bolstered by palm kernels, coconuts and bananas.

Equatorial Guinea, Cameroon and a number of other countries also poor in mineral resources have always relied on agriculture as their main source of income. As well as coffee and cocoa, cash crops include cotton, sugar and tobacco, and some processing industries have grown up around them. Heavily forested countries produce timber and rubber as well as palm oil and bananas, usually exported as raw materials for use in manufacturing and food-processing industries across the world. In Cameroon, for example, four-fifths of the workforce is employed in growing or processing crops, with only a small additional manufacturing industry assembling imported raw materials and components. The manufacturing sectors of Ivory Coast and Senegal are both dominated by food processing. In Senegal most industry is linked to the extraction of groundnut (peanut) oil and fish processing. Local cotton also supports four cotton-ginning mills, spinning, weaving, printing and dyeing plants. An integrated textile complex at Kaolack in western Senegal was opened in 1987 to boost production, and much of its output is exported. Ivory Coast concentrates on processing primary products such as cereals, cotton, fruit and sugar for the export market.

Local success stories

Kenya has had more success than most countries in the region in diversifying away from cash crops. Its large manufacturing sector produces paper, cement, vehicles, electrical equipment, tires, batteries, paper, ceramics, chemicals and petroleum products. Three vehicle assembly plants – Associated Vehicle Assemblers, General Motors Kenya and Leyland Kenya – produce commercial and four-wheel drive vehicles from kits supplied by major American and European manufacturers. New factories are now being planned or built, often in conjunction with overseas companies, to manufacture glassware, shoes and machine tools.

Nigeria, too, has begun to manufacture goods that were formerly imported from the West. An integrated iron and steel industry is planned around three steel-rolling mills opened in the late 1980s at Oshogbo in the southwest, Katsina on the central northern border with Niger, and Jos in the center of the country. An even larger complex has already been set up at Abeokuta in the southwest, financed by a Russian company.

Although manufacturing remains a relatively minor industrial sector in Central Africa, the service sector – from the selling of kola nuts in the street by children to banking and insurance – has developed rapidly to earn considerable revenue for the region. Cottage industries also flourish. Traditional crafts in northern Nigeria are particularly well developed. Tourism has also increased dramatically in recent years with the development of cheaper world travel and some basic marketing of the region's attractions. These include magnificent beaches along the west coast, the spectacular Rift Valley in western Kenya, Mount Kilimanjaro in northeastern Tanzania, and game parks, particularly in eastern Africa, that protect wildlife.

A dyeing art Much of the industry in the region is on a local scale using traditional methods. Dye pits in northern Nigeria have been used for centuries to color locally produced cloth with extracts of plants and tree barks.

A NEW GENERATION OF INDUSTRIALISTS

Since the 1960s, many industrial projects using local resources have been established throughout Central Africa, but comparatively few are African owned. Both Africans and foreigners saw the West as the model for the future, and most projects were set up under Western guidance. Although African ownership has become more common in recent years, foreign companies still have vested interests in the region.

Plantations, irrigated agriculture and mining are prime examples of foreign-owned industries in Central Africa. Palm oil plantations in Zaire are run by Unilever and produce oil for export, as well as soaps and detergents for the domestic market. Fyffes Bananas, a British company in Cameroon, transformed banana growing on plantations into a major export industry, and sugar production in northern Senegal is Lebanese owned and British managed.

It is becoming accepted among international investors that promoting self-sufficiency among the African workforce should be an integral part of resource development. Training for managerial posts is increasing, but although the number of Africans in middle management has risen, senior management is still made up largely of foreigners. The African population provides the labor.

A thriving informal sector

The importance of the informal sector in African industry has only recently begun to be appreciated for its harmony with a traditional way of life as well as for its commercial value. For instance, bark and leaves collected by both men and women are processed into medicines or dyes and sold locally, usually by women. In the savanna, animal hair and wool from the herds are collected by women and children, dyed and then woven by men into elaborate blankets. Mats and fences are made from local grasses, and firewood is collected and sold or made into charcoal. In recent years there has been a big increase in the number of carved wooden and metal objects in the markets, made to meet the demand for souvenirs from tourists throughout the region.

TRADITIONAL METALSMITHS

Abundant mineral resources have encouraged the development of a traditional jewelry industry throughout Central Africa. In Senegambia, the region of the Senegal and Gambia rivers, jewelry making is restricted to one particular caste and remains a family business with traditional practices handed down over generations. This, together with limited investment capital, has meant that methods of manufacture have remained basic and usually very small in scale.

Metal is imported from a wide variety of sources: gold from Ghana, silver from Sierra Leone and other metals from Europe, where supplies are considered more reliable. It is melted in small, locally made ceramic crucibles no bigger than a teacup, on small charcoal furnaces worked by a hand-powered fan. The molten metal is then poured into molds, hand beaten, filed and polished into the final article.

Jewelry making is slow and labor intensive as metal is usually melted for the production of one item at a time. This is crucial as the metal can only be fashioned when it is at precisely the correct temperature – a decision made according to the jeweler's experience and skill. Styles and patterns of jewelry vary considerably, and regional designs are very distinctive. Marketing the product is far from sophisticated. A selection of perhaps a dozen articles is usually displayed in small glass cases on the roadside, often at the same site as the workshop, where tourists and locals haggle over prices.

Gold digging *(right)* Prospectors in Burkina search for elusive gold nuggets to supply the age-old metalsmithing trades in the region. Successive generations carry on the trade, partly out of tradition and partly for tourist dollars.

There is no European involvement in the informal sector. The work itself is labor-intensive, investment capital is negligible, and management is frequently highly sophisticated. Unlike many large-scale schemes, this method of working provides significant opportunities for women. It represents resource use by Africans for Africans, and provides an alternative to subsistence farming or manual labor. Women vendors in Ghana and Nigeria do not just survive, they prosper. Regular trading also provides a social network between rural areas, where goods are grown or gathered, and the urban areas where they are sold.

Smallscale flexibility

The demands that these locally run industries make on the workforce are very different from the demands made in the formal sector. The production and sale of goods are done when people have time and when they need to make money. Working in this way enables people to perform many other activities such as farming, herding, fishing, growing vegetables, housework and childcare.

In urban areas women vendors buy and sell small quantities of perfume or cosmetics at a small profit. Taxes must be paid on a market stall, so word-of-mouth advertising is both attractive and effective. Vendors are often self-employed, especially if they are selling something they have made, such as peanut butter. Often several people sell for a single employer, each making a small amount of commission on individual items. Many enterprises are confined to families, though development schemes encourage cooperatives. Entrepreneurs in urban areas are generally more sophisticated. Some jewelry makers in Banjul, on the island of St Mary on the Gambia river, have imported machinery, and the sewing machine has revolutionized small and medium-sized tailoring enterprises all over Central Africa.

The forest-based industries

Rainforest lumberjacks *(above)* Forests cover half to three-quarters of the total land area in the region, making wooden-related products a valuable resource. Highly desirable hardwood, used in construction and furniture making, is exported worldwide.

Keeping the economy buoyant *(right)* Felled trees are floated down river to processing and distribution plants. Although the majority of trees are used for lumber products, the forests also support food, energy and scientific industries.

Most of the region, from Guinea in the west to the Rift Valley in the east, supports dense rainforest. The forests provide an enormous variety of raw materials for commercial exploitation, but they are themselves a particularly fragile resource. In many parts overuse has destroyed the forests. Yet demand for tropical hardwoods continues to rise. Although some of the market has recently been lost to Asia, Africa still remains the main supplier of hardwood logs to Europe. While imports into the United Kingdom, Belgium, Luxembourg and the Netherlands are largely of sawn wood, imports into France and Germany are predominantly logs needed to supply the domestic plywood industry and sawmills. In France, African hardwoods are traditionally used for joinery, while in the Netherlands and the United Kingdom softwoods are normally used instead.

Although most of the primary forest of western Africa has now been removed, considerable resources still remain in Congo, Gabon, Zaire, Cameroon and the Central African Republic. Their potential value has not been fully realized, as multinational drug companies, among others, are now beginning to appreciate. While many drugs can be synthesized in Western laboratories, the tropical forests are yielding drugs that as yet cannot be manufactured.

Meeting a range of local needs

Although Western markets value African hardwoods for their visual appeal and are just beginning to explore their wider chemical potential, in local use the forests are still valued most for fuel and food. Wood and charcoal are often the main sources of energy for households and industries throughout the region. Traditional methods of curing tobacco use huge quantities of wood. Waste is enormous in the traditional curing ovens. Some tobacco companies in Kenya are encouraging local smallholders to grow trees as well as tobacco, but awareness of the danger of burning the forests too quickly is not widespread.

Other diverse industries based on tree

products exist throughout the region. Plywood and matches are by-products of the logging industry, and gum can be harvested from the forests as a crop, with ever increasing local involvement in the various stages of production. Uganda and Congo both manufacture paper, and Kenya is nearly self-sufficient in paper products. Various other countries have built huge paper mills, often with government sponsorship.

In addition, local populations often depend on the forests to provide a source of food and income; they collect fruit, roots and leaves to eat and to sell in the marketplace. Fabric dyes for locally produced cloth are also manufactured from tree barks and roots. Tropical fruits, grown as plantation crops, are also important sources of revenue in the region. As well as being prepared for export, they are processed by local manufacturers, and some are used in small-scale manufacturing. Palm oil and palm kernel processing are established in virtually every Central African country regardless of natural resources or level of development. One of the largest mills was built by the Ghana Oil Palm Development Corporation. Much of the palm oil is exported and finds its way into sweets for the American and European market. The abundance of copra (made from sun-dried coconut meat) has led to soap manufacturing in Liberia, Burundi, Rwanda and elsewhere. Soap manufacturing is well suited to fledgling industry, as small firms can use the old-fashioned labor-intensive boiling method of production.

THE WORLD'S TREASURE HOUSE

MINERAL WEALTH · INDUSTRIAL POTENTIAL · DIVISION OF LABOR

Southern Africa contains some of the richest deposits of minerals in the world. Mining is a major activity in the region, though many areas are hampered by shortages of skilled labor, water, transportation and power. South Africa has exploited its mineral wealth most successfully, and on the strength of this has developed a strong industrial base employing migrant workers from neighboring countries. Widespread coal reserves, particularly in South Africa, provide much of the energy needed for mining and industry; the region has very little oil, gas or nuclear power. Several countries, including Comoros, Mauritius and Madagascar, have little industry except commercial farming. Cash crops are also grown on a huge scale in South Africa, providing the raw materials for food processing.

COUNTRIES IN THE REGION

Angola, Botswana, Comoros, Lesotho, Madagascar, Malawi, Mauritius, Mozambique, Namibia, South Africa, Swaziland, Zambia, Zimbabwe

INDUSTRIAL OUTPUT (US $ billion)

Total	Mining	Manufacturing	Average annual change since 1960
50.9	13.9	30.6	+2.3%

INDUSTRIAL WORKERS (millions)
(figures in brackets are percentages of total labor force)

Total	Mining	Manufacturing	Construction
4.7	1.0 (3.3%)	2.95 (9.8%)	0.73 (2.4%)

MAJOR PRODUCTS (figures in brackets are percentages of world production)

Energy and minerals	Output	Change since 1960
Bituminous coal (mill tonnes)	183.3 (5.3%)	+358%
Oil (mill barrels)	169.3 (0.7%)	N/A
Copper (1,000 tonnes)	770.0 (9.0%)	-22.3%
Nickel (1,000 tonnes)	69.5 (8.6%)	+124%
Chrome (mill tonnes)	2.17 (51.5%)	+153%
Vanadium (1,000 tonnes)	16.4 (53%)	N/A
Zirconium (1,000 tonnes)	154.5 (21.1%)	N/A
Gold (tonnes)	639.0 (35.9%)	-18%
Diamonds (mill carats)	25.6 (43.1%)	+86.2%
Manufactures		
Canned fruits (1,000 tonnes)	264.2 (5.1%)	No data
Ladies' dresses (mill)	19.1 (4.1%)	No data
Ferro-alloys and chrome (mill tonnes)	3.4 (26.6%)	N/A
Household hardware (1,000 units)	10.3 (4.6%)	No data

N/A means production had not begun in 1960

MINERAL WEALTH

Mining for metallic minerals probably first began in the region in the 1st millennium AD. By the end of the millennium copper and gold had become established as part of Iron Age culture. In Zimbabwe alone, archaeologists have discovered more than 1,100 prehistoric gold mines and 150 copper mines. Today mining is significant in all the mainland countries of southern Africa with the exception of Malawi.

South Africa has by far the greatest range of minerals – approximately 70 exploitable substances. European mining started in 1685, but exploitation for industrial development did not come until the late 19th century with the diamond rush of 1870 and the gold rush of 1886. The discovery of these valuable minerals provided the impetus for the country's rapid economic development, transforming it from a near-subsistence agricultural region to an industrial power. Growth in the workforce and improvements in transportation followed quickly. Before the end of the 19th century the number of white settlers had soared to about 1 million, and there was a massive expansion in railroad construction.

Today South Africa has the world's largest reserves of gold, chromium, and the "platinum group" of metals (platinum, palladium, rhodium, ruthenium, iridium and osmium). It is also a major producer of both gem and industrial diamonds, uranium and vanadium. Other important minerals include copper, iron ore, asbestos, chrome, silver, beryllium (used mainly as a hardening agent in alloys) and tin. Of these, gold is by far the most important metal, and accounts for more than half of all mineral sales.

The golden arc

The South African gold fields, which form a 300-km (185-mi) long curved area running through Johannesburg and into the northern Orange Free State, are thought by geologists to have been laid down some 2,700 million years ago. The gold occurs in sheets of low-grade ore, anything up to 4 km (2.5 mi) below ground. More than 40,000 tonnes (40 percent of the world's total gold stocks) have been recovered from this part of South Africa, and mining still continues.

Gold is mainly used as bullion – kept in reserve for bank notes that have been issued – but industry also makes high demands on gold reserves. Excluding official coins (such as the South African Krugerrand), by far the greatest proportion of gold produced is used in jewelry, followed by electronics, dentistry and medals. During the 1970s, the dramatic increase in the price of gold on world markets revitalized the industry, and veins that were previously considered uneconomic to mine were opened up.

Outside South Africa, gold is mined principally in Zambia and Zimbabwe; copper in Zambia and Namibia; cobalt – derived from copper ore and widely used in industry for the production of hard-wearing alloys, turbine blades and permanent magnets – in Zambia; chromium in Madagascar and Zimbabwe; uranium in Namibia; tin in Namibia and Zimbabwe; diamonds in Angola, Namibia and Botswana, and to a lesser extent in Lesotho and Swaziland; and asbestos in Zimbabwe. Malawi has begun to exploit its bauxite potential.

Fueling industry

Coal is the major energy resource in the region. In South Africa it is the only locally produced fossil fuel and meets 80 percent of the country's energy needs. It is also the country's second largest export earner, despite the fact that in recent years many countries have refused to trade with South Africa for political reasons. Coal burnt in South African power stations provides energy – at some of the lowest prices in the world – for large areas of the region.

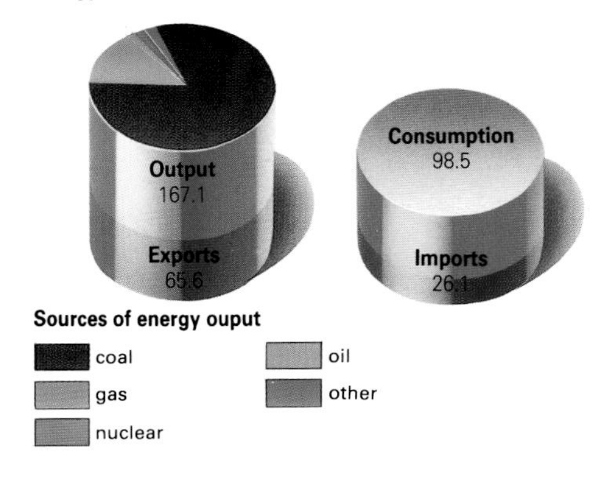

Energy production and consumption *(above)* The region, especially South Africa, is rich in coal reserves. Coal and its derivitive fuels make South Africa 80 percent self-sufficient in energy. Oil comes from the Angolan enclave of Cabinda north of the border.

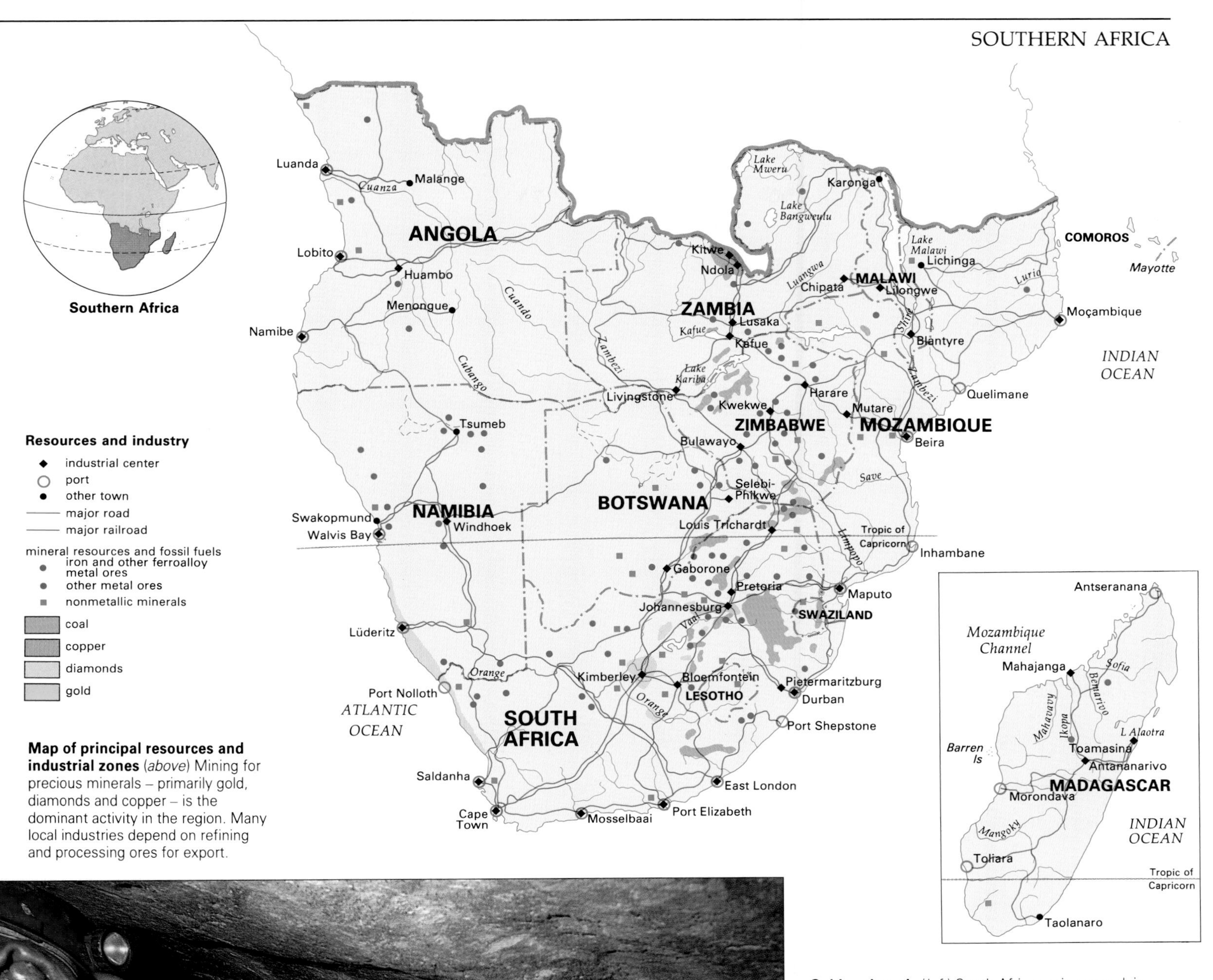

Map of principal resources and industrial zones *(above)* Mining for precious minerals – primarily gold, diamonds and copper – is the dominant activity in the region. Many local industries depend on refining and processing ores for export.

Golden threads *(left)* South African miners work in dangerous conditions to produce the majority of the world's gold. At the end of their day's work the miners are searched for any hidden nuggets, and their clothes are vacuumed for gold dust.

Lacking any reserves of petroleum, South Africa has developed synthetic fuel plants that produce fuel oil by distilling coal. The development of this process formed a major part of the government's plan to make the country self-sufficient in basic commodities, thereby reducing its vulnerability to economic sanctions.

Angola is the only significant petroleum producer in the region. Reserves of oil have also been discovered in Swaziland and Mozambique, but these have yet to be exploited. Limited amounts of oil are extracted from shale in Madagascar.

South Africa has developed nuclear power from its extensive uranium reserves, while in Zambia most industry is driven by hydroelectric power from the Kariba Dam on the Victoria Falls, and the Kafue river project.

Copper furnace *(above)* The region's copper refineries and processing plants, such as this one in Zambia, provide southern Africa's growing manufacturing sectors with the raw material to make alloys and electronic components that are sold abroad.

Nuclear jets *(right)* Southern Africa has a growing uranium industry and has joined the United States and Australia as a significant producer and exporter. This mine in Namibia uses a high-powered water jet to extract the radioactive ore.

INDUSTRIAL POTENTIAL

Although southern Africa has diverse and abundant resources, industry remains second to agriculture as the dominant activity in the region, except in South Africa. The lack of industrial development is largely attributable to the European colonial powers that dominated the region, except for South Africa, until the 1960s and 1970s. Under their control, the policy of most countries in the region was to supply cheap raw materials to the colonizing nations and to import manufactured goods, a pattern that still persists to some extent.

The workshop of the region

Industrial development has been exceptional in South Africa: the country produces just under half of all the manufactured goods in the whole of Africa. These include iron and steel products, machinery, automobiles, transportation equipment, chemicals, foodstuffs, clothing, electrical goods, plastics and a range of consumer goods. Defense and high-technology industries have also been developed, due largely to the country's recent history of political isolation.

The stimulus and capital for industrial expansion followed the discovery and exploitation of the country's considerable mineral reserves, notably gold and diamonds, in the late 19th century. South Africa also has the advantages of a large workforce, plentiful skilled labor, a well-developed transportation system, reliable power supplies and growing markets, both domestic and regional.

Industrial activity in South Africa is concentrated in the main urban centers. The PWV (Pretoria–Witwatersrand–Vereeniging) complex, which also includes Johannesburg, is the largest industrial complex and accounts for over half of South Africa's output. The government has made repeated attempts to decentralize industry to the poorer rural parts of the country, but so far with little success.

Most of the other countries in the region have developed limited manufacturing industries, usually based on their natural resources. In Zambia, for example, the copper industry has expanded

URANIUM IN NAMIBIA AND SOUTH AFRICA

South Africa and Namibia (under South African rule until 1989) both have deposits of uranium, but because of international sanctions, in force until 1990, neither country was able to exploit them fully. Namibia's deposits are dotted all over the country, but uranium is extracted at only one site: the Rossing mine near Swakopmund halfway up the coast. When it opened in 1976, the mine – operated by the Rio Tinto Zinc company, but under South African control – was the world's largest open-cast uranium mine. It can produce up to 4,250 tonnes per year, but, despite contracts with Japan, Britain, Taiwan, France and Germany, has only been working at two-thirds capacity. With 2,000 employees, it is one of Namibia's largest employers. The low-grade ores are processed on site. Uranium has no associated domestic industry.

In South Africa uranium is produced at 15 sites, largely as a byproduct of gold mining. The ore is low grade, but profitable because gold covers the costs of extraction. Even during the period of international economic sanctions, the country was the world's third largest producer of uranium, with export markets in Western Europe and Israel. Some ore goes to Valindaba near Pretoria for enrichment, and in 1984 the country's first commercial nuclear power station – at Koeberg near Duynefontein – became operational. A second reactor has since come on line.

from supplying the raw material to making finished goods from the metal, and Lesotho now supplies polished diamonds to the world market, rather than exporting them in their raw state.

The agricultural sector also supports a wide range of industrial activities: South Africa, for example, produces wine; timber goods and paper are exported from Swaziland; Angola produces coffee and processes tobacco into cigarettes; Zimbabwe manufactures leather footwear and textiles; agriculture in Botswana supplies its leather-processing and meat-canning factories; and the Comoros Islands, off the northeast coast of Malawi, specialize in perfume-oil extraction.

Constraints on industrialization

The availability of raw materials does not in itself guarantee the viability of an industrial sector. Shortage of water, for example, particularly in the southwest of the region, places severe constraints on industrial development. In Botswana and Namibia this hampers diamond mining as abundant water supplies are needed to separate the diamonds from the waste material in which they are found. In South Africa, the Vaal river – the main source of water for a large part of the country's industrial complex – is nearing the limits of its capacity. Plans have been made to import water from Lesotho via the Orange river.

In the landlocked countries of the region, industrialization also has to contend with transportation problems. Although many countries have railroad systems for carrying raw materials and manufactured goods to ports for export, problems arise when goods need to be carried through several countries using different track gauges. Political hostilities between states also disrupt long-distance traffic, while poor power supplies add to any existing difficulties.

Furthermore, the recession in much of the Western world during the 1980s caused a fall in demand for African raw materials, and an inevitable drop in export revenues. This reduced the capital available for industrialization. Another effect of Western recession has been a reduction in financial aid, which for the poorer countries in the region provides much needed funds for industrial development and the construction of large-scale, highcost projects such as dams and hydroelectric plants.

DIVISION OF LABOR

The structure of the southern African workforce still bears signs of the region's colonial past. Before independence, industry in most of the region was controlled by Europeans, usually from one of the colonial powers. Black Africans – about 90 percent of the population – were generally employed as cheap, unskilled labor. Rather than teaching European skills to the indigenous population, the colonialists imported people from Europe to do the skilled jobs.

Since independence, Africans have taken control of much of the region's production, though "neocolonial" ties still exist in many areas. The South African economy – which dominates the region – remains largely controlled and run by the minority white population, but nevertheless depends almost totally on a black African workforce.

Commercial enterprises, including cash crop production, are often owned by foreign-owned multinational companies. Foreign ownership can bring advantages: increased employment, for example, and the introduction of new skills, technology, training, capital and markets. There are, however, major disadvantages. Foreign owners are often regarded as exploitative and profiteering, and some governments – notably those of Mozambique, Angola and Zimbabwe – have attempted to remove this dependency, which they regard as offering unfair terms of trade, by nationalizing many private sectors of the economy.

Sewing together *(above)* Women's sewing cooperatives, such as this one in Botswana, help to provide clothes for many of the people in the region. Most consumer goods bought in southern Africa are made by small, local producers.

Building model boats *(below)* At the Beira Shipyard in Mozambique, the locals are rebuilding part of their Portuguese colonial legacy. With traditional tools and using plans that were left behind, a Portuguese trading vessel is being recreated.

Lonrho textile mill Southern Africa's largescale industry is based on production for huge foreign-owned multinationals. The finished products are usually for export and the benefits of manufacturing are not felt by the local people.

The migrant workforce

The mining industries, vital to the region's economy, employ a significant part of the industrial workforce. In South Africa, thousands of employees – mainly men – are migrant workers who have left their home countries to find work in South African mines and factories. Many are driven there by the lack of employment opportunities in their own countries. Population pressure and the shortage of land aggravate the situation. Some are officially contracted and others come without any guarantee of employment. Every year in South Africa the mining and manufacturing sectors demand additional labor. It is estimated that at any one time there are over 1 million of these "guest migrant workers" in the country.

There are some benefits to this system. People are employed, often housed, fares are sometimes paid, and they are able to send back their wages to their home country. At present about 15 million people in the region depend on wages earned in South Africa. The money sent home is an invaluable help to their families, and is also an important source of revenue for other countries, particularly Botswana, Mozambique, Zimbabwe, Lesotho, Malawi, and Swaziland. Half of Lesotho's earnings comes from the "export" of labor to South Africa.

A drain on development

On the negative side, most of the migrant workers are employed as unskilled or semiskilled labor and do not receive any training. There is little incentive for the employer to train workers who are only there on a temporary basis. In addition, family life and personal life suffer. The major disadvantage of migrant labor, however, is that the pool of able-bodied workers in the home countries is reduced, handicapping the development potential of poorer areas.

Industrial development is also hampered by high levels of unemployment, especially in urbanized areas where populations are expanding much faster than the rate of economic growth. Political instability in the region has affected the productivity of the workforce, and put pressure on scarce resources. War in Mozambique and Angola has destabilized the economy preventing development of an industrial base.

In South Africa increased industrial action and union activity in the mining and manufacturing sectors have improved working conditions. Nevertheless wages remain very low in comparison with those of the white workforce.

INDUSTRY AND APARTHEID IN SOUTH AFRICA

At the turn of the century, the development of the Witwatersrand gold fields in northeastern South Africa created job opportunities for huge numbers of unskilled workers. Black people from all over the region flocked to the area to find work in the mines, and the industry rapidly expanded. As it grew, the South African government systematically implemented a policy of "apartheid". This entailed the segregation of different racial groups, state-protected job discrimination and the exploitation of black labor. During the 1930s, "homelands" were established by the government – areas designated specifically for the settlement of South Africa's black population.

The homelands were situated in places that had limited agricultural potential and no industry or mineral deposits. Economic opportunities for people living in these areas were therefore severely limited, making it easy for the government to exploit the new pools of labor. In 1950 the Population Registration Act and Group Areas Act gave legal force to apartheid and legitimized the homelands system.

The gold mines alone employ 485,000 people, a large proportion of whom come from the homelands. Manufacturing industry also provides employment for thousands of homeland residents: for example, the South African metal and engineering industry employs 400,000 people in more than 9,000 companies. Many workers commute on a daily basis; others migrate for several months, if not years, at a time. Although the laws that created apartheid were abolished in South Africa in the early 1990s, the homelands still exist, and the male inhabitants of these areas continue to migrate to the Witwatersrand gold mines in search of work.

Glittering prizes

Diamonds are a valuable resource in 6 of the 10 mainland countries of the region and form a vital part of the southern African economy. Extracting and processing diamonds provides a great many jobs and earns valuable foreign exchange.

The primary source of diamonds is a rock called kimberlite, which occurs in pipes and fissures in volcanic rock. The ore is mined, usually by open-cast methods, and then gently crushed. A variety of techniques are used to separate the diamonds – the hardest naturally occurring substance in the world – from the waste material that surrounds them. Repeated washing of the crushed ore causes diamonds – which are denser than the rock in which they are found – to collect at the bottom of the container. The most modern method involves x-raying the ore. This causes the diamonds, a pure form of carbon, to fluoresce, making them easy to pick out.

Most diamonds mined from the interior of southern Africa are obtained from volcanic sites, the most famous of which is the "Big Hole" in Kimberley, Cape Province, South Africa. In coastal areas, particularly in Namibia and Angola, ancient river beds, beaches and river mouths are mined for "alluvial" diamonds, which have been washed down from the interior of the continent.

Controlling the market

Diamonds have provided the means for rapid economic growth for a number of southern African countries. In 1968, for example, exploitation began at Orapa in east-central Botswana, at what proved to be one of the world's richest sources. By 1990 diamonds accounted for 60 percent of exports and one-third of government revenue. There is, however, a potential risk in reliance on one primary resource: market recessions, price fluctuations and the danger of a nonrenewable resource running out could destabilize the country's economy.

Furthermore the diamond market is controlled by De Beers Consolidated Mines; a South African company, registered in Switzerland and operating out of London. De Beers has a virtual monopoly in diamond trading – 85 percent of the world market. It exerts control through supply, not price, so when the market is weak, De Beers has to buy up

Alternative methods (*above*) Independent and, for the moment, illegal mining operations do exist. Often using primitive processing, smallscale mining offers local people a profitable though potentially dangerous job.

Diamonds in the rough (*left*) The newest South African diamond pipe is this one at Finch Mine to the west of Kimberley. Extracting diamonds from ore is a long process, adding to the resource's high value.

and store excess production in order to keep control. Southern African governments do not therefore have control over the prices of one of their major resources.

In 1991 the close-knit diamond world was shaken by the news that new legislation was permitting Angolan citizens to possess and deal in diamonds (and Angolan diamonds are reputed to be among the finest in the world). The rush of illegal digging that followed resulted in enough diamonds to rival the legal production figures. If the market continues to be flooded by diamonds that have been smuggled out, it could ruin Angola's chances of becoming a major producer, while depriving the government of much needed revenue.

Diamonds for industry

There are two main types of diamonds: about 80 percent by weight are industrial quality; the other 20 percent are gem quality, used for jewelry. Industrial diamonds are either shaped into cutting tools for high-precision machining, or ground up into an abrasive powder known as bort. As the demand for industrial diamonds grows, the region's mines face competition from plants producing synthetic diamonds, which now account for about 30 percent of the world market.

Although gem quality diamonds attract much public interest and are commonly bought for investment, the lesser quality diamonds are of far greater importance to world industry, which uses them for a variety of machining processes such as grinding, sawing, drilling and polishing.

COAL, CRAFTS AND CASTE

RAW ENERGY · GROWTH AND MODERNIZATION · EXPLOITING THE POPULATION

Fossil fuels, mineral ores, jute and a vast workforce are among the principal industrial resources of the Indian subcontinent, primarily an agricultural region trying to develop its industrial sector. Electricity and other kinds of energy to feed large-scale manufacturing plants are more limited. India itself has large reserves of coal that can be used to generate electricity, and is developing a number of oil and gas fields as well as nuclear power. Other countries in the region, especially Pakistan and Bangladesh, lack these advantages and a reliable national power supply is still a problem. The complicated social structure, famine and environmental disasters combine to work against rapid industrial development. The potential of the region is enormous but any resolution of its many difficulties is likely to be painfully slow.

COUNTRIES IN THE REGION

Bangladesh, Bhutan, India, Maldives, Nepal, Pakistan, Sri Lanka

INDUSTRIAL OUTPUT (US $ billion)

Total	Mining	Manufacturing	Average annual change since 1960
86.33	7.53	68.0	+5%

INDUSTRIAL WORKERS (millions)
(figures in brackets are percentages of total labor force)

Total	Mining	Manufacturing	Construction
43.2	2.4 (0.7%)	34.2 (10.2%)	6.6 (2.0%)

MAJOR PRODUCTS (figures in brackets are percentages of world production)

Energy and minerals	Output	Change since 1960
Coal (mill tonnes)	205.3 (4.3%)	+271%
Oil (mill barrels)	266.8 (1.2%)	+7900%
Iron Ore (mill tonnes)	32.8 (5.8%)	+77%
Bauxite (mill tonnes)	4.0 (4.1%)	+255%
Manufactures		
Refined sugar (mill tonnes)	10.5 (16.4%)	+193%
Cotton woven fabrics (mill sq. meters)	16969 (14.2%)	+263%
Jute fabrics (mill sq. meters)	3735 (92%)	+48%
Footwear (mill pairs)	197.4 (4.4%)	+1368%
Broadleaved sawnwood (mill cu. meters)	15.1 (12.1%)	+647%
Nitrogenous fertilizer (mill tonnes)	7.9 (8.5%)	+652%
Cement (mill tonnes)	45.5 (4.1%)	+415%
Transistors (mill)	3560 (5.2%)	N/A
Bicycles (mill)	7.4 (7.3%)	+662%

N/A means production had not begun in 1960

A long-established iron and steel plant (*above*) at Jamshedpur, in southern Bihar, which is one of the chief centers of India's metal industries. Women working as laborers are engaged in rebuilding one of the site's dilapidated structures.

Energy balance (mill. tonnes coal equivalent)

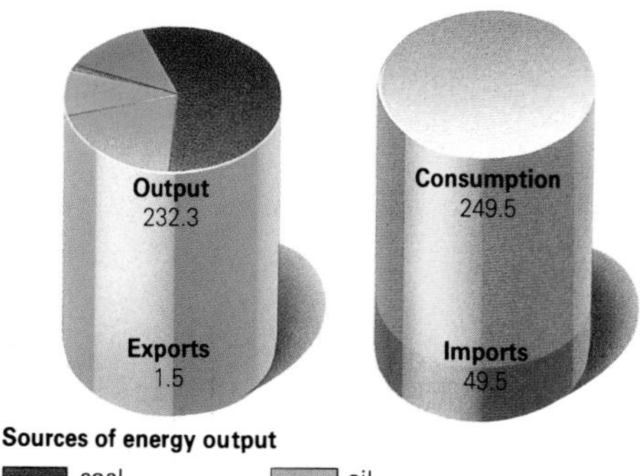

Sources of energy output

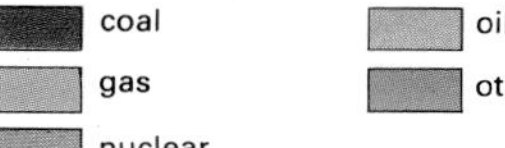

RAW ENERGY

India has sizable amounts of natural resources including coal, iron and manganese ore. Coal supplies are vast. Reserves of lignite (brown coal) are sufficient to meet the country's needs, but those of bituminous coal, used to make iron, are too low to match iron-ore resources, so much of the ore is processed in other parts of the world. Coal reserves are not evenly distributed. They occur in West Bengal and Bihar in the northeast and in Madhya Pradesh and Andhra Pradesh in the center and southeast, and have to be transported hundreds of kilo-

Energy production and consumption (*left*) Although expansion of the region's energy resources, including coal, natural gas (which has seen enormous growth in recent years) and hydroelectricity has been rapid, industry is dependent on petroleum imports.

Map of principal resources and industrial zones The mineral resources of the Indian subcontinent are extensive, but patchily distributed. Coal, including lignite, and iron ore are the most widespread, but even so large areas lack significant deposits of any kind. The region has great potential for resource development. Many deposits are not fully exploited, and mineral surveys of some areas are still to be made.

meters. In Pakistan the best reserves are in Kalabagh, West Punjab. India has lignite deposits at Tamil Nadu in the southeast, where they are used to produce fertilizers, power and briquettes, and at Rajasthan and Gujarat in the northwest. Lignite is also found in Bangladesh.

Singhbhum and areas of Orissa state in northeast and east India contain some of the Earth's richest iron-ore deposits, making the region one of the world's largest exporters of iron ore. Small but concentrated deposits of high-grade iron ore also occur in Pakistan and in Sri Lanka, where it is used for electrosmelting.

Chromite and titanium are found as well in India but full exploitation of these resources has not yet been achieved. The country is short of other metals, but the Geological Survey of India has revealed deposits in several areas, none of which has yet been mined. Sulfur can be extracted from pyrites at Amjore, Bihar.

Natural gas is present in huge quantities in Pakistan, especially in the northeast at Siu between Baluchistan and Punjab. A network of gas pipelines links gasfields with the main consumption areas. Bangladesh has natural gas at Comilla and Sylhet, used primarily to produce fertilizers and thermal power.

Pakistan has 20 different types of minerals including chromite, celestite, bauxite and barite. However, many of its mineral resources are underexploited or low grade. It does have high-quality, easily extractable limestone deposits, including marble, which form the basis of an important cement and construction industry. Sri Lanka has a seemingly inexhaustible supply of graphite (its major export), used in nuclear reactors and for making pencils. The island is also rich in many kinds of gemstones.

Nepal, Bhutan and the Maldives are comparatively poor in natural resources. Nepal has coal, iron ore, copper and a handful of other minerals in small scattered sites. Very little has been done to exploit them. Bhutan has not been surveyed and there is little or no industry there. The same is true of the Maldives, one of the poorest countries in the world.

Postcolonial industrialization

Traditionally India has a strong manufacturing base in fine cotton and silk clothing, exporting it all over the world. Jewelry is also a long-established craft in the region, combining the finishing and mounting of local precious stones with metalworking skills. The development of modern industries in the region began on a modest scale before independence in 1947. New cotton mills were built in Bombay and Tamil Nadu, though only one steel plant, Jamshedpur in Bihar, was constructed despite the availability of coal and iron ore. India's industrial development since independence has followed a

Goggles and turbans Workers at a cement plant in Gujarat, northwestern India, protected against irritant dusts, resemble extras from a space movie. Not all workers are so lucky. In the rush to industrialize, safety measures may be disregarded.

variety of routes. The government has created numerous public sector industries, including iron and steel plants, engineering enterprises, aerospace factories, and electrical and petrochemical concerns. Many dams have been built to provide hydroelectric power (HEP).

At the same time, however, the government has restricted many private plants to producing items considered to be socially useful, such as pharmaceuticals. The output of "luxuries" such as private cars was for many years strictly limited by quota, though after the 1980s there was a significant increase in the production of consumer goods, much of it due to the growth of largescale manufacturing. Private enterprise on a smaller scale also makes a significant contribution.

GROWTH AND MODERNIZATION

New industrial growth across the subcontinent has been generated by a mixture of internal expansion, foreign investment and the exploitation of new resources. Investment in the private sector, usually by large combines, has been a force for change in the petrochemicals industry and in the automobile trade. Public-sector investment has been concentrated in basic industries such as coal, iron, steel and heavy engineering, as well as the defense and aerospace industries. In India, central government funding has created a string of new cities in the northeastern and eastern states of Bihar, West Bengal and Orissa. These are the areas that form the heart of the steel and heavy engineering industries since the deposits of coal, iron ore and other minerals used in processing and production are concentrated there.

The success of some agricultural projects in the region has enabled farmers and cooperatives to invest in smallscale engineering projects. In the Punjab, for example, mechanization on the farms and the development of irrigation techniques have generated considerable growth in medium-sized engineering industries. These changes have also given a boost to the transportation sector that moves the goods from one place to another.

Looking for expansion

The Pakistani government has also attempted to encourage industry, and to reduce the overwhelming reliance on agriculture that existed at independence. Shortages of electricity and fuel in Pakistan have hindered industrial development, and hydroelectric schemes have been developed on the Indus and Jhelum rivers in an attempt to meet demand. Agriculture now accounts for less than a quarter of national income, and the manufacturing sector has increased to contribute one-fifth.

During Pakistan's initial efforts to industrialize in the early 1950s, manufacturing expansion concentrated on processing agricultural products. However, industry in the area soon became dominated by cotton textile mills. These are still significant employers in the sector. Woolen textiles, sugar, paper and tobacco industries developed alongside. In the mid 1950s imports were reduced, so foreign companies began to establish branches within Pakistan to produce consumer goods and eventually items such as light engineering components, fertilizers and chemicals. In the last decades of the 20th century there was a move into steel products but, despite attempts to diversify, cotton textiles continue to be the major export.

Bangladesh is still dominated by agriculture, which causes a regular pattern of seasonal unemployment. Repeated natural disasters such as devastating

Celluloid success (*right*) India ranks above the United States and Japan as the largest producer of movies in the world; more than 70 million people visit the cinema each year. Movies are made in all the major languages, and romance and social themes are the most popular. The movie industry is subject to government censorship.

Turning wheels (*below*) Making bicycle parts at a factory in Lahore, Pakistan. There has been considerable expansion in light engineering, but much of Pakistan's manufacturing machinery, such as the lathes and presses seen here, has to be imported, as do many of industry's raw materials.

floods have undermined much of the country's attempts to establish other enterprise. As a result, little heavy industry has developed, though the economic climate is able to sustain a wide range of cottage industries. The government has tried to tackle the country's endemic problems with an industrialization policy, but lack of mineral resources has restricted its success. Jute, cotton, hides and skins traditionally form the basis of processing industries; jute and tea are still major exports. Bamboo plants provide excellent raw materials for papermaking and there are four mills in the country. Bangladesh also has sugar factories, glass and aluminum works, a shipyard at Khulna in the south and a steel mill farther west at Chittagong.

Where the old meets the new

In many parts of the region traditional industries are still practiced. Bangladesh's cottage industries produce yarn, textile fabrics, carpets, ceramics and cane furniture. Modern largescale counterparts have grown alongside them, and sometimes a modern industry can grow from the roots of a traditional one. India has, for example, a number of shipyards that construct vessels both for defense and for the country's large merchant fleet. At the same time, scattered along the coastline, particularly on the coast of Gujarat in the northwest, a number of small yards still produce wooden cargo ships (dhows). These craft – manufactured almost entirely by hand from local timber – ply across the Arabian Ocean to the Gulf states and eastern Africa.

The region is, potentially, a huge market for consumer goods. In recent years there has been a growth in production of items such as refrigerators, televisions and air-conditioning units. Modern office equipment is also increasingly popular – photocopiers are built in India under license from, and with the cooperation of, Western or Japanese companies. The automobile industry has also changed drastically in the last decade – again with Japanese cooperation – and a number of relatively modern and lower cost vehicles were manufactured in India during the 1980s.

Many of the cars and motorcycles produced in the region are no longer even made in the country that developed them. The original tools and jigs have simply been transported to create a new production line in India. In Madras the Indian Enfield company still manufactures a motorcycle that originated in Britain in the early 1950s. In the 1980s the company began to export these machines back to Britain, another twist in the complex import–export relationship that has long linked the two nations.

Exporting machinery and equipment to the rest of the world is, relatively speaking, still in its infancy. India has a very wide-ranging manufacturing capacity – covering the spectrum from sewing-machines and bicycles to nuclear reactors and rockets. Much of what is produced, however, is of low quality and relatively costly in world terms, which makes it difficult for Indian goods to compete successfully in international markets. Some progress has been made in exporting railroad rolling stock and trucks, and the export of handmade cotton goods has also thrived in recent years.

POWERING INDIAN INDUSTRY

A great obstacle to the development of manufacturing in the region is the lack of a reliable supply of power, even in the larger cities. Although many parts are rich in coal, using it to generate electricity has been a slow and difficult process. One of the problems is that the coal, especially in India, has a high ash content, making it a poor quality fuel. Another is that, in the days of the British empire, much of the coal mined in the region was exported back to Britain, and India has made full use of its own resources for only a relatively short period.

Since independence, boosting coal production in India and investing in new power stations to supply the demand for electricity has been a priority for every administration. Now coal-fired thermal power stations provide two-thirds of the country's electricity, and can meet the needs of a rapidly expanding iron and steel industry. India's urban population has more than quadrupled since independence, and this expansion has led to a vast increase in the need for electric power. The demand is not only to feed domestic services – lighting and water pumps – but also for new factories making the electrical and consumer goods that are becoming an everyday part of life for the burgeoning middle classes.

EXPLOITING THE POPULATION

The vast population of the region could be a major industrial resource both as a workforce and as a market. However, a number of economic and social considerations have prevented the situation developing the way it has in Europe and the United States. At independence the Indian government believed that the lack of industrialization was partly due to the reluctance of Indian entrepreneurs to invest. Other, perhaps more significant, factors were that the British had given no encouragement to local industry, which eventually would have come into competition with their own exporting industries. They also neglected to develop a higher education system that would, in the future, meet the needs of a new technological society.

The local culture, too, and the values instilled by religion, played a part in the way the workforce reacted to possible change. India's bureaucracy before independence was dominated by Hindus, notably of the educated higher castes. When the region was partitioned in 1947, the new Muslim state of Pakistan found itself with a severe shortage of educated clerical workers and government officials.

Religious divisions and social barriers

Within mainstream Hindu India, the persistent legacy of the caste structure is another reason for the lack of an entrepreneurial industrial class. One group of castes – called collectively the Vaisya – have traditionally controlled the business sector, by which is meant trade and moneylending, or banking, rather than investment in production. It is significant that one of India's most famous business houses, which embraces a wide range of industries from trucks to hotels, is dominated by the Tatas, a family from the Parsi community of Bombay. The Sikhs too are noted for their industrial dynamism. Their energy has turned the Punjab into one of the most affluent states in India, with highly developed agriculture and manufacturing plants. In the Hindu community caste barriers can still cause resistance to certain kinds of "demeaning" jobs.

A shortage of management skills is also partly responsible for the inefficiencies of government enterprises. The region possesses very few management training schools, and those that do exist tend to be exclusive – the preserve of those who can afford the high fees. There is no shortage of educated manpower at some levels – India has an excess of trained engineers of every type – but career aspirations are often limited to obtaining a safe, salaried, permanent job in a large corporation or in government service. Such jobs are seen as a sure way to avoid sliding into the widespread poverty that afflicts so many people in the country, but it is not an attitude that encourages individual initiative and enterprise.

A ship dismembered A giant boiler lies in pieces in a shipbreaker's yard in Chittagong, Bangladesh, as a worker tackles the task of dismantling its boiler rods for scrap. Throughout the region, nothing is too large or too small for industrial reuse.

Organized labor

The Indian government distinguishes between the "organized" and the "informal" sector. The informal sector consists of factories with less than 12 employees, or less than 50 if electric power is not used. Large numbers of people work in the informal sector, either making completed products or manufacturing parts for larger firms. Industries in the two sectors often rely on each other: small-scale dyers, for instance, may use factory-produced cloth and then sell it on to established retail firms. Within the organized sector, legislation ensures some employment protection, and there are limited regulations regarding health and safety at work. Many workers in smaller concerns do not even have these benefits

– it is simply beyond the scope of any administration to monitor the huge variety of small enterprises. In any event, many operate on such small profit margins that they cannot bear the cost of additional overheads.

Another aspect of uncontrolled activity in the informal sector is the impossibility of imposing environmental controls. Waste substances – often highly toxic – are simply dumped as quickly and cheaply as possible. In the competitive world of the small business the expense of ensuring proper disposal would seriously undermine profitability.

Many workers in the informal sector are short-term migrants to the cities, working there to earn a cash income to supplement the meager living that prevails in rural areas. It is therefore very difficult to organize labor, and unions are restricted almost entirely to the organized sector. In some cities rickshaw drivers have formed unions, but attempts to organize the poor in this way are liable to bring a violent backlash from employers. In both Pakistan and India unions in larger private or public-sector concerns are quite well established, and have been involved in political struggles. A strike by Indian railroad workers in the 1970s and disputes in the Pakistani cotton trade were typical incidents that have slowed development.

Gradually the region, particularly India, has seen the development of a new urban middle-class with buying power. Often they are professionals and skilled workers employed in a government scheme. Bangalore in southern India has benefited from central government investment in hi-tech aerospace and electronics industries and an expansion of state government employment. These have combined to give the city the fastest-growing economy in India, making it the "Silicon Valley" of the region. At the other end of the scale, many people, including children, work in poorly paid jobs in factories making items such as local hand-rolled cigarettes (bidis). In northern India and Pakistan children are commonly found in sweated workshops producing rugs and carpets.

WHERE NOTHING GOES TO WASTE

One characteristic feature of industry in the subcontinent is the vast number of smallscale enterprises, many of which are involved in reconditioning or recycling goods. Oil drums, for example, can be beaten out flat and used to make steel trunks. Sometimes this is done by leaving them on the road for passing trucks to run over. In the West a burnt-out electric motor is often simply thrown away, but in India small electrical workshops rewind armatures and reconstruct bearings – cheap labor makes it cost effective.

Workshops will typically mix technologies and techniques; the very old with the very new, the very cheap with the more expensive. Ropemakers use wooden spindles to weave nylon fibers bought from a mill into modern synthetic rope. Quartz clock movements are installed in handmade wooden cases, and weaving techniques, once employed to make raffia furniture, are now used with plastic threads on steel-framed chairs.

Working for eight rupees a day Child labor is a fact of life in a region where poverty is widespread, labor is cheap, and children are regarded as potential wage earners from an early age. This 10-year-old boy is employed at a carpet loom in Pakistan – a craft in which children traditionally work beside their fathers.

Sackcloth to silks

A cottage industry *(left)* Throughout the region, but especially in Bangladesh, where textiles are a vital constituent of the country's economy, most textile factories are small, family-owned workshops that employ only a handful of people. In this factory in Dhaka meters of yellow cotton fabric are being made up into garments to be sold locally.

Fashion articles for the West *(right)* In the garment export-processing zone of Chittagong, in southeastern Bangladesh, denim jeans are laid out in the sun to fade into the light hues demanded by fashion-conscious Western consumers. Finished garments now constitute the single most important element in Bangladesh's exports.

The region's textile industry uses a wide range of technologies, from smallscale local production to large modern mills. In Bengal before partition the dominant textile industry was the production of jute to make sacking. Bangladesh has managed to rebuild its textile industry, which was curtailed when, as East Pakistan, it was severed from India. Jute processing now contributes significantly to the economy.

Locally produced cotton for home use – a form of subsistence industry – is still found in some rural areas. Cotton is collected from the seeds of certain plants in the genus *Gossypium* and then spun into thread using a simple weighted bob or a spinning wheel. The thread can then be woven into cloth. Although the fabric would traditionally have been colored with vegetable dyes, these are seldom used today. Even the smallest production units are likely to use synthetic dyes produced by the chemical industry. The Indian government has encouraged the local production of handloom cloth, and restrictions on the output of modern mills through quotas have helped to sustain this cottage industry.

Silk – spun from the protective cocoon of silkworms – is a more specialized process and thrives mostly in southern India, near Bangalore in Karnataka and the towns of western Tamil Nadu. The quality can be extremely high, and a silk with fine gold filigree weaving is a prized possession. Almost all of these garments are sold in the domestic market, though India exports silk for clothing, furnishings and upholstery.

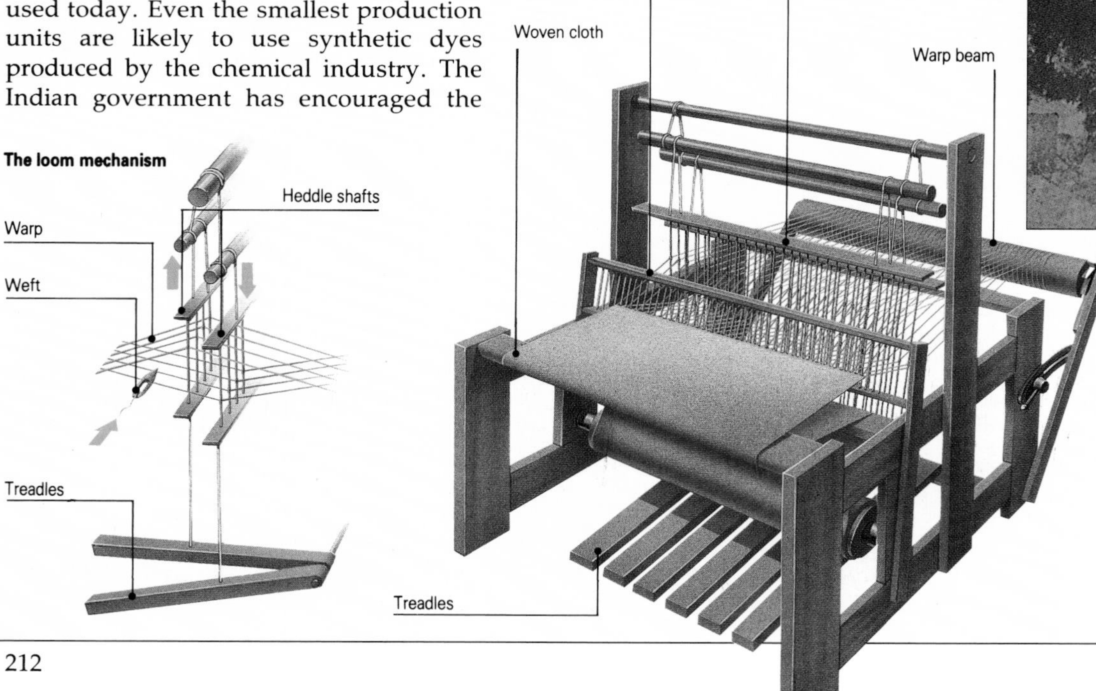

Traditional technology Handlooms are still used for weaving cotton and woolen thread throughout the subcontinent, though many factories are now fully mechanized. All looms work on much the same principles. The warp yarn is strung longways. The heddle shafts raise and lower the warp threads to create a "shed," through which the weft (the horizontal thread) passes in the shuttle. The reed moves back to press each new line of weave in place. On mechanized looms a motor replaces the treadle.

Modernizing the industry
Modern textile factories are located throughout India, but are concentrated around the cities of Bombay and Ahmadabad in the northwest and in the inland towns of Tamil Nadu.

In the 1960s modern synthetic fabrics became highly fashionable and caused a minor revolution in the textile industry. Supply depended on licensing from Western companies and relied on the petrochemical industry for raw materials – factors that made nylon and terylene expensive compared with cotton, hence something of a status symbol. In the 1980s and 1990s, however, people rediscovered the comfort of natural fibers, and synthetics are now more likely to be used sparingly in blends as reinforcing agents.

The production of finished garments is an important industry throughout the subcontinent. Although in urban areas there is a trend toward selling off-the-peg clothing, the vast majority of people wear clothes that need no tailoring – a lungi or dhoti for a man, or a sari for a woman, is simply a length of cloth wound in a particular way. Many people buy cloth and take it to one of the thousands of tailors that work in towns and cities throughout the region. New clothes can be made quickly and efficiently by these men, and as labor is cheap the cost is low. The people who produce finished clothing for export work to the standards laid down by the overseas importers.

LOW-TECH MEETS HIGH-TECH

UNTAPPED RESOURCES · CHANGING INDUSTRIAL STRATEGIES · WORKERS TURNED CONSUMERS

China and its neighbors present a fascinating paradox. China, though slow in moving into consumer manufacturing, has impressive mineral and energy resources to draw on. In addition, it contains over one-fifth of the world's population, a vast potential workforce. Since 1978, when liberal economic and trade policies were instituted after the death of Mao Zedong, China's economy has grown significantly, and the standard of living has soared. Yet China remains among the world's poorer countries, with much scope for developing more efficient modern industry. Taiwan, Hong Kong and Macao, on the other hand, have very little in the way of natural resources, and they import raw materials for a wide range of successful manufacturing industries including fashion, toys and miniaturized electronics.

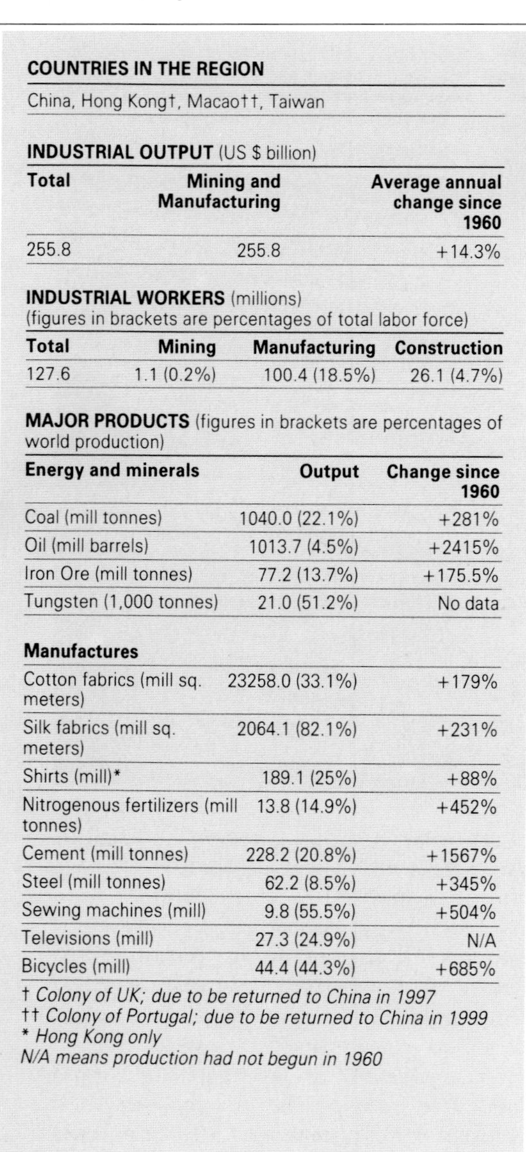

COUNTRIES IN THE REGION

China, Hong Kong†, Macao††, Taiwan

INDUSTRIAL OUTPUT (US $ billion)

Total	Mining and Manufacturing	Average annual change since 1960
255.8	255.8	+14.3%

INDUSTRIAL WORKERS (millions)
(figures in brackets are percentages of total labor force)

Total	Mining	Manufacturing	Construction
127.6	1.1 (0.2%)	100.4 (18.5%)	26.1 (4.7%)

MAJOR PRODUCTS (figures in brackets are percentages of world production)

Energy and minerals	Output	Change since 1960
Coal (mill tonnes)	1040.0 (22.1%)	+281%
Oil (mill barrels)	1013.7 (4.5%)	+2415%
Iron Ore (mill tonnes)	77.2 (13.7%)	+175.5%
Tungsten (1,000 tonnes)	21.0 (51.2%)	No data
Manufactures		
Cotton fabrics (mill sq. meters)	23258.0 (33.1%)	+179%
Silk fabrics (mill sq. meters)	2064.1 (82.1%)	+231%
Shirts (mill)*	189.1 (25%)	+88%
Nitrogenous fertilizers (mill tonnes)	13.8 (14.9%)	+452%
Cement (mill tonnes)	228.2 (20.8%)	+1567%
Steel (mill tonnes)	62.2 (8.5%)	+345%
Sewing machines (mill)	9.8 (55.5%)	+504%
Televisions (mill)	27.3 (24.9%)	N/A
Bicycles (mill)	44.4 (44.3%)	+685%

† Colony of UK; due to be returned to China in 1997
†† Colony of Portugal; due to be returned to China in 1999
** Hong Kong only*
N/A means production had not begun in 1960

UNTAPPED RESOURCES

China possesses a wealth of natural resources that could be used in industry. Significant metallic deposits include iron ore, bauxite, gold, lead, zinc, copper, manganese, mercury, nickel, tin, tungsten and vanadium. Among the country's nonmetallic minerals are asbestos, diamonds, fluorspar, gypsum, phosphates, potash, salt, sulfur and talc.

China also has plentiful energy resources, but difficulties in exploiting them (the remoteness of sites combined with poor service facilities) mean that the country only has sufficient energy for 80 percent of its industrial power needs. Coal is the most important energy resource. China is the world's leading producer, with reserves estimated at 770 billion tons (enough to last 400 years). Oil and natural gas reserves are also immense, and China is currently the world's sixth largest producer. In addition, the region has enormous potential for developing hydroelectric power, especially

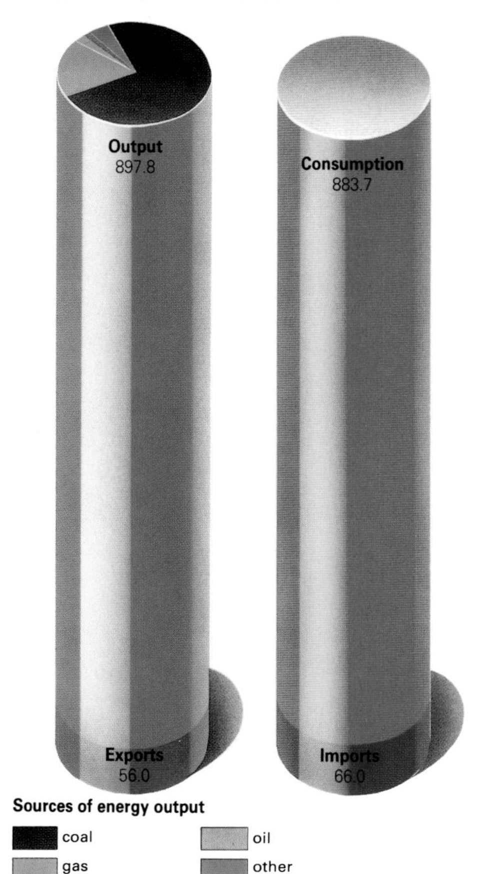

Energy production and consumption Coal dominates in China, which also has oil and natural gas reserves in unknown quantities. Oil is already produced in sufficient quantities to be exported. Both hydroelectric and nuclear power are being developed.

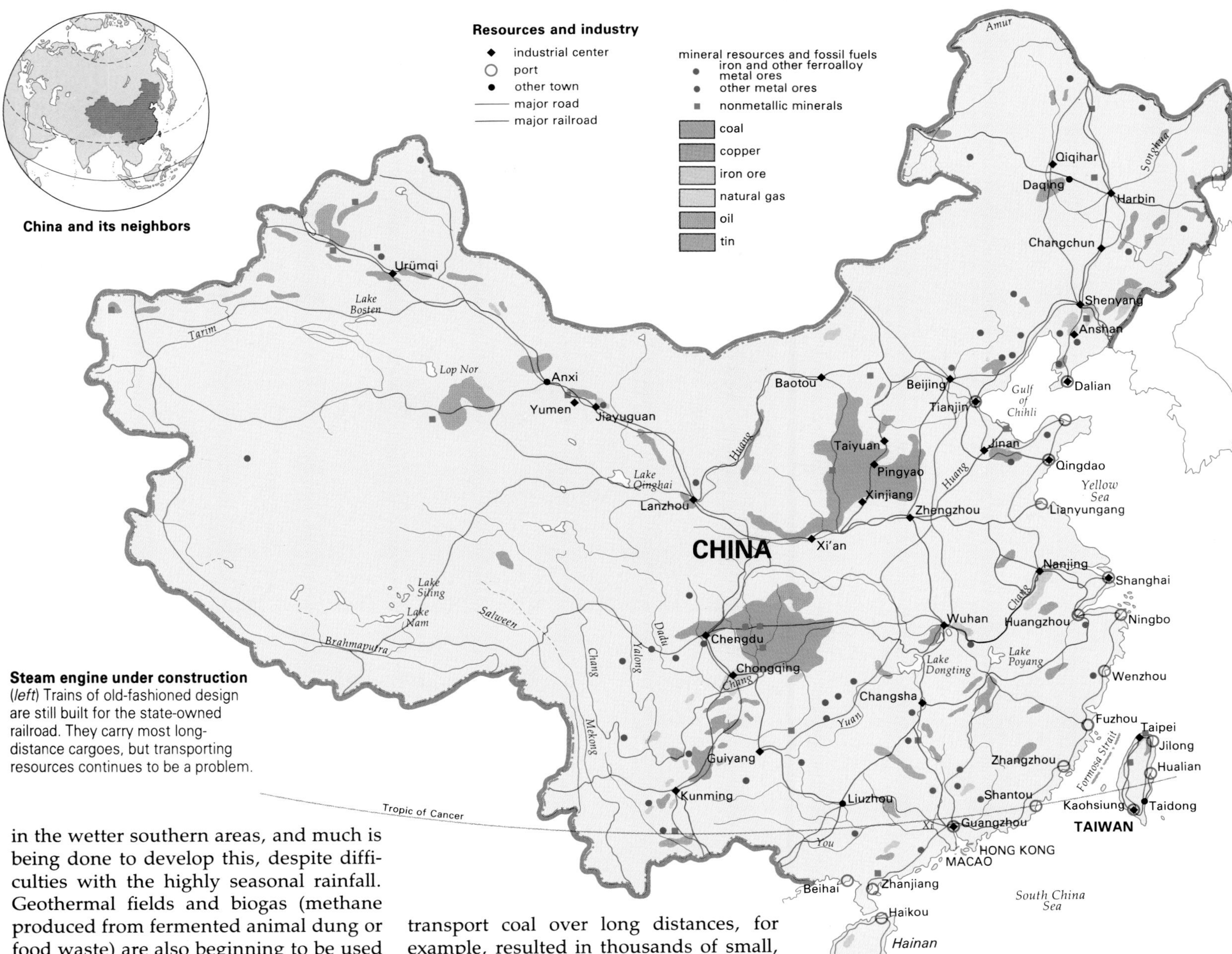

Steam engine under construction (*left*) Trains of old-fashioned design are still built for the state-owned railroad. They carry most long-distance cargoes, but transporting resources continues to be a problem.

Map of principal resources and industrial zones China's vast and diverse resources are scattered throughout the country. Coal and heavy industry are found mostly in the east. Trade and export industries are confined to the eastern coastal areas.

in the wetter southern areas, and much is being done to develop this, despite difficulties with the highly seasonal rainfall. Geothermal fields and biogas (methane produced from fermented animal dung or food waste) are also beginning to be used as sources of power. There is also enormous potential in the region for generating nuclear energy.

The other, much smaller territories of the region, by contrast, have very limited resources. Hong Kong has some feldspar and kaolin, while small deposits of kaolin, coal and marble are found in Taiwan, which also has hydroelectric potential and is now developing nuclear power. In both territories – and in Macao too – mining is of little significance, accounting for less than one percent of national production.

Despite the wealth of China's industrial resources, their huge potential has never been fully exploited. The sheer size of the country is partly responsible, linked to its lack of an adequate transportation network, especially the scarcity of good roads. Attempts were made in the 1950s and 1960s to find a way round this problem by creating localized industrial bases, though these had mixed success. One attempt to cut out the need to transport coal over long distances, for example, resulted in thousands of small, locally run mines being developed, but their relative inefficiency has meant that the coal they produce is expensive.

The uneven spread of rich resources makes solving the transportation problem a key factor in exploiting them efficiently. Some important coalfields lie close to the industrial heartland of Manchuria in the northeast, but the main reserves are found in distant Xinjiang in the far northwest, accessible only by means of the already busy railroad system. Many of the country's oilfields are also in the northwest (Xinjiang and Yumen), though others are more conveniently placed: for example, around Daqing in the northeast, offshore in the gulfs of Chihli and Tonkin, and in the estuary of the Zhu (Pearl) river in southeastern China.

Modernizing with foreign money

After the communist revolution in 1949 most industrial resources were developed by the state, with little private or foreign participation. Since 1978 a more liberal strategy has been pursued, involving both state-run industries and joint ventures with foreign companies. Traditional labor-intensive methods are still practiced, especially in southern China, where there are many small mines and hydroelectric power stations. Foreign investment is now much more common – especially in the north – with many joint ventures, such as the American-backed open-cast coal mine at Pingyao in Shanxi Province. Such projects were jeopardized by the crushing of the student demonstration in Beijing's Tiananmen Square in 1989, which led to withdrawals of foreign investment. By the early 1990s, however, foreign capital and technology were beginning to flow into the country again.

CHANGING INDUSTRIAL STRATEGIES

China's new communist leaders in 1949 adopted the Soviet Union's approach to manufacturing, concentrating on heavy industry, with rigidly centralized state control. After 1978, however, the government economic reforms brought a shift of emphasis to light industry. Industrial enterprises were able to form direct connections with one another, without interference by a higher state authority. At the same time China began to import manufacturing techniques and the equipment to make them work from Japan and the West, vastly improving the efficiency and quality of its manufactured products.

The 1978 economic reforms led to rapid industrial growth. After years of restrictions there was an unleashing of consumer spending, and local industries boomed to meet the new demand. Manufacturing expansion was so rapid that it threatened to get out of control, and in 1988 with the country suffering from soaring inflation, the government introduced an austerity program. Curbing the consumer boom proved no easy matter, however, partly because of the government's policy of continuing to pay wages to workers who had been laid off.

Balancing heavy and light industry

The 1980s saw a significant shift in the pattern of China's manufacturing. There was a pronounced move from heavy to light industry, as well as from low-technology, labor-intensive methods to a more modern, high-technology and capital-intensive approach.

Heavy industry is concentrated in the northeastern provinces, a location that reflects its development there before and during World War II by the Japanese, and later by the Soviet Union. Steel-rolling mills and tubing plants around Anshan and Shenyang in the northeast supply steel for bridges, oil pipelines, machinery and machine tools. Harbin, farther north, produces agricultural machinery as well as heavy pumps and turbines. Smaller centers of heavy industry – especially steelmaking – are found in other regions of the country, including Baotai and Taiyuan in the north, Liuzhou and Guiyang in the south, Chengdu in the center of the country, Jiayuguan in the north and Urümqi in the far northwest.

HONG KONG: DOORWAY TO THE WESTERN WORLD

Hong Kong, a British-owned territory on the south coast of China, is a thriving center of light manufacturing industry, especially plastics, toys and electronic gadgets, including watches and calculators. The colony consists of Kowloon and the New Territories on the mainland, and islands just off the coast of which Hong Kong Island is the largest. In particular, Hong Kong sits at the hub of the international rag trade, and often copies designer clothing at a fraction of the price for the mass market. With its accumulated experience in manufacturing for markets across the world, it is well placed to provide an example and an export route for China's developing light industry.

Hong Kong occupies a unique and powerful position in the plans for the industrial expansion of China, especially when the territory reverts to Chinese rule in 1997. It is one of the world's busiest trading ports, and a key point of contact between China and the countries of Southeast Asia. Its political status as a British colony has allowed it to establish essential longstanding links with the external commercial world, which will be vital to Chinese industry in the future. It is also an important training ground for the financial and export services essential to successful business cooperation. In the late 1980s and 1990s Hong Kong conformed well to China's open-door policy and the development of the Special Economic Zones. But the question of the colony's trading future after 1997 was by no means assured.

Glassblowing in Chengdu (*above*) The capital of Sichuan province in central China is a flourishing industrial area. Local mineral resources include lead, zinc, marble and salt. The bright colors worn by the workers indicate significant social changes since the death of Mao, who favored drab colors.

Joint venture in the northeast (*right*) This Chinese–Czech factory is in Shenyang, Liaoning province. The area produces pig iron and steel, and has the highest concentration of heavy industry in the country. Joint ventures these days are likely to involve countries outside the former Soviet bloc.

Computer factory, Shanghai China's principal commercial center since modern times, Shanghai is now the base of the growing high-tech sector. Many companies sponsor workers' studies at the city's numerous institutes of higher learning. Computers present a particular challenge: the Chinese language has no alphabet, and several thousand characters representing individual words must be fitted onto an extra-large keyboard.

Although heavy industry has generally developed close to the resources that it needs, there are a few exceptions. In particular, Shanghai's industry is supplied from distant sources with coal carried by sea from the northern ports, and iron ore brought along the Chang river. This pattern resembles the way industry is organized in Taiwan and Hong Kong.

Light industry is a rapidly growing sector in China, with output fast approaching that of heavy industry. Distribution is relatively even around the country, reflecting the attempt in the 1950s and 1960s to encourage cities to be as industrially self-sufficient as possible. There are, however, increasingly important concentrations in the coastal provinces, especially close to Hong Kong. Light industry is found in and around most major cities, producing a range of items including radios, bicycles, sewing machines, cotton yarn, cloth, cassettes and cameras, as well as pharmaceutical and food products.

Special Economic Zones

Another important change in China's manufacturing since 1978 was caused by the government's "Open Door Policy", designed to bring China more fully into global trade. As part of this strategy Special Economic Zones (SEZs) were set up in Fujian and Guangdong Provinces, on Hainan island, and in 14 "designated" ports. Foreign firms were encouraged to establish new ventures in these areas, using both Chinese and foreign materials. The plan was to boost high-technology manufacturing while at the same time carefully controlling contact with the West. This policy was a reversal of government ambitions during the 1950s and 1960s, when support was given to a more egalitarian distribution of industry and wealth across the country, but was a return to the idea of treaty ports established by imperial China for trade with the West.

The zones have already begun to play an important role in shaping the development of China's industry. In particular Shenzhen, adjoining Hong Kong's New Territories, has been successful in attracting industrial investment and joint manufacturing ventures from Japan, the United States and Hong Kong. The zones have encouraged new concentrations of industry in the eastern, northeastern and coastal provinces, much of it light, high-technology and capital-intensive manufacturing. Alongside the foreign ventures a small Chinese private-industry sector has also begun to emerge. However, the Tiananmen Square episode, and the possibility of future political disruption, cast some doubt on the future of the whole Special Economic Zone policy, especially in the inland provinces.

WORKERS TURNED CONSUMERS

People have always been the cheapest resource in China, and saving time or labor costs has never been a consideration in any industrial program. This was particularly true during the "Great Leap Forward" (1958–60) when Mao Zedong tried to give the country an overnight manufacturing base by encouraging decentralized cottage-style industries right across China. Cities and villages were encouraged to become self-sufficient. Farmers were ordered to seek for coal beneath the fields that they tended, and thousands of small agricultural machinery factories were set up, as well as backyard blast furnaces. The experiment was not a success, however, as lack of planning caused great problems. The locally made agricultural machinery was often of poor quality, as was the pig iron produced in the local blast furnaces, which themselves often crumbled in the rain as they were made from soft adobe brick. Dams for hydroelectricity and coal pits caused great harm to farming land, contributing to disastrous famines (1960–61) in which it has been estimated that some 25 to 30 million people died.

Spools of silk in Hangzhou *(above)* The city of Hangzhou in Zhejiang province in the southeast is an ancient and renowned center of silk manufacturing. Zheijang produces more silk than any other province except Sichuan, and is a leading producer of tapestry, embroidery and silk umbrellas for domestic and export markets.

Scrap metal, a cottage industry *(right)* Backyard industry was encouraged in the early 1960s as a way of accelerating China's development. Now, however, these Dai minority people, like many of their fellow Chinese, are likely to be in business for themselves.

TRADITIONAL TEXTILE INDUSTRIES

Local textile craft industries are widespread throughout the region, in both rural and urban areas, and even where modern factory production has become dominant. Manufacturing is by individuals or small groups, dependent for success on their ability to create styles and patterns that are valued for their distinctiveness. Silk, of which China is the world's leading producer, is of very high quality. In a number of cases the craft products have become associated with a particular area or city. Shanghai, for example, has long been known for its textile crafts, and indeed it was immigrants from this city who took their skills with them when they fled to Hong Kong, Taiwan and Macao during the 1949 revolution.

It is impossible to travel around the region without seeing evidence of a local craft textile industry. Craftspeople are sometimes organized together in centers for tourists who wish to examine production methods and buy cloth. Traditional local designs and styles are also increasingly incorporated into the textile industry at factory level.

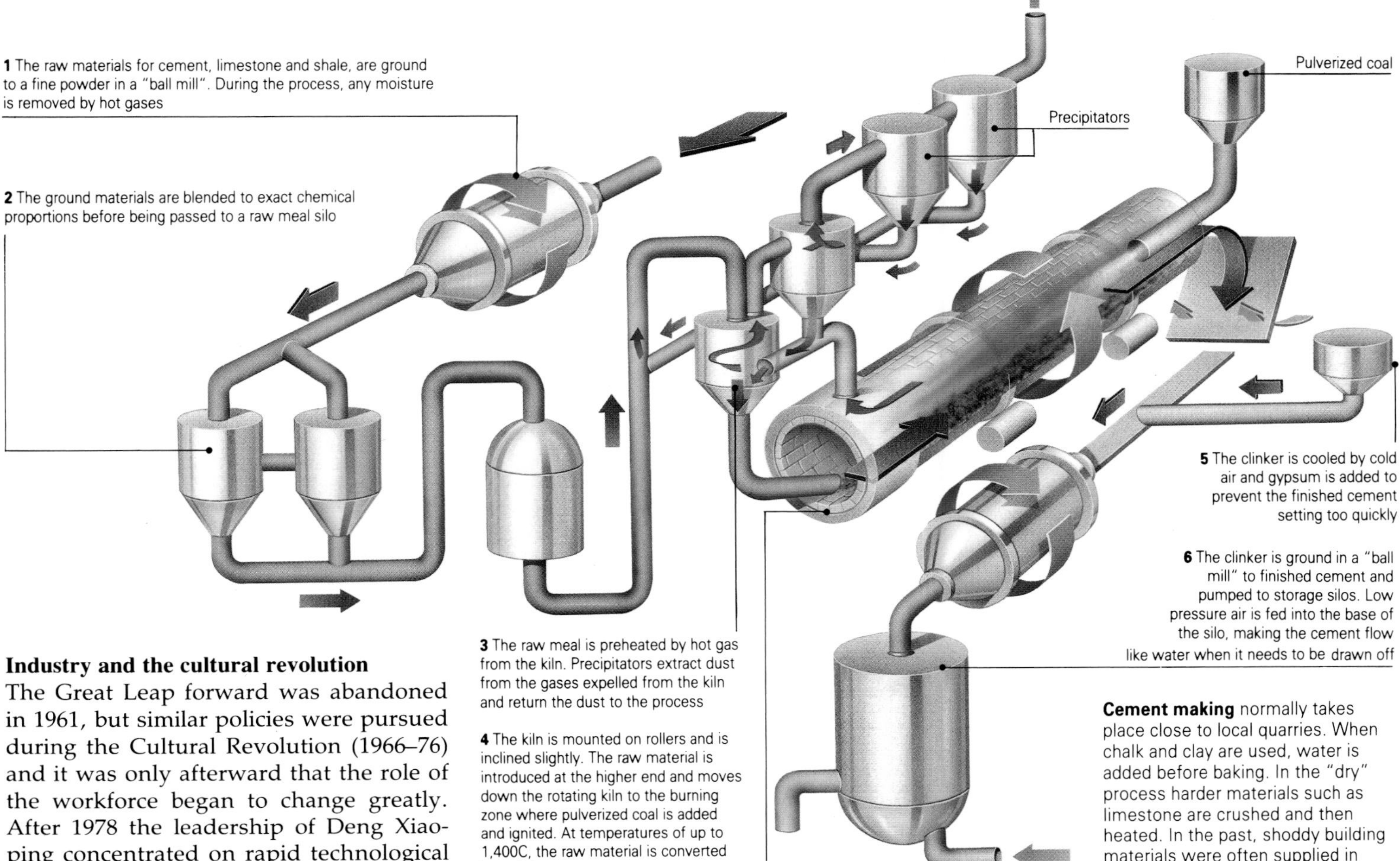

Cement making normally takes place close to local quarries. When chalk and clay are used, water is added before baking. In the "dry" process harder materials such as limestone are crushed and then heated. In the past, shoddy building materials were often supplied in China's rush to build factories.

Industry and the cultural revolution

The Great Leap forward was abandoned in 1961, but similar policies were pursued during the Cultural Revolution (1966–76) and it was only afterward that the role of the workforce began to change greatly. After 1978 the leadership of Deng Xiaoping concentrated on rapid technological change – to produce high-quality goods at low cost – rather than on mass participation in industry. Lack of capital and labor skills were highlighted, and new policies introduced to deal with these. Foreign investment was encouraged, as well as foreign joint ventures and the importing of new technology and training. At the same time an attempt was made to loosen state control of manufacturing, so firms could act with some degree of freedom, and the government encouraged the growth of a private sector of industry.

Although much has been achieved, China's manufacturing is still dominated by large – and often inefficient – establishments, with three-quarters of output coming from the more than 100,000 state-owned enterprises. Collectives and co-operatives (of which there are about 300,000) account for most of the remaining production, while private firms are only responsible for one percent of industrial output.

The view from offshore

Unlike China, Taiwan and Hong Kong have both pursued a free-market approach to their industry. Although the role of government intervention has been very much greater in Taiwan than in Hong Kong, both have seen a flourishing of private entrepreneurial activity. While foreign multinationals play an important role in both territories, the number of locally based companies is also significant. Hong Kong has no fewer than 50,000 firms with 10 or fewer employees, many of them family concerns. Although small, these enterprises are frequently modern and use technologically advanced methods of production. The abundant cheap labor in Hong Kong combined with traditional skills in textiles has helped to make the territory the world's leading exporter of clothing.

Labor difficulties provide further insight into the very different state of industry in China and the other territories of the region. In Hong Kong and Taiwan labor problems derive largely from straightforward shortages. Taiwan has grown increasingly dependent upon an immigrant workforce, while in Hong Kong, where immigration is strictly controlled, entrepreneurs have begun to look outside the colony for cheaper labor, setting up factories in Macao, and Shenzhen in mainland China. Both Hong Kong and Taiwan are well supplied with skilled labor, and both have experienced the growing power of trade unions.

China suffers no labor shortages. Indeed, the country has a surfeit of manpower, with policies of overemployment frequently undermining attempts to encourage greater efficiency on the production line. Nor are there union problems, as all such organizations are tightly controlled by the state. Skill shortages, however, have dogged industrial development, preventing manufacturing from rising to meet the needs of the vast potential market that is offered by China's almost limitless domestic market.

China's leadership has recognized that foreign participation and decentralization are essential to achieve modernization. Such changes, however, also cause a dilemma, as they involve some loss of political control over industry, which the government is reluctant to permit. This great contradiction is one of many facing a regime that is struggling both to control a state-run industrial sector and to reap the benefits of a free market.

Breaking and building ships in Taiwan

Shipbreaking – industrial recycling (*left*) Natural resources on the small, mountainous island of Taiwan are both limited and inaccessible. Since 1949 the island has had to develop its industrial base rapidly to support a large urban population; about a third of the workforce is employed in manufacturing. Shipbreaking enables Taiwan to acquire raw materials through recycling. Today a universe-class oiltanker measures 345 m (1,130 ft), providing vast quantities of scrap. With more than 300 million tonnes currently afloat, the shipbreaking industry has huge potential to expand.

In the dry dock at Kaohsiung (*right*) Located on Taiwan's southwest coast, Kaohsiung is one of two deepwater ports and a shipbuilding center. The huge vessels in use today require extensive facilities for building and maintenance. Typical facilities at a major port or shipyard include dry docks, cranes and workshops to perform welding, rigging, carpentry, machining and electrical work. New ships are becoming more specialized to handle specific kinds of cargo, and there is constant pressure in the industry to increase speed, economy, and environmental safety.

Taiwan, lacking significant resources of its own, has had to build up its industrial base using imported materials. The shortage of resources is at the heart of the shipbreaking industry. In the 1960s the government became aware of a growing demand from the manufacturing sector for scrap iron. Taiwanese industrial policies have generally included little state involvement, but in this and a few other cases, the government decided to play a leading role. In 1965 it set up a state-owned shipbreaking enterprise. A purpose-built site was established at Kaohsiung and, with the cheap labor then available, the country soon came to dominate this industry worldwide. At the same time shipbreaking also provided both the iron and the capital for a shipbuilding industry, also state owned.

By 1973 a dockyard with a capacity of 1.6 million tons of repair was in operation, and ship construction was soon well under way. This development in turn encouraged the growth of Taiwan's steel industry in the late 1970s, with the first blast furnace starting up in 1977. For some time the China Shipbuilding Corporation of Taiwan was the new steel sector's only customer, though other sectors – including vehicle and heavy machinery production – have now added to demand. In Taiwan, the role of the state in shipbuilding contrasts with government policies in Hong Kong, where purely laissez-faire strategies have been pursued.

WORLD LEADERS IN SHIPBUILDING AND BREAKING

Shipbuilding and shipbreaking are distinct but related industries. Lacking the resources to feed heavy industry, shipbreakers perform the increasingly crucial function of "recycling" vast tonnage. The resulting labor cost would be prohibitive in the shipbuilding countries, which find it more cost-effective to utilize domestic resources or to import them. Taiwan and South Korea participate in both industries on a major scale, placing them in the top five of each sector as measured by annual output in deadweight tonnage (dwt).

Shipbuilding	('000 dwt)
Japan	8,178
South Korea	3,642
Germany	515
Brazil	430
Taiwan	383

Shipbreaking	('000 dwt)
Taiwan	17,592
China	6,086
South Korea	5,461
Pakistan	1,887
Bangladesh	1,517

Competition and declining markets

Shipbuilding in Taiwan has had to face a number of difficulties. One particular

problem was that development occurred at a time when there was a surplus of tankers and other ships on the world market. The industry has also been hampered by lack of capital and technology, both of which have had to be imported largely from Japan and the United States. The raw materials too are mostly imported. Skilled labor has proved an additional problem, though technical training has done much to overcome this. A further hurdle has been the fierceness of competition, especially within the Far East, from Japan and South Korea. It is largely as a result of this competition that the China Shipbuilding Corporation of Taiwan has not flourished to the extent predicted in the 1970s. Although fluctuations in the exchange rate have reduced the competitiveness of Japanese shipbuilding, both Japan and South Korea remain dominant in East Asia, while Germany and Brazil also rank above Taiwan as world shipbuilding countries.

The China Shipbuilding Corporation of Taiwan is currently trying to overcome financial problems by diversifying. To this end it has developed production of offshore drilling rigs, as well as nuclear plant equipment and other steel structures. Taiwan now has a total shipbuilding capacity of one million tonnes per year and a repair capacity of three million tonnes, but its survival may well depend on expansion of the country's commercial fleet, and on production of weapons-related items. The future of the industry will also largely hinge on decisions made by the government.

HIGH-TECH BOOM IN THE PACIFIC

SUPPLYING THE WORLD WITH TIN · ADDING VALUE TO RESOURCES · VERSATILITY AND AMBITION

Southeast Asia has been a rich source of precious metals, minerals and gemstones for hundreds of years. The region is famous for its jade (mostly from Burma) and for its hardwood, made into intricately carved furniture and boxes. More recently Southeast Asia has become a leading exporter of tin, particularly from Thailand, and has significant reserves of oil. Both resources help to fund new industrial growth. Since the 1960s, a young labor force has provided the impetus for expansion into high-technology electronics. However, progress has been affected by political events. Vietnam, Laos, Cambodia and Burma are isolated from their neighbors, and internal wars during the past 50 years have forced a decline in their traditional mining and manufacturing, preventing new opportunities to diversify.

COUNTRIES IN THE REGION

Brunei, Burma, Cambodia, Indonesia, Laos, Malaysia, Philippines, Singapore, Thailand, Vietnam

INDUSTRIAL OUTPUT (US $ billion)

Total	Mining	Manufacturing	Average annual change since 1960
89.5	16.7	68.4	+7.2%

INDUSTRIAL WORKERS (millions)
(figures in brackets are percentages of total labor force)

Total	Mining	Manufacturing	Construction
18.5	0.4 (0.15%)	15.1 (6.3%)	3.0 (1.3%)

MAJOR PRODUCTS (figures in brackets are percentages of world production)

Energy and minerals	Output	Change since 1960
Coal (mill tonnes)	14.3 (0.3%)	+610%
Oil (mill barrels)	785.8 (3.5%)	+312%
Natural gas (billion cu meters)	78.9 (4.1%)	+1073%
Tin (1,000 tonnes)	92.3 (46.0%)	-22.4%
Nickel (1,000 tonnes)	63.4 (7.8%)	+199%
Manufactures		
Processed palm and coconut oil (mill tonnes)	6.5 (35.3%)	+413%
Canned fruits (1,000 tonnes)	665.0 (12.9%)	+302%
Sawnwood (mill cu. meters)*	19.5 (15.6%)	+123%
Natural and synthetic rubber (mill tonnes)	3.9 (27.1%)	+54%
Rubber footwear (mill pairs)	47.2 (19.9%)	N/A
Jet fuels (mill tonnes)**	7.8 (5.5%)	N/A
Cement (mill tonnes)	36.3 (3.3%)	+1628%
Radios and sound recorders (mill)	57.7 (20.3%)	N/A

** Broadleaved timber only (coniferous excluded)*
*** Mainly Singapore*
N/A means production had not begun in 1960

SUPPLYING THE WORLD WITH TIN

The countries that make up the region of Southeast Asia are richly endowed with metals and minerals, especially in the older mountain chain running south from the border between China and mainland Southeast Asia to the Indonesian islands around Bangka, north of Java. Rubies, sapphires, gold, silver, and other valuable resources have been mined extensively in past centuries, but as it enters a new industrial era the region has made its fortune extracting modern riches: for example, tin, oil and gas.

About 200 years ago Chinese settlers began to mine tin in parts of the peninsula. By the second half of the 19th century they had developed the richest tin fields in the world in the Kinta Valley in southeast Malaysia. Largescale European interest was not awakened until the early 20th century when dredgers – machines that suck metal from the mine floor – were imported to boost mining activities. Until World War II Malay was the dominant source of tin, but afterward there was rapid expansion in Indonesia and Thailand as world demand strengthened. Malaysia's output continues to fall gradually as the richest fields become depleted and only 25 percent of the mines operating in 1985 were still working in 1990. The expensive large dredgers of the 1960s and 1970s are less successful today and cheaper gravel pumps, which are more flexible in poor mining conditions, are being reintroduced.

In Indonesia, where tin resources are still relatively plentiful, 28 state-owned dredgers contribute to 60 percent of the country's tin production. Thailand is seeking to increase offshore output from extensive undeveloped tin deposits near Phuket, an island lying off the southwest coast of Thailand. Even so, production is dropping due to dwindling resources. In 1990 Indonesia, Malaysia and Thailand together accounted for 35–40 percent of

Tin bath (*below*) Plentiful in Malaysia, but scarce worldwide, tin is usually found mixed with mud in alluvial deposits in streams, rivers or oceans. Sucked up from the water bed, the tin is laundered – washed and separated – and the concentrate collected.

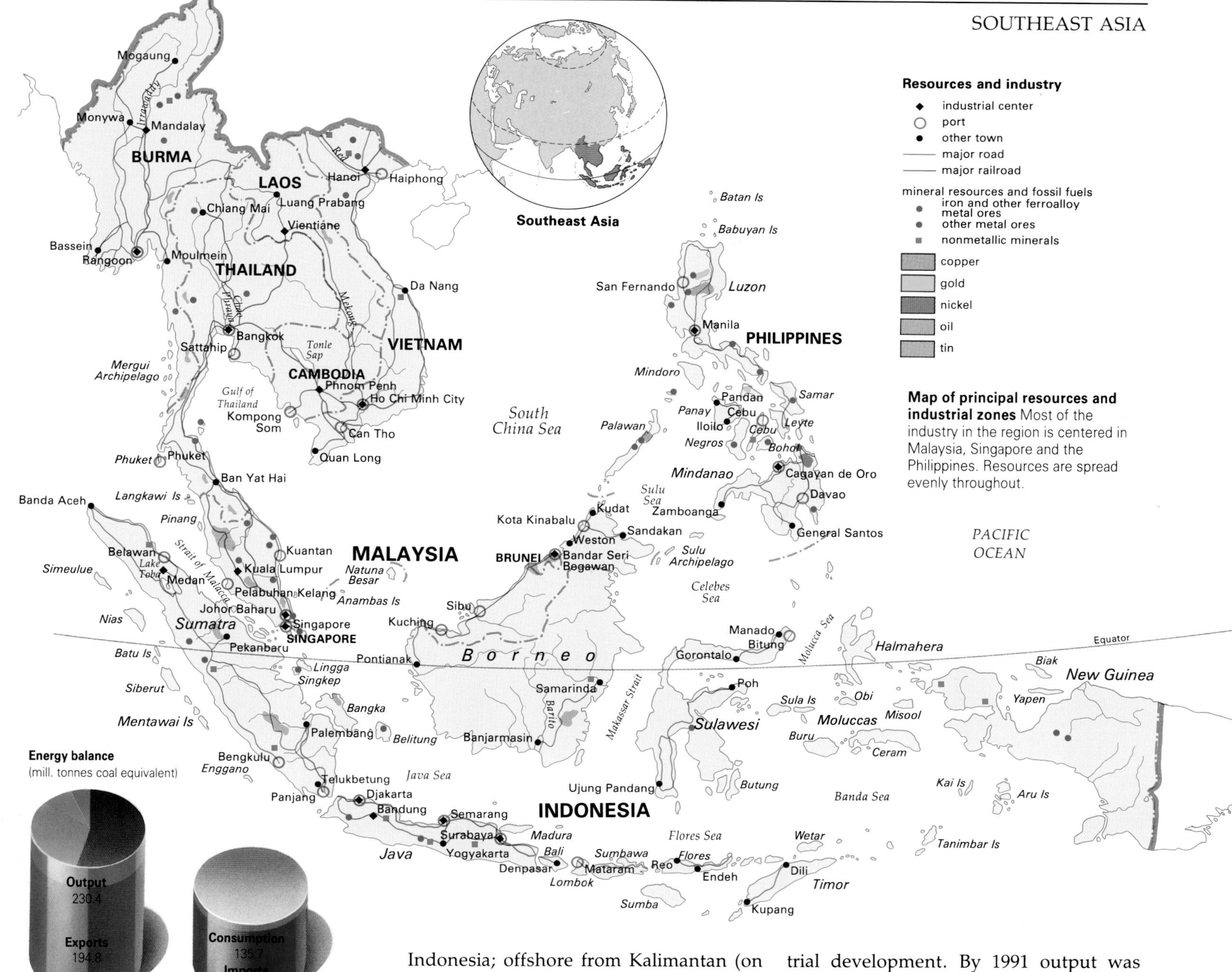

Map of principal resources and industrial zones Most of the industry in the region is centered in Malaysia, Singapore and the Philippines. Resources are spread evenly throughout.

Energy production and consumption *(above)* Not only is the region a net exporter of oil and natural gas, it also has the third largest oil refinery complex in the world. Coal is the only other major fuel. The region has no nuclear capability.

world output compared with over 60 percent in the 1960s, and the output of tin in Burma has also fallen.

The fossil fuel boom

During the second half of the 20th century Southeast Asia has become a significant producer of oil and natural gas. Extensive deposits, and a small variety of other minerals, are found in central and lowland Burma; eastern Malaysia; Sarawak and Sabah on the northwest coast of Borneo; eastern Sumatra; northern Java in Indonesia; offshore from Kalimantan (on the island of Borneo); and along the coast of western Malaysia and Thailand. The giant oil field at Seria in Brunei (northeastern Borneo) has passed its peak of production after nearly 60 years of mining, but has turned the tiny kingdom of Brunei into one of the richest countries in the world. Indonesia is fast becoming a major player in this field: in 1990 it was producing over 425 million barrels of oil per year more than Brunei. Malaysia's output has been rising as new fields have been opened up, particularly off the coast of the state of Terengganu in western Malaysia. These are more than replacing declining oil output in Sarawak. Oil production in Thailand should rise as new fields at Nuang Nuan in the Gulf of Thailand are brought into production.

While still surveying for further oil supplies, Thailand is developing extensive offshore gas fields. These feed the increasing energy demands of new industrial development. By 1991 output was meeting 70 percent of the country's needs. Urbanization, too, boosts the domestic demand for energy, and there is a large prospective market for any surplus to the north of the region, in densely populated areas such as Japan.

More than 45 percent of Indonesia's gas comes from Mobil's Arun fields in Aceh, in northern Sumatra, much of the output being exported to Japan and South Korea. Malaysia and Brunei have more recently developed fields along the coast of Brunei, Sarawak, Sabah and off Terengganu. Production has increased from 4 billion cu m in 1983 to approaching 20 billion cu m, 80 percent of which goes to Japan. In addition, an important potential oilfield is to be found beneath the waters around the Spratly Islands in the South China Sea. However, six surrounding nations, including China, claim the islands so development has been halted until territorial disputes are resolved.

ADDING VALUE TO RESOURCES

In the 1950s and early 1960s a number of countries in Southeast Asia built up manufacturing industries to exploit their natural resources. Their aim was to establish processing industries, particularly oil refining, metalworking, food packaging, textiles and electrical assembly to add value to local raw materials, and to reduce the region's reliance on imports.

Oil refining is by far the most significant processing industry in the region. Singapore, which was a British colonial trading center from the early 19th century until just after World War II, has transformed itself since independence into a leading processing center for oil, as well as rubber, tin and oil palm. The tiny island state is now the third most important center for oil refining in the world after Houston (Texas, the United States) and Rotterdam (the Netherlands), and has developed rapidly since 1961. It has five major refining complexes with a total capacity of 43 million tonnes. Singapore can more than fulfill the needs of the region, and has enough capacity to export processed petroleum across the world. Recently the industry has become more sophisticated and also produces chemical byproducts, plastics and synthetic fibers.

Following Singapore's lead, other oil producers in the region are anxious to develop their own refining capacity to serve at least their domestic needs for fuel oil, gasoline and kerosene. Pertamina, the Indonesian state-owned company, has six refineries in Sumatra and one in Kalimantan with a similar capacity to Singapore. It is also developing liquid propane gas (LPG) plants with Japanese investment, particularly in Bontung and Bodak in eastern Kalimantan. Malaysia, with 10 million tonnes of oil-refining capacity, has gone a step further by building an export-oriented refinery at Lutong in Sarawak to compete with Singapore. Continuing high demand for plastic, a byproduct of the refining process, attracts further investment in the industry from multinational giants including Kellog-Thyssen and Rheinstahl.

Stagnation caused by war

Many of the poorer countries in the region, including Laos and Cambodia, have very little largescale manufacturing industry, and war has prevented any effective development. Vietnam's industrial economy has regressed in the second half of the 20th century as a result of the Vietnam war (1955–75). Some manufacturing industries, such as cement, paper, metals, glass, cotton textiles, food, drink and tobacco, had been developed in the north (mostly in Hanoi, Haiphong and Cholon) before hostilities commenced.

The written word *(above)* Southeast Asia, especially Singapore, is quickly becoming a major center for printing and color reproduction. The combination of cheap labor, good transportation and vast improvements in printing technology has fueled this growth.

Economic driving forces *(left)* Although some United States' automobile manufacturers are well established in the region, the past two decades have seen a big increase in Japanese investment and development of joint venture and licensee automobile factories. Companies such as Toyota are able to take advantage of cheap labor and tax incentives.

There were also some specialized craft-based industries in the Tonkin delta, but more recently Vietnam has lacked sufficient funds to modernize them. After 1976, reconstruction efforts created a revival in the cement, steel and power-production industries, mostly in the north. They are still dogged by shortages of both raw materials and skilled labor.

Burma, too, has lagged in industrial development in the late 20th century. Its inward-looking military government has halted the potential growth of the tourist industry, which in turn stunted development of the country's traditional craft industries, particularly jewelry (made from local gemstones, gold and silver), woodcarving and textiles.

Fostering joint ventures

In the countries belonging to the Association of Southeast Asian Nations (ASEAN) – Brunei, Indonesia, Malaysia, the Philippines, Singapore and Thailand – industries processing the area's rich resources now attract international attention. Shell, Dunlop, MB Caradon (formerly Metal Box) and Unilever have developed manufacturing bases in Singapore, because of the oil and rubber refining there. Similarly, the palm oil industry in the Philippines is bringing in considerable investment from abroad. A joint venture between Shell and Mitsubishi has invested US $1.3 billion in an olefin plant in Indonesia making textile fibers.

Japanese and American investment is very important in maintaining copper smelting in the Philippines and aluminum smelting in northern Sumatra using local hydroelectric power. However, it is less vital to the older tin-smelting industries in Pinang, Phuket and Pelim near Bangkok, where local Chinese capital is used. A new development has been the expansion of Krakatau Steel at Cilegon near Djakarta in Indonesia; it produces nearly 5 million tonnes of steel products mainly for export to Japan.

High-tech and electronics industries are also beginning to transform several manufacturing sectors across the region. Malaysia and Thailand have been most successful in this field to date, Malaysia in particular having developed a semiconductor and integrated circuit (IC) chip industry, exporting products all around the world. Manufacturing automobiles, trucks and motorcycles is also growing in importance, with help from abroad. Expansion in this area reflects increased domestic demand and international confidence in manufacturing in the region.

MICROCHIP TECHNOLOGY FROM PINANG

The tiny Malaysian island of Pinang is known as the Silicon Island of the Pacific basin due to its rapid and successful development in the microelectronics industry. A cheap labor force and good communications with the mainland – the island has a port, an international airport and a road link – have made it an attractive prospect to multinational companies.

It has particular expertise in the production of semi-conductor devices (so-called because they are neither good conductors nor good insulators) used in electronic circuits as switches, amplifiers, light-detectors and for numerous other functions. They can be put into complex but easily manufactured microelectronic circuits and are the key elements for the majority of electronic systems from computers to communications equipment. Their versatility owes much to their compact size, reliable performance and low cost.

Originally Pinang imported most of the components, and was responsible for only the last two stages of chip making, the labor-intensive assembly and final testing and sorting. However, the island was able to attract investment to upgrade production, recruit skilled workers and go into silicon wafer processing. It has also developed local support for the industry that has encouraged continuing investment. The Malaysian-owned company National Semiconductor has established two subsidiaries producing high quality products that are exported to various other Asian countries and also to the United States.

VERSATILITY AND AMBITION

Southeast Asia has a population of 450 million people, the majority of whom are young. The potential workforce is therefore enormous, and this wealth of human resources has already begun to play an important part in the region's development. Levels of urbanization are still low – below 30 percent in Thailand, Vietnam, Indonesia and Burma – and a large proportion of the population works in agriculture and fishing (more than 45 percent in most countries). There are few very large industrial cities in the region given the overall size of the population. Bangkok and Djakarta, with populations of more than seven million, stand out as being exceptionally large.

Recently, education has become more readily available to the younger generations and literacy has increased, except in Cambodia, Laos and the remoter islands of Indonesia. A young, educated workforce is essential to expansion in the emerging labor-intensive light industries such as garment manufacture and electronics, particularly as the region is becoming increasingly competitive with the West. Whereas Malaysia, Singapore and, to a lesser extent, Thailand have been relatively successful in meeting the more sophisticated aspirations of these educated workers, in Indonesia and the Philippines they have struggled to find jobs. This has led to an explosion of people going overseas to areas of shortage, most significantly to the Gulf areas in the Middle East, though this may change.

Filipinos are the most numerous migrant workers; because English is one of the country's two official languages, it makes it easy for them to travel to the United Kingdom, continental Europe, Australia, Hong Kong and the United States. With the development of inter-ASEAN economic zones such as the one based in Singapore, the region should be able to provide employment for most of its own workforce. In the last few years Singapore has even begun to take in guest workers to cover its own shortages.

Shifting patterns of employment

As the region has developed modern manufacturing industries, a noticeable divide has grown up between a commercial, capitalized and technologically-oriented sector and one that is often family oriented and more traditional, using a low level of technology, and with informal labor and wage arrangements. Over time, the traditional sector has been drawn into providing products and services for the commercial sector, and the high-tech sector has come to specialize in labor-intensive products. Traditional wood-carving skills are used in the manufacture of furniture and plywood, for example, and long-standing expertise in paper and metal processing has been turned toward high-quality color reproduction and printing for Western and Japanese markets.

One consequence of this shift in emphasis is that the traditional resource-based industries are becoming less attractive areas for employment, and generally offer wages that support a much lower standard of living. The mining sector now accounts for about half of one percent of the labor force in all countries where figures are available, except Brunei. Even the plantations (producing rubber, sugar and coconuts) that, historically, were very large employers, operate with many fewer people today. Throughout the 1980s and early 1990s the workforce began to move away from subsistence farming and was increasingly absorbed into the manufacturing and service sectors.

Following the pattern *(above)* The majority of industry in the region is in the middle ground between cottage and high-tech industries. Manufactured goods, usually clothes and household items, are either for domestic consumption or limited export.

Like father, like son *(below)* Very smallscale manufacturing, such as this explosives workshop in Indonesia, is being squeezed out of the region. Except in Thailand, Burma and Vietnam, family-owned workshops mainly produce trinkets for tourists.

Low levels of unionization

In spite of the spread of education among the growing young adult urban population, there has been little union activity in manufacturing centers. Demands for improved wage levels and conditions of work are usually the result of rising standards of living, though improvements have sometimes been introduced by paternalistic governments. There is often a powerful work ethic among the population, particularly driven by the motivation to acquire enough wealth to marry, buy homes or consumer goods, fund further education or set up small businesses. The lack of union activity is one of the features that attracts foreign companies seeking new locations for factories to satisfy their global markets. Most of the governments in the region actively encourage foreign investment and frown on trade unionism. In the socialist republic of Burma, for example, trade unions are not permitted, although centrally

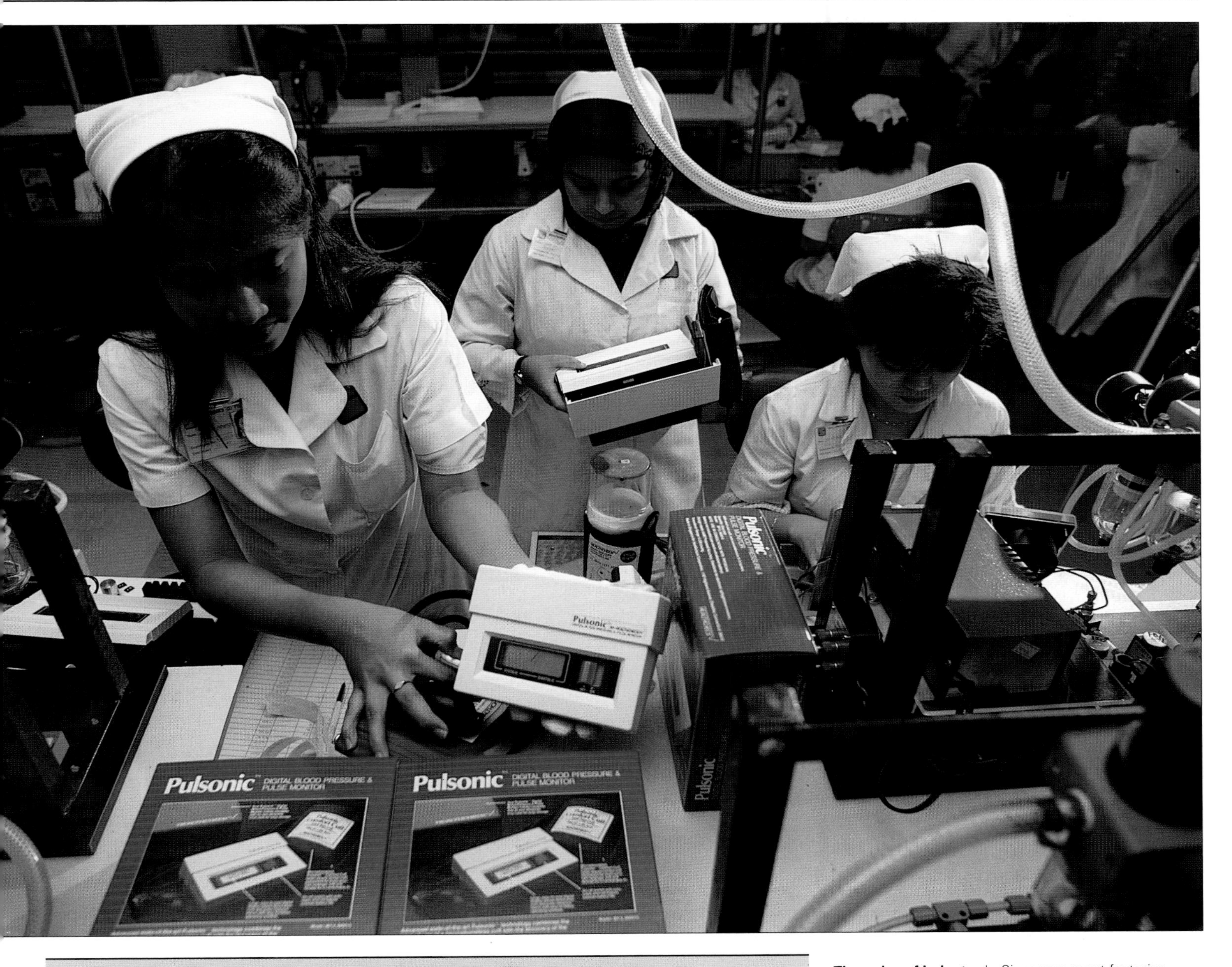

The pulse of industry In Singapore most factories, such as this medical supplies company, are manufacturing outposts of multinational organizations, specializing in the production of high-tech electronic and miniaturized equipment.

controlled workers' councils and local conciliation committees mediate in labor disputes. Northern Vietnam has a similar government-controlled system to mediate in labor disputes.

Among the foreign investors in many of these countries are the Chinese, who act as a local source of wealth to fund entrepreneurial ventures. This is important in countries such as Thailand and Indonesia that particularly favor joint industrial ventures. The few individuals who have become very wealthy in business have often expanded from a modest beginning, and such people act as a motivating example to others. The ultimate wish of most working Asians is to improve on their traditional low-waged employment status in order to own and run their own small business.

SINGAPORE – CROSSROADS OF THE EAST

A tiny island at the meeting points of the Indian and Pacific oceans, northern Asia and Australasia, Singapore is one of the world's most exotic commercial centers. Colonial traditions – including the widespread use of English as a first language – and a long history of trade have helped Singapore become a magnet for multinational companies. Singapore's commercial significance is due largely to the free port status it had in colonial times. At its height in the mid 20th century, Singapore was the largest port in Asia and the fifth largest in the world. Although shipping is still a leading industry, it is no longer the most important.

The island is heavily and efficiently promoted as a tourist and conference destination. Tourists go there for the superb food, luxury hotels, and duty-free electrical and photographic equipment. They also use it as an attractive and convenient gateway to or from Australasia or the rest of Asia. Alongside tourism, international business flourishes. Branches of multinational firms and a thriving workforce provide the capital and service industries to make Singapore a significant commercial center. The island has been particularly successful in establishing oil and petrochemical industries as well as electrical and electronics manufacture, industrial chemicals, printing and publishing. Metalworking, especially for ships and oil-rig repair, is a more recent spinoff from the old shipbreaking industries that were set up on the Jurong industrial estate in the 1970s.

Space is a recurring problem in such a heavily urbanized country and Singapore has joined forces with its neighbors, Johore to the north and Riau islands to the south, in developing a growth triangle. In addition, the free trade economic zone on Batam Island, 12 miles away, has attracted industrial development by international giants including Philips, Sony and Toshiba.

Carving up the forests

Timber is undoubtedly one of the most important exports of the region, especially to Japan where it is a mainstay of the construction industry. Hardwood from Southeast Asia is used to make wooden houses, or as shuttering in the manufacture of concrete slabs. The more valuable woods are exported in large quantities to Europe and Japan to be used as high-quality veneers or in cabinet making. Teak, prized for its visual appeal and durability, is frequently used for interior fittings in ships or luxury cars or for domestic utensils, particularly wooden bowls. Its exceptional resistance to acid makes it popular for laboratory benches and other fittings. At the other end of the spectrum, kempas and kapur are heavy-duty timbers used extensively in the construction of bridges, farm buildings, wharf decking, industrial flooring and in railroad sleepers.

Meeting insatiable demand

Southeast Asia was originally covered with extensive tropical rain forest, but exploitation of hardwood as an industrial resource has accelerated in recent years, causing extensive depletion. Much of the wood has been used to produce furniture, timber for the building trade, paneling, flooring, plywood, paper and other commercial products.

During the 1980s output of processed wood more than doubled in most of the region except Vietnam and the Philippines, and increased tenfold in Indonesia. The development of the printing industry in the region has meant that the consumption of pulp and paper is growing twice as fast in Asia as in Europe and North America, and the product is seriously underpriced in terms of the real costs of soil depletion, the drain on the water supply, atmospheric pollution and global warming. While the scale of forest depletion is not comparable to the much more extensive areas disappearing in South America, Southeast Asia is still selling a valuable resource for shortterm gains and longterm losses.

Demands for timber and for its by-products, woodchip and paper goods,

Awaiting the chop *(above)* Log felling and the timber trade is an important part of the region's economy. Nearly half of the labor force in the area is involved in forestry or some aspect of wood processing. After petroleum and rubber, timber is Malaysia's biggest foreign export earner.

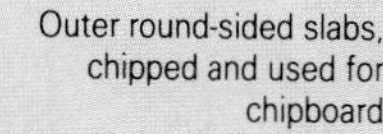

From tree to toothpick *(left)* There is little wastage in wood processing. The bark is used for compost and fuel, and the rounded outer portion is processed into chipboard. The inner sections of wood are cut into either planks or beams depending on the quality of the wood.

Carving a niche There is a long tradition of wood carving and furniture making in the region, but most of it is small scale. Attempts are being made to develop this industry to benefit both the timber trade and local skilled craftsmen.

have encouraged tree-felling practices in the region that prevent woodland from renewing itself. Thailand finally banned logging after discovering that forest cover in the country fell from 29 to 19 percent between 1985 and 1988. Furthermore, the high level of demand for this versatile resource has forced the lumber industry to expand into the neighboring countries. In consequence, there is a considerable amount of illegal cutting and smuggling from both Burma and Laos.

Less wood, more income

One possible way to slow down the rate of tree felling would be to generate a higher level of income from less of the raw product. In effect, this would mean adding value to the wood while it is still in the region, by processing it or using it in manufacturing industries and then exporting the finished goods. Manufacture of wooden furniture in the region remains small scale, but is an obvious candidate for expansion.

The traditional craft of decorative carving is currently only a very small part of the modern wood products industry since the vast majority of timber is exported in its raw state to be worked on by European and Japanese craftsmen. Manufacturers specializing in furniture are scattered in hundreds of small factories, often employing a large, skilled workforce but making slim profits. In terms of productivity these firms rank about 20th in the region compared with other ventures, though they have a much higher position in terms of numbers of employees.

At the beginning of the 1990s, the wood-processing industries making plywood, veneer sheets, sleepers, chipboard and fiberboard were still relatively small, but the emphasis for the future is on growth. A program to promote this sector has already had some success in Indonesia. Exports of raw timber have been banned, and as a result plywood has recently become the country's main non-oil export. With average prices for logs and sawn timber staying at a standstill during the 1980s, it is doubly important that the industry finds other ways to expand apart from increasing the volume of cutting.

INTENSIVE MANUFACTURING

A CULTURE DEPENDENT ON IMPORTS · EXPANSION IN A LITTLE SPACE · THE JAPANESE WORK ETHIC

Korea and Japan have limited natural resources but have become world leaders in manufacturing and technical expertise through the energy and acquired skills of the people. Since World War II the Japanese have expanded and diversified manufacturing very rapidly by importing foreign technology and fostering intense rivalry between firms. Shipbuilding, the steel industry, vehicle and machinery manufacture, chemical works and miniaturized high-technology equipment have all contributed to economic growth. While South Korea has followed the Japanese route, North Korea has pursued the Soviet model of a centrally planned industrial program. The dependence on imported raw materials and energy throughout the region has led to growth being concentrated near the major ports.

COUNTRIES IN THE REGION

Japan, North Korea, South Korea

INDUSTRIAL OUTPUT (US $ billion)

Total	Mining	Manufacturing	Average annual change since 1960
1100.9	19.4	1081.5	+8.0%

INDUSTRIAL WORKERS (millions)
(figures in brackets are percentages of total labor force)

Total	Mining	Manufacturing	Construction
29.6	0.2 (0.23%)	22.6 (25.6%)	6.8 (7.7%)

MAJOR PRODUCTS (figures in brackets are percentages of world production)

Energy and minerals	Output	Change since 1960
Coal (mill tonnes)	86.9 (1.8%)	+29.5%
Limestone (mill tonnes)	231.0 (9%)	+407%
Graphite (1,000 tonnes)	129.8 (12.7%)	+36%
Manufactures		
Steel (mill tonnes)	129.1 (17.7%)	+481.3%
Automobiles (mill)	17.2 (36.2%)	+2163%
Ships (mill gross tons ordered)	24.1 (57.7%)	+590%
Excavators (1,000)	115.3 (65.2%)	No data
Clocks and watches (mill)	382.9 (31.8%)	No data
Semiconductors/transistors (mill)	60455.0 (88.9%)	N/A
Calculators (mill)	72.3 (85%)	N/A
Televisions (mill)	28.1 (25.7%)	+540%
Radios and sound recorders/players (mill)	88.4 (29.5%)	+534%
Pianos and organs (1,000)	543.0 (53.8%)	No data
Industrial robots (1,000)	55.8 (70%)	N/A

N/A means production had not begun in 1960

A CULTURE DEPENDENT ON IMPORTS

The ingenuity and work ethic of the people of Japan and Korea are the key resources of the region. Minerals and fossil fuels are limited, generally occurring in remote and scattered deposits that are not economic to work.

Japan, in particular, imports huge quantities of fossil fuels. It has some stocks of coal, but is dependent on other nations for oil and natural gas. About 70 percent of Japan's oil comes from the Middle East – making it vulnerable to political instability in that region. The remainder comes from Indonesia, China, Brunei and Mexico. Liquid petroleum gas (LPG) is supplied by Indonesia and Australia. Japan substantially reduced its dependency after the 1973 oil crisis, cutting imports from over 2 billion barrels annually in the late 1970s to 1.5 billion in 1985. Most of these savings were achieved by fuel-economy measures in manufacturing, by thermal-electric power stations switching to gas, and by steel furnaces changing to coal. Soaring car ownership, however, has raised consumption to 1.7 billion barrels per year in the early 1990s.

Two-thirds of the region's substantial bituminous coal reserves of 1.2 billion tonnes are located in Japan, but its quality as a fuel is poor. Output from fields in the extreme north near Kushiro on Hokkaido and south near Kitakyushu on Kyushu island has fallen dramatically over the last 30 years, mostly for economic reasons. Mining is an expensive operation due to difficult geological conditions, and the deposits tend to be far away from major industrial centers. For all these reasons, it is more economical for Japanese industry to import better-quality Australian, Indian or Chinese coal. North and South Korea, by contrast, are able to meet most of their energy needs with local coal and hydroelectric power (HEP).

Natural hazards and energy issues

The combination of a mountainous landscape and a wet climate gives the region

Propelling the economy Despite having to import almost all of its iron ore, the heavily steel-dependent shipbuilding industry in Japan leads the world. Japanese shipyards have the capacity to build the world's largest supertankers.

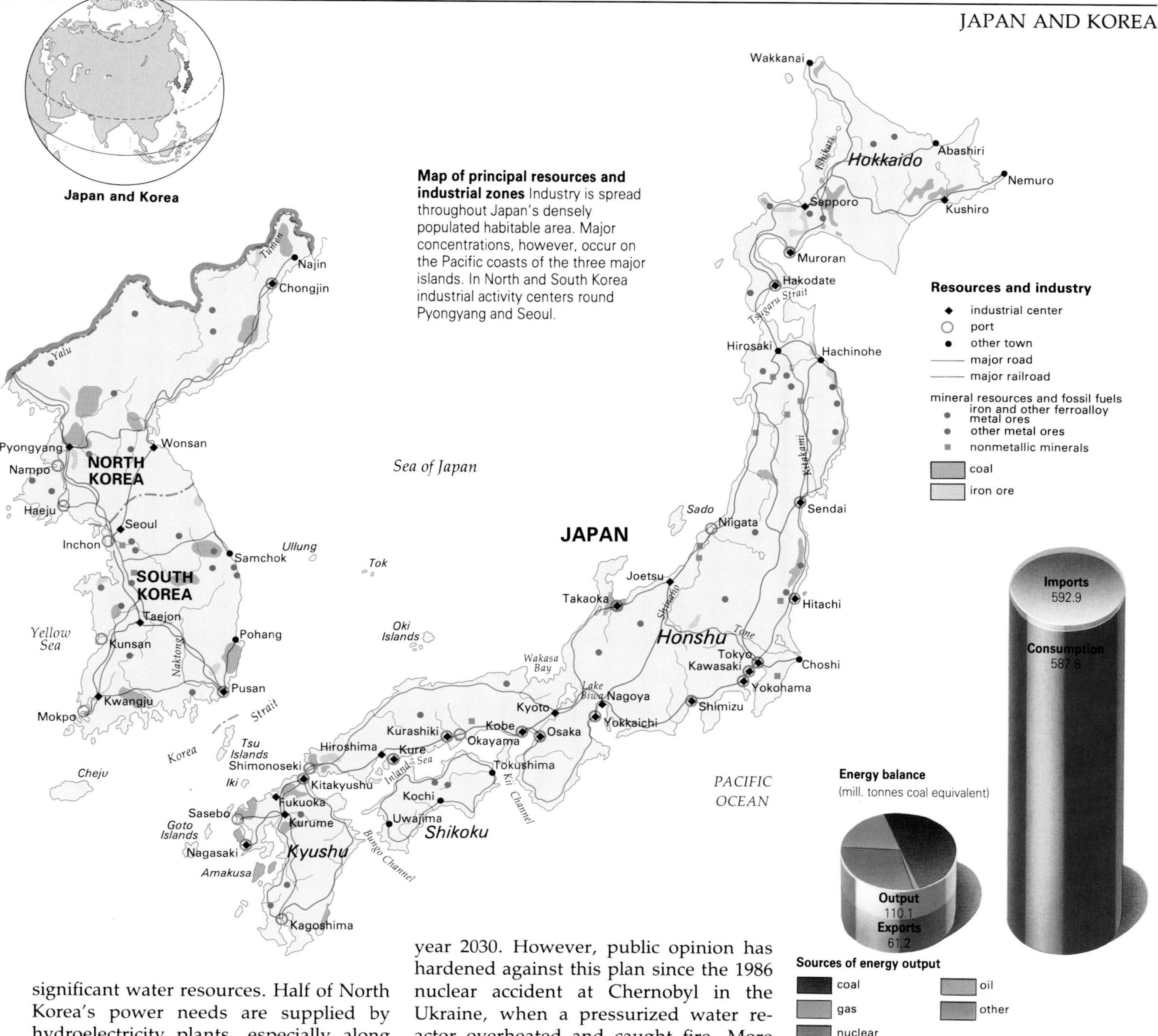

Map of principal resources and industrial zones Industry is spread throughout Japan's densely populated habitable area. Major concentrations, however, occur on the Pacific coasts of the three major islands. In North and South Korea industrial activity centers round Pyongyang and Seoul.

Energy production and consumption North and South Korea rely almost totally on local coal and hydroelectric power for their energy needs. Japan, lacking significant energy sources, has to import energy – its imports exceed consumption.

significant water resources. Half of North Korea's power needs are supplied by hydroelectricity plants, especially along the Yalu river, on the border with northeast China. Power generated in this area is also exported to China. HEP in Japan and South Korea supplies 7 and 12 percent of electricity respectively from stations located on rivers near major urban-industrial centers in central Honshu and Seoul. Potential for hydroelectric power plants remains underdeveloped in remoter areas of Japan, partly because of potential damage to dams from earthquakes, typhoons and volcanic eruptions.

These natural hazards raise concerns about nuclear power programs, which now yield half of South Korea's electricity and 27 percent of Japan's. Uranium enrichment and fast-breeder reactor plants have recently been developed in Japan and the nation aims to produce 60 percent of its power from nuclear sources by the year 2030. However, public opinion has hardened against this plan since the 1986 nuclear accident at Chernobyl in the Ukraine, when a pressurized water reactor overheated and caught fire. More effort is now going into developing geothermal and solar energy sources.

Minerals too expensive to mine

Japanese output of iron ore, copper, lead, tin, manganese, chrome and tungsten fell sharply or stopped after 1979 as metal-processing firms found it cheaper to import raw materials from Australia, India or Korea. Only the production of zinc and limestone flux – used for metal smelting – remains significant.

North Korea has substantial reserves of iron ore. Both North and South Korea are major producers of tungsten, and of a significant amount of the world supply of graphite. Most of the region's graphite is exported to the United States, Japan and the former Soviet Union for the manufacture of heat-resistant metal-smelting containers, brushes for electric motors and rods for nuclear reactors.

The mountainous landscape of the archipelago means that Japan's Pacific coastal belt in particular has an extraordinarily high density of people, cities, industries and transportation systems. This causes an acute shortage of land along the coast where industries, heavily dependent on imports, are also concentrated. Flat land suitable for factory building has had to be found in more distant coastal areas, or by reclaiming extensive areas from the sea, particularly in the bays of Tokyo, Ise (Nagoya) and Osaka.

In the picture (*above*) Japanese camera companies – for example, Nikon, Sony, Canon and Minolta – dominate the world market. The industry has grown spectacularly since it began during World War II, when German Leica lenses were copied by the Japanese for military viewfinders and periscopes.

Kim Chaek iron and steel complex (*above*) North Korea, one of the last communist countries, still subscribes to Marxist industrial theories. This plant in Chongjin city still uses labor-intensive work practices.

EXPANSION IN A LITTLE SPACE

Significant industrialization first began in Japan under the Meiji government (1868–1911) and in Korea after the 1950–53 war. In 1945, after the end of World War II, Japan began a remarkable program of reindustrialization that led to massive and rapid growth in its manufacturing industry. In the early 1950s the labor-intensive textile and clothing industries were the leading employers in Japan. The main concentrations of mills and textile manufacturers were found in the prefectures of Chubu, Kinki and Kanto in the central part of Honshu island. However, competition from new industries for land and labor drove much of the production into cheaper rural areas, though synthetic fibers continue to be largely tied to coastal petrochemical complexes. Strong competition from Asia has caused the Japanese textile industry to decline and emphasis has shifted to fashion, clothing design and automated production.

Metallurgy, metalworking and chemicals became the other key sectors after 1920, especially around Osaka, Kobe, north Kyushu and Tokyo. The militarization of Japan provided a major stimulus for these industries, and in the postwar period they became, in turn, the focal points of reindustrialization.

The period from 1950 to 1973 also saw the rapid growth of shipbuilding, automobile manufacture, and the petrochemical industry. Large plants and huge integrated complexes spread around the major port facilities. To avoid bad congestion and pollution around these ports, Japanese companies began to site new enterprises in smaller coastal and inland cities – often at great environmental cost to rural and fishing areas. Seven of the nine steelworks opened between 1960 and 1974 were built on newly reclaimed coastal strips in areas that previously had little industry. Together in that 14-year period they raised Japanese output from 22 to 117 million tonnes of steel, enabling significant growth to take place in shipbuilding and automobile manufacture.

Adapting to circumstances

The energy crisis in 1973 triggered another set of major changes in Japanese industry which have had longterm consequences. The growth industries of the 1960s, especially metallurgy, shipbuilding, chemicals and automobiles, entered a period of adjustment and streamlining which emphasized reducing capacity, saving energy, labor and materials and manufacturing more sophisticated products. In areas such as Tokyo, which had high growth and a shortage of labor, the steelworks and shipyards could be

Space-age technology (*below*) Japanese industry is closely integrated. Merging electronics, optical research and computer software gave rise to a new field of expertise, the aerospace industry. Here, scientists are fine-tuning an observation satellite.

rationalized or radically modernized. The smaller rural cities that lacked a broader industrial base and were dependent on these heavy industries, however, found it harder to adjust to the changes.

A notable feature of Japanese industry has been its investment in research and development, leading to an enormous expansion of the "knowledge-intensive industries". The most high-profile of these are electronics (semiconductors, computers, telecommunications, video recorders and televisions), optical products (cameras and fiber optics), ceramics, antipollution devices, biotechnology and robotics for automated production.

Innovation was encouraged by intense rivalry between the large Japanese industrial corporations. Firms often transferred technological expertise acquired in "old" product lines to "new" ones, and the practice of using established clusters of subcontractors enabled this to be done quickly. Camera firms such as Canon, Konica, Minolta and Ricoh expanded into manufacturing photocopiers for the office boom, then into facsimile (fax) machines, while Matsushita (incorporating National, Panasonic, Technics and JVC), Sharp and Toshiba diversified from audiovisual consumer electronics into office machinery and facsimiles. Brother shifted from sewing machines to typewriters and Toray, originally a maker of synthetic textile fibers, has developed into a major manufacturer of carbon-fiber materials used in a wide range of manufacturing from textiles to aerospace.

A disproportionate growth of these new industries has occurred in the outer Tokyo Metropolitan region where a cleaner environment, large numbers of universities, national research institutes, and a potential workforce of engineering graduates attract the research and development units of major companies as well as their most advanced manufacturing facilities. Tokyo's stock market and financial services provide the funding for expansion and diversification. In an attempt to spread some of these facilities to other areas, the government has, since 1979, promoted policies to disperse these activities among the 25 "technopolises" located throughout Japan.

Korea – in Japan's footsteps?

Political division after the Korean War created the conditions for divergent industrialization in both North and South Korea. Both countries, however, developed their steel and defense sectors, generally concentrated around Pyongyang and Seoul because of the poor internal transportation facilities in the rest of the country. Since the 1960s, South Korea has set out to challenge Japan with a vigorous program to manufacture goods for export, drawing on their cheap labor resources and on technological research. A government policy of spreading industrial centers evenly throughout the region has helped to keep costs low, and South Korea has succeeded in establishing itself as a significant presence in world markets in clothing, sports footwear, consumer electronics, ships, automobiles and, since the 1980s, microchips. Recent rises in labor costs and fluctuating currency-exchange rates have exposed industrial weaknesses. In spite of the efforts of the South Korean manufacturers, Japanese-made products are still competitive with about half of their goods.

GROUPING FOR POWER

Industry in the region is dominated by groups of firms known as "*chaebol*" in Korea and "*keiretsu*" in Japan. Six *chaebol* produce about 40 percent of South Korean manufactures and services. *Keiretsu* are estimated to employ over 15 million Japanese, a quarter of all workers. Among them are giants such as Mitsubishi, Mitsui, Fuji, Sumitomo and Dai-Ichi Kangyo, which are double the size of the largest United States or European companies.

Chaebol are family-run conglomerates. Each has a bank at its center and comprises firms operating over a wide spectrum of industries and services. For example, Samsung has 24 companies that work in advertising, aerospace, chemicals, computers, consumer electronics, engineering, foods, glass, insurance, medical services and diagnostic systems, newspapers, paper, property, synthetic fibers, petrochemicals, shipbuilding, semiconductors, watches and retailing.

Both *chaebol* and *keiretsu* operate systems to pool members' resources to initiate research, raise finance, improve design, increase their buying power, improve transportation facilities and sell products more efficiently within the group. Intense rivalry between different *chaebol* or *keiretsu* stimulates their individual diversification, restructuring, innovation and labor redeployment among group companies during periods of change.

Training to work as a team The Japanese work ethic and culture in large corporations are both powerfully motivating forces. This management school near Mount Fuji promotes teamwork and solidarity rather than individual achievement as a way of increasing productivity, cooperation and job satisfaction.

THE JAPANESE WORK ETHIC

Japan has the world's most highly skilled industrial labor force. Twice as many engineering graduates leave its universities each year than engineers from universities in the United States – a country with twice Japan's population. Schools in Japan emphasize discipline and intensive teaching methods ensure, high academic achievement. The literacy rate is 100 percent and high standards are reached in a wide range of subjects. Great emphasis – encouraged by strong parental commitment – is placed on success in exams, the reward for which is a place at university and the possibility of a secure job afterward. The high esteem in which science and technology is held is another factor that encourages the best students to follow a technical career.

Jobs for the highly skilled, however, are concentrated in Kanto prefecture, particularly in Tokyo, where 45–65 percent of all the researchers, scientists, engineers and data-processing specialists work. Only the manufacture of semiconductors – a more standardized industry – is dispersed, with Kyushu, "Silicon Island", as the major center, followed by the Tohoku region in Honshu.

In an effort to spread skill resources more evenly and decongest the Tokaido megalopolis, several interlinking policies are being applied by the Japanese government. The Tokyo urban area is being restructured to concentrate new industrial research and technology functions in Chubu, centered on Nagoya, as Tokyo specializes more in the finance and information-based services. High-technology industry is being fostered in other areas, and more manufacturing companies are being encouraged to set up their operations in northern Honshu, Kyushu, Shikoku and Hokkaido.

Culture and technology

Japanese cultural attitudes strongly influenced the working practices of the people during the period of economic reconstruction following World War II. The legacy of Confucianism can be seen in the emphasis that companies place on teamwork and commitment to the organization, as well as in meticulous business etiquette. The importance of individual success, vital to workers in other countries, runs counter to Japanese

COMPANY MEN AND WOMEN

Much of the success of Japanese industry has been attributed to the attitude of the workforce, and the relations between employer and employee. A Japanese worker recruited by a large company expects to stay with that company for his or her entire career. In return, the company offers guaranteed employment for life. As a result of this longterm commitment, companies have every incentive to ensure that their workers receive training, and they are prepared to invest a good deal of time and effort to make sure that the workforce is equipped with the right skills for the job. The knowledge of a secure future on both sides makes this kind of investment feasible.

Another benefit of this system is improved quality control of the product. Much of the responsibility for manufacturing standards lies in the hands of individual workers, who have a great incentive to make certain that a high standard of quality is maintained. This is a major factor in contributing to the longterm success of their company, which, in turn, helps to ensure their continued employment.

Close quarters drill These workers processing bamboo shoots operate in a very crowded environment without accident or incident. Densely populated industrial regions containing a workforce well-schooled in teamwork limit the need for personal space in the work area.

cultural expectations. The Japanese ideal is of a harmonious team, and cohesiveness between team members is reinforced by group activities such as daily exercise sessions, in which all employees are expected to participate. Although companies have staff hierarchies, many firms make workers of all levels wear uniforms to foster a sense of corporate identity. At one time Japanese methods of management were seen as old-fashioned and inefficient, but the success of Japan's economy has led many Western companies to investigate whether they could benefit from these management techniques – though the cultural exchange is far from straightforward, given the vastly different ways of life in East and West.

Although imitation of foreign technology has been important to Japanese manufacturing, the nation has never lost sight of its own traditions. Inspiration from cultural and artistic traditions has helped the Japanese to meet the challenges of modern urban-industrial society through technological creativity. Miniaturization, for instance, is of special significance in a crowded region where space is at a premium. The Japanese art of paper folding (*origami*), with its complex three-dimensional forms, contributed to the development of computer-aided design (CAD). Folding fans (*sensu*) were the germ of an idea for the design of laptop computers, and from lengths of cloth (*furoshiki*), used for wrapping and carrying things, came the inspiration for folding solar panels. The complexity of pictorial characters (*kanji*) in the Japanese language made it necessary for word-processing software to be developed to a very sophisticated level, and prompted the development of optical scanners and aspects of artificial intelligence.

Breeding hybrids

The Japanese innovate rapidly, often by recycling existing technology with new ideas. This is partly a response to the dearth of natural resources, and is partly due to intense rivalry between companies and growing Korean and Taiwanese competition, especially in electronics. The leading manufacturers, including Matsushita, Toshiba and Sony, are making videophones and stereo televisions – hybrids of existing television, video recorder, compact disk, telephone and stereo technologies. The science of robotics is also a hybrid, resulting from the fusion of electronics and machine tools. Other aspects of "crossover" technology are illustrated by the use of compact disks in top-of-the-range automobiles, or by Sony's use in security systems of light-sensitive sensors originally developed for home video cameras.

Robots in the workshop

The phenomenal postwar development of the Japanese motor industry illustrates many of the factors that have brought success to Japanese industry as a whole. Automobile output expanded dramatically during the 1960s and 1970s, and by 1980 Japan was the world's largest producer, accounting for 28 percent of world production.

Most of today's market leaders were already making vehicles in the 1930s, often imitating American products. The government, keen to foster local production, promoted mergers between small firms to pool working capital and technical skills in larger, more stable, companies. Later, to support expansion in east Asia, it encouraged firms manufacturing other kinds of machinery – such as Toyota and Mitsubishi – to begin making military vehicles. In the aftermath of World War II a number of firms were able to transfer technology acquired in wartime industries, such as building fighter aircraft, to automobile manufacture.

In the 1950s Japanese companies made major gains through the study of mass-production methods and the acquisition of equipment from Detroit, the home of vehicle manufacture in the United States. At this time a number of firms established formal technical links with Western automobile manufacturers.

THE TOP MOTOR VEHICLE PRODUCING COUNTRIES

Japan is the world's most prolific producer of automobiles and commercial vehicles (*italics*). More spectacular than the sheer volume of automobiles produced is the rate of industrial growth in this area. From 1960 to 1990 Japanese auto production increased by almost 5000 percent.

Country	Vehicle production (,000)			
	1960	1970	1980	1990
Japan	165 *595*	3,179 *2,124*	7,038 *4,006*	8,198 *4,488*
United States	6,675 *1,194*	6,542 *1692*	6,376 *1,667*	7,105 *3,856*
Germany	1,817 *239*	3,528 *318*	3,530 *380*	4,312 *270*
France	1,1136 *234*	2,458 *292*	3,488 *505*	3,414 *538*
Soviet Union	139 *360*	344 *572*	1,327 *872*	1,262 *991*
Italy	596 *49*	1,720 *134*	1,445 *167*	1883 *225*
Canada	326 *72*	923 *231*	847 *528*	1,008 *1,018*
Spain	42 *17*	455 *77*	1,048 *131*	1,498 *315*
United Kingdom	1,353 *458*	1642 *472*	924 *389*	1,227 *311*

Assembly robot

The many joints and pivots of the robot allow its "hand" to be manipulated to almost any angle

Robots can be manufactured to perform different tasks. The "hand" can be a welder, or can pick things up with grippers or a suction pad

The stages of automobile assembly

4 With all the panels correctly aligned, a series of robots at several stages make the final welds to the body. The completed body moves to the next stage of production

Basic industrial robot *(left)* Most are made up of several joints or axes that rotate or extend components to the right position. A robot such as this one is programmed to perform a series of standard movements using up to six major axes.

Mass-producing small automobiles

At the core of the motor manufacturing boom in the 1960s and 1970s was a remarkably rapid expansion in the production of small automobiles, light commercial vehicles and pickups. These were particularly suited to the growing markets in Japan and the developing east and Southeast Asian countries, which were demanding a supply of inexpensive, durable vehicles with excellent fuel economy. The difficulties of parking in urban centers meant that drivers began to place a premium on smaller vehicles for their maneuverability.

Several forces combined to enable Japan to satisfy these needs in both domestic and export markets. The expanding motorcycle industry created new engineering expertise in light, small, fuel-efficient motors. Between 1950 and 1963 the number of manufacturers increased, intensifying rivalry. This, in turn, forced the whole industry to lower costs, while at the same time improving the quality and reliability of the product.

A rapidly increasing, well-educated, youthful Japanese workforce enabled the motor manufacturers, who became increasingly prestigious employers, to

No time off for lunch The Japanese have the largest share of the world automobile market and they are expanding all the time. Before automation, manufacturers could not make vehicles fast enough. Now almost all production is by robots.

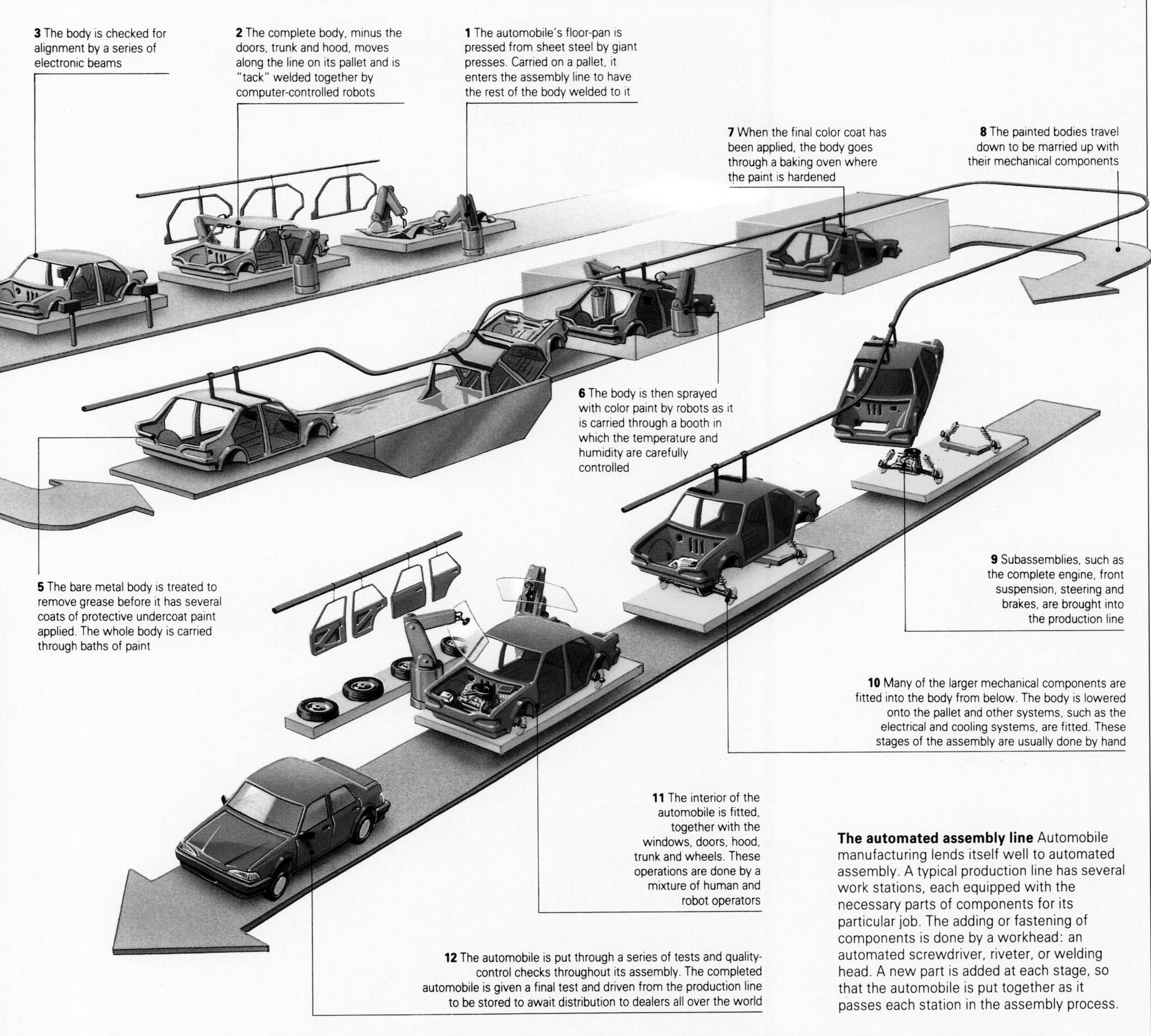

The automated assembly line Automobile manufacturing lends itself well to automated assembly. A typical production line has several work stations, each equipped with the necessary parts of components for its particular job. The adding or fastening of components is done by a workhead: an automated screwdriver, riveter, or welding head. A new part is added at each stage, so that the automobile is put together as it passes each station in the assembly process.

attract rural labor willing to work hard for low wages. The expanding steel industry provided cheap raw materials and the construction of state-of-the-art mass-production factories, such as Toyota's 1959 Motomachi plant, offered huge economies of scale in the manufacture and assembly of automobiles.

These factories were closely linked with clusters of hundreds of competing parts-producers. The use of large numbers of specialized subcontractors offers flexibility to the major manufacturers, as well as economies of scale. These clusters gave birth to the "*kanban*" or "just-in-time" technique, which means organizing parts deliveries so that materials arrive as they are wanted, and companies are saved the trouble and expense of transporting huge stocks of components. The *kanban* system was pioneered at Toyota, and greatly reduced overheads in a manufacturing culture where the scarcity and cost of land is a major difficulty. During the same period manufacturers began to develop automated production lines with specific manufacturing stages (especially welding and spraying) being performed by robots. These innovations helped the manufacturers to build on their strengths – highly efficient and cost-effective mass-production of vehicles that they could market intensively and successfully all over the world.

Diversification and sophistication

To remain competitive, Japanese firms are expanding their model ranges to cater for more varied markets at home and abroad. Total dependency on imported oil is a major incentive for the Japanese to focus on improved fuel economy, achieved through the use of lightweight materials and better aerodynamics. At the same time, the need to combat urban air pollution has led to the development of clean-burning engines. Some years ago the public perception of the Japanese auto industry was that it was following in the footsteps of its Western competitors. More recently, though, it is regarded as the technological pace-setter.

Living national treasures

Japan's American-led postwar Westernization and the phenomenal growth of its industrial and technological bases have reduced the role of the country's ancient traditional crafts and customs. The Japanese government, worried that the old ways of life would be completely replaced by the new, set up an Agency for Cultural Affairs in the 1970s within the Ministry of Education.

The agency began to seek out traditional craftsmen preserving ancient techniques in time-honoured professions. The most skilled among them were created Living National Treasures. A person chosen for this honor receives a yearly stipend from the government and is expected to promote his or her art through educational programs and personal demonstrations.

The classification covers most of the fine arts, theater and crafts. The number of Treasures is limited in each category and once honored, the Treasure retains the title until death. Groups can be selected by the agency as well, and several *Kabuki* and *No* theater groups have held the title.

Through the Agency for Cultural Affairs and the Living National Treasures, the Japanese government hopes that the price of continued economic success will not mean the loss of traditional values and customs.

A Living National Treasure carries on the traditions of the past. This ceremonial sword maker demonstrates the time-consuming and demanding ancient art.

MINING AND STEELWORKING

AGRICULTURAL DECLINE, MINERAL BOOM · THE RISE AND FALL OF MANUFACTURING · TRADE UNION DOMINATION

Extracting and processing the natural resources of the region account for most of the industry in Australasia and Oceania. Production of iron and steel – together with the exploitation of Australia's rich mineral reserves – has recently overtaken agriculture as the dominant industry in the area since the imposition of European Community (EC) limits on meat and dairy imports. New Zealand extracts coal, natural gas and petrochemicals. Nauru produces phosphates, and both Fiji and the Solomon Islands mine gold and silver. Fruit canning for export is important in tropical Oceania; wood and copper in Papua New Guinea. Antarctica is also rich in mineral deposits, but mining there is banned for ecological reasons. Tourism is a fast-developing industry across the whole of the region, causing a small boom in local crafts.

COUNTRIES IN THE REGION

Australia, Fiji, Kiribati, Nauru, New Zealand, Papua New Guinea, Solomon Islands, Tonga, Tuvalu, Vanuatu, Western Samoa

INDUSTRIAL OUTPUT (US $ billion)

Total	Mining	Manufacturing	Average annual change since 1960
87.8	12.5	61.6	+3.0%

INDUSTRIAL WORKERS (millions)
(figures in brackets are percentages of total labor force)

Total	Mining	Manufacturing	Construction
2.55	0.23 (2.1%)	1.6 (15.1%)	0.7 (6.5%)

MAJOR PRODUCTS (figures in brackets are percentages of world production)

Energy and minerals	Output	Change since 1960
Coal (mill tonnes)	180.3 (3.8%)	+390%
Iron Ore (mill tonnes)	97.8 (17.3%)	+376%
Copper (1,000 tonnes)	437.0 (5.1%)	+71%
Nickel (1,000 tonnes)	133.6 (16.5%)	+16%
Zirconium (1,000 tonnes)	490.0 (66.9%)	+65%
Uranium (1,000 tonnes)	3.5 (10.0%)	+1300%
Gold (tonnes)	196.0 (11.1%)	+326%
Diamonds (mill carats)	36.3 (37.8%)	No data
Manufactures		
Knitted woolen underwear (mill)	94.7 (3.9%)	No data
Woolen carpets and rugs (mill sq. meters)	44.9 (17.1%)	No data
Cement (mill tonnes)	7.1 (0.6%)	+82%
Steel (mill tonnes)	6.9 (0.9%)	+88%
Aluminum (mill tonnes)	1.3 (5.9%)	+525%
Trailers (1,000)	22.4 (2.2%)	No data

AGRICULTURAL DECLINE, MINERAL BOOM

Australia – rich in copper, gold, lead, zinc, nickel, manganese, uranium, diamonds, opals, tungsten and rutile (titanium) – has traditionally been an enormous treasure trove of valuable resources for other nations of the world. Major industrial growth came after World War II with the exploitation of its high-quality coal, oil, bauxite and iron-ore deposits.

In particular, the continent has become an important supplier of minerals and energy to Japanese manufacturing and service industries. Coal reserves in Australia are estimated to be about 500,000 million tonnes, of which about two-thirds is recoverable, and the country is now the world's biggest exporter. Australia has good reserves of oil and gas and more has been discovered in sedimentary basins, particularly in the Timor Sea and off Western Australia. The country is a net exporter of energy.

Precious and semi-precious stones contribute significantly to the country's mineral wealth. Coober Pedy, an isolated spot in the centre of South Australia, supplies 90 percent of the world's opals, and stocks are by no means exhausted. In spite of continuous mining since 1916, less than 2 percent of natural opals have been brought to the surface.

Industry in New Zealand has long been

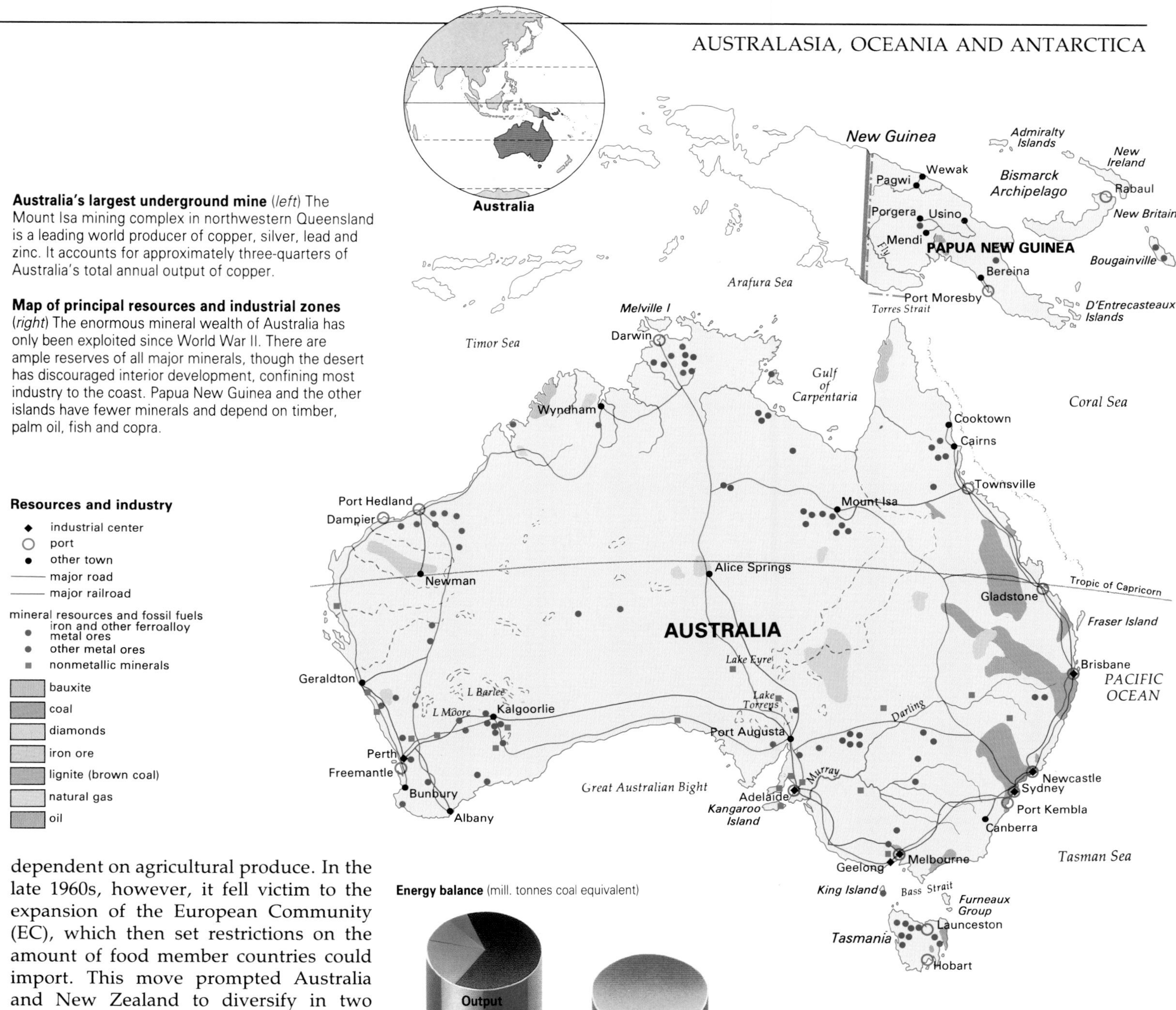

Australia's largest underground mine (*left*) The Mount Isa mining complex in northwestern Queensland is a leading world producer of copper, silver, lead and zinc. It accounts for approximately three-quarters of Australia's total annual output of copper.

Map of principal resources and industrial zones (*right*) The enormous mineral wealth of Australia has only been exploited since World War II. There are ample reserves of all major minerals, though the desert has discouraged interior development, confining most industry to the coast. Papua New Guinea and the other islands have fewer minerals and depend on timber, palm oil, fish and copra.

dependent on agricultural produce. In the late 1960s, however, it fell victim to the expansion of the European Community (EC), which then set restrictions on the amount of food member countries could import. This move prompted Australia and New Zealand to diversify in two main areas. One strategy was to move into mining natural ores and developing mineral-related industries. By the 1970s the agricultural industry, despite the growth of output and exports, had been eclipsed by the phenomenal success of this venture. Another strategy was to think of new ways of processing food to find a way round the restrictions. The Te Rapa Cooperative Dairy Company in New Zealand, for example, blasts milk into huge evaporators to turn it into convenient powder form with a longer shelf life. It is able to sell this processed milk alongside the fresh product.

The 1980s were also boom years for mineral-related industries in the island states around Australia and New Zealand. In Papua New Guinea, for example, mining and agriculture were equally important mainstays by 1989. Copper is the major export, followed by gold, coffee, cocoa, copra (dried coconut meat yielding

Energy balance (mill. tonnes coal equivalent)

Output 205.8
Exports 90.2
Consumption 130.3
Imports 22.1

Sources of energy output

coal
oil
gas
other

Energy production and consumption Huge reserves of coal dominate Australia's energy industry. Commercial gas production began in the 1960s, and Australia now supplies more than 70 percent of the oil it consumes. New Zealand also has coal, but its chief energy resources are hydroelectric and geothermal.

coconut oil) and palm oil. The French overseas territory of New Caledonia has even richer mineral resources, including nickel (the main export) chrome, iron, manganese, mercury, silver and lead.

The main resources on Fiji are sugar, gold and copra, though tourism is probably the island's most profitable industry and fishing brings in significant revenue.

Antarctica's protected environment

Antarctica has a plentiful supply of valuable minerals including gold, silver, beryl and graphite. Extraction, however, is banned by an international treaty because of the environmental sensitivity of the area. Quite apart from that major consideration, any attempt to mine the area would be difficult and expensive in view of the permanently frozen conditions, the special equipment required and the transportation problems involved. Antarctica also possesses the largest coalfield in the world, but the high ash content of the coal makes it poor quality. The development of power stations able to burn low-grade coal, together with the world's increasing demand for energy, may see this giant coalfield being tapped in the 21st century if international legislation allows it.

THE RISE AND FALL OF MANUFACTURING

During the 1950s and 1960s, Australia used much of its earnings from agricultural industries to develop a considerably wider manufacturing base. In Sydney, Melbourne and Brisbane production of motor vehicles, textiles, clothing, footwear and food products peaked in the early 1950s, but by the 1970s this industry was already in rapid decline. In order to revitalize it, both the Australian and New Zealand governments imposed taxes on imported goods as a means of protecting domestic production, and capital investment in manufacturing – much of it from abroad – boomed under this protective umbrella. For instance, Australia's two largest automobile companies, Ford Australia and General Motors Holden, are American owned.

The world recession in the early 1980s caused a great change in these economic policies. Important industries such as steel and motor vehicles were subsidized so that exports could increase, and once-protected industries, such as New Zealand television assembly and shipbuilding, disappeared.

Three sectors dominated Australian manufacturing at the end of the 1980s: food (including drinks) and tobacco; machinery and equipment (including vehicles and public transportation); and paper products and publishing. In New Zealand food processing was the biggest sector of the manufacturing industry in 1984, and by 1989 it had increased its output by 40 percent. Over the same period automobile assembly was cut back; one third of New Zealand's 16,000 factories shut down during the 1980s.

Innovation has been the key to recovery in manufacturing industries during the late 1980s and early 1990s. Newer industries in Australia include aerospace, especially major engine repairs, avionics and airframes. Australia has also developed an electronics and information-based industry that supplies domestic telecommunications equipment and also produces computer software, medical, scientific and defense equipment and specialized electrical components.

Island-based industries

Papua New Guinea, much like the rest of the region, is economically dependent on extracting and exporting its natural resources in a relatively raw state (resource-based or primary industry). The main activities are producing plywood and wood veneers, and processing copra and oil palm, mainly for the food industry. Manufacturing (product-based industry) is much less developed. In recent years the government, trying to protect the country's natural resources, has set up a Department of Primary Industry project. This is intended to encourage smallholders to restock their land with high-yielding hybrid coconut and cocoa plants, to be used for further processing.

Copra is an important raw material for many Pacific Island industries, including French Polynesia, Western Samoa, the Solomon Islands, Tonga and Tuvalu. Most copra for export in Papua New Guinea is dried in hot-air mechanical driers that produce a higher-quality product than by the traditional method of drying in the sun. Pressing and treatment by solvents yields coconut oil used in edible fats and confectionery as well as in the manufacture of soaps, detergents, shampoos, synthetic rubber and glycerin. The remains of the pressed copra are used as fertilizer and animal feed. About half of the production comes from village-owned plantations, the rest from land owned by individuals or private companies. A large number of plantations were taken over by the government in an attempt to control the entire coconut industry from within the country itself.

Geothermal power in New Zealand (*above*) New Zealand has an abundance of hot springs and geysers that can be tapped for geothermal power. Lake Taupo on North Island is one such source – a huge natural lake in a volcanic crater.

Phosphate mining on Nauru (*below*) Deposits on this tiny coral island have the richest concentration of phosphates found anywhere on Earth, and the extraction industry totally dominates the economy.

Map of principal resources and industrial zones New Zealand's commercially mined mineral resources include coal, iron sands, gold, natural gas and quarry rock. Forestry and farming, especially on North Island, are also important resources for industry, as they are on the Pacific island states where industrial capacity is small.

Palm oil has been exported from Papua New Guinea since 1972. Oil from the outer layer of the palm fruit is used for making soap, candles and lubricating grease, and in processing tin plate and coating iron plate. Oil from the kernel is mostly used in food processing, especially in margarine, chocolate products and in some pharmaceuticals. The cake residue after processing is used as cattle feed. 1972 is a significant date in the industrial history of New Guinea because the government and a British company set up equal ownership of the New Britain Oil Palm Development Company, which enabled local smallholders as well as company plantations to have their crops processed by the same factory.

On New Caledonia various forms of nickel are extracted in great quantity for the export market, and this is the country's chief resource-based industry. Brewing, the cement works and production of paper goods are the principal industries; factories on the island also produce goods ranging from soft drinks and foods to automobile accessories, industrial chemicals and clothing. Trade in leather goods and clothing also began to prosper on Fiji in the mid 1980s. Tax-free zones on the island employ 9,000 people in 96 factories producing leather garments, shoes, food products and furniture.

TOURISM – A NEW RESOURCE

Throughout the region a new resource is coming to prominence: the spectacular scenery now attracts large numbers of foreign visitors and is creating a profitable tourist industry. The variety of things to see is enormous: ranging from Australia's major cities, its lush rainforests, arid desert, miles of coast and coral reefs to the beautiful landscape of New Zealand; from the tropical beaches of the Pacific islands to the frozen wilderness of Antarctica. For the foreseeable future tourism should continue to be a major export earner, in spite of the fact that the region is a long flight away from Europe and the United States. In 1990 two million tourists visited Australia, spending $1 billion and contributing to the employment of 400,000 people. In the same year major resort or hotel development projects worth $16 billion were under construction.

The South Pacific islands, once they have invested in the basic necessities (hotels, ports, airports and internal transportation systems), will be able to exploit a major source of future income. Tourism generates more profit for the small islands than commodity exports, and the poorest countries of the region such as Tuvalu and Kiribati probably have the most to gain. Its main benefit is to provide employment for the local workforce, though wages are generally low. However, if food and other goods have to be imported for the tourists then potential income is drastically reduced. Sensitivity is essential in developing tourism, which can damage fragile environments and alter the character of unspoiled landscapes.

Making use of the oceans

The fishing industry and related products dominate in the Solomon Islands, Vanuatu, Kiribati and American Samoa. On American Samoa canned tuna is the major export, and fish meal is produced as a by-product of the industry. In the island's government industrial park 20 companies produce machinery, roofing iron, sandals, drink bottles and recycled aluminum. On the island of Tonga, the preparation of bananas and copra for the domestic and export markets forms the basis of most industrial activity.

Whaling was, until recently, an important industry in the Antarctic islands. Whale products were used for many manufacturing processes including making cosmetics, and oil from the baleen whale was made into glycerin and used in the manufacture of explosives. However, pressure from worldwide conservation groups encouraged countries such as the Soviet Union, Japan, Norway and Iceland to stop commercial whaling in the 1980s.

TRADE UNION DOMINATION

Australian state initiatives in building, transportation, factories and the services meant that, in the late 19th century, jobs were plentiful. Gold was discovered in 1851 in New South Wales and the sudden boom in mining brought immense wealth to the colonies as well as a vast number of immigrants from other parts of the world. Whether or not they were successful, they tended to stay on and turned to other ways of making a living. Factories sprang up, especially around Victoria, which badly needed new sources of employment after the gold rush ended. Trade unions became very influential and almost completely organized the labor market. The years between 1872 and 1893 were said to be the "Golden Age" for wage earners, and trade union dominance has persisted to the present day.

During the next century some sectors of the manufacturing industry received

Manufacturing specialized machinery (*left*) in New South Wales. At the center of the Australian iron and steel industry, the province is well positioned to house giant engineering projects like this.

Supplying the world with opals (*right*) Coober Pedy in South Australia is the source of most of the world's opals. Apart from their use as gemstones, opals are used in making abrasives.

Traditional crafts on Tonga (*below*) This enormous hand-decorated tapa cloth is not for sale; hundreds of hours of labor have created an heirloom piece that will be saved for a family wedding, birth or funeral.

government protection in the form of tariffs and bonuses. This policy was intended to provide employment and raise incomes, but in the mid 1960s doubts were raised about its effectiveness. It had led to an increase in the price of products, effectively taxed other sectors, and also reduced the quantity of goods for export. Management became complacent and reluctant to expand or innovate, and the unionized workers demanded higher wages and benefits. Consequently the cost of manufacturing was completely out of line with productivity.

In some industries workers were getting three or four months' annual vacation, and strikes called for trivial reasons had a devastating effect on essential services. In late 1982 Australia was paralyzed by strikes. Laden cargo ships blocked the harbors as dock workers refused to unload them; raw sewage floated in the sea and garbage piled up in the streets; communication and travel were continually disrupted.

Management overhaul

During the 1990s, Australia and New Zealand saw the urgent need to improve managerial efficiency and work practices. They made tax concessions to boost research and development, and encouraged the transfer of new technology among firms. Venture capital was made available for new developments, and training programs were revitalized: 1.6 million new jobs were added to the 6 million available in 1983, the majority in community services; finance, property and business services; the wholesale and retail trade and recreational and personal services. Manufacturing lost 58,000 jobs.

Employment on the islands

Industry on the islands of Oceania is labor-intensive and wages tend to be low. Generally it is the foreign-based owners who reap the revenue. Resource-based industries have suffered, among other things, from a shortage of skilled labor. In American Samoa the tuna-canning factories depend on workers coming from neighboring Western Samoa.

A principal feature of the employment scene has been the expansion of part-time jobs and the numbers of working women. By the end of the 1980s 20 percent of the population was in part-time work and almost half of them were female employees. When unemployment rocketed in the 1990s more men than women lost their jobs, and levels of young people out of work soared to three times the national average.

Migrants in the Australian workforce live in close-knit groups in the cities. Italians, Greeks, Lebanese, Chinese and Vietnamese form a rich multi-cultural society that is particularly influential in running shops and restaurants. Migrant families are often self-employed in the retail or catering trades, or work in semi-skilled jobs in transportation, manufacturing or other largescale industries.

MEDIA MAGNATES – FAIRFAX AND MURDOCH

Australia, the land of opportunity, has produced an impressive number of entrepreneurs who have made their mark on the world stage, particularly in the publishing and media industries. Among the first was John Fairfax (1805–1877), who bought Australia's most prestigious newspaper, the *Sydney Morning Herald*, in 1838. He went on to acquire other newspapers, becoming the most successful publisher of his time and the founder of the longest surviving family-owned newspaper group in Australian journalism.

The John Fairfax Group went public in 1956, but family members always had a controlling interest. It expanded into magazines, television studios and radio stations throughout Australia and gained interests in telecommunications and property. The unbroken line of succession proved too much to sustain in 1987 when Warwick Fairfax became involved in the group at the age of 26. Three years later the group made losses of $73.5 million and most of the board resigned. Warwick Fairfax left the company in 1991.

Another media magnate, Rupert Murdoch, bought up Australian regional papers and in 1969 was wealthy enough to buy the British tabloid *The Sun*, which subsequently became the highest selling paper in Great Britain. In 1981 he took over *The London Times* and sacked the editor, amid much controversy. By this time Murdoch had financial interests in American publishing and in television. He had revamped the Twentieth Century Fox film company and in 1984, as one of Australia's most dynamic businessmen, he surrendered his Australian citizenship and became a United States citizen in order to satisfy the media-ownership laws.

The Australian steel industry

Australia's mineral resources include all the raw materials for the production of iron and steel. There are rich supplies of limestone and iron ore in South Australia, chromite (an oxide of iron and chromium) in Western Australia, limestone in New South Wales, coking coal and manganese along the east coast. The industry began in the 19th century, and gave birth to the almost monopolistic Broken Hill Proprietary Limited (BHP), now the country's biggest company. BHP began by smelting silver and lead before moving into steel, backed by its good reputation for dynamic leadership and profitability. Two-thirds of the total iron and steel is produced at Port Kembla, New South Wales, almost one-third in Newcastle, NSW, and the remainder in South Australia at the port city of Whyalla.

During the first half of the 20th century, the development of new technology in the steel industry led to radical improvements in efficiency. In 1915, for example, 3.5 tonnes of coal were needed to make 1 tonne of steel, but by the 1940s this had been improved to a ratio of 1.5 tonnes of coal for every 1 tonne of steel.

The blast furnace *(below)* A modern blast furnace stands between 30–40 m high (100–130 ft), and is 9 m (30 ft) in diameter, and can produce around 2,000 tonnes of hot metal daily. A combination of iron ore, coke and limestone is used to feed into the furnace to produce molten iron. The iron collects at the base of the furnace and slag is removed. Most iron ores smelted today contain 60–80 percent iron oxide, with the rest a mixture of earth, clay, sand and stone. To produce steel, the carbon content of crude iron must be reduced from about 4 percent to less than 1 percent, making the metal less brittle.

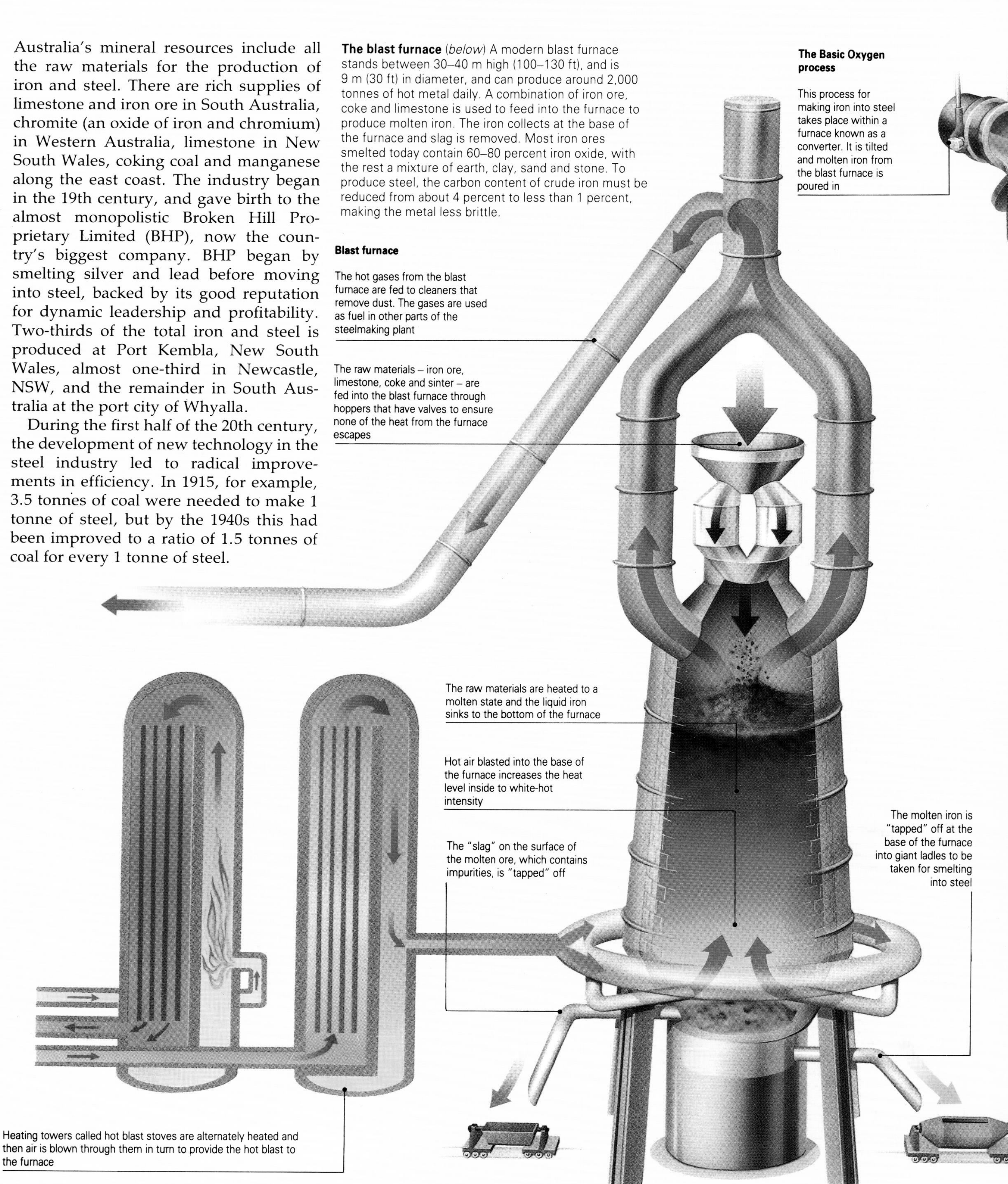

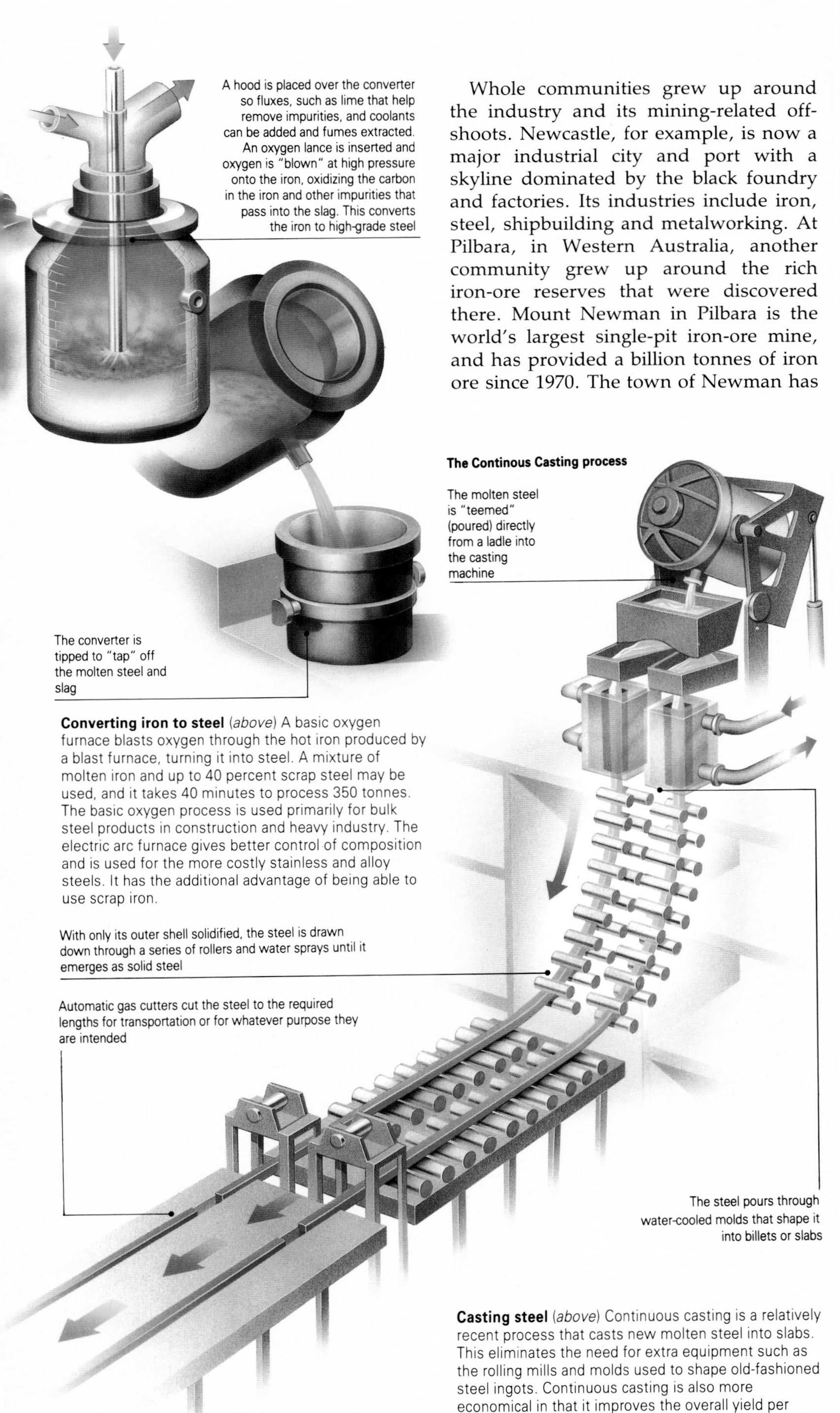

Converting iron to steel *(above)* A basic oxygen furnace blasts oxygen through the hot iron produced by a blast furnace, turning it into steel. A mixture of molten iron and up to 40 percent scrap steel may be used, and it takes 40 minutes to process 350 tonnes. The basic oxygen process is used primarily for bulk steel products in construction and heavy industry. The electric arc furnace gives better control of composition and is used for the more costly stainless and alloy steels. It has the additional advantage of being able to use scrap iron.

Casting steel *(above)* Continuous casting is a relatively recent process that casts new molten steel into slabs. This eliminates the need for extra equipment such as the rolling mills and molds used to shape old-fashioned steel ingots. Continuous casting is also more economical in that it improves the overall yield per tonne, and the quality is better.

Whole communities grew up around the industry and its mining-related offshoots. Newcastle, for example, is now a major industrial city and port with a skyline dominated by the black foundry and factories. Its industries include iron, steel, shipbuilding and metalworking. At Pilbara, in Western Australia, another community grew up around the rich iron-ore reserves that were discovered there. Mount Newman in Pilbara is the world's largest single-pit iron-ore mine, and has provided a billion tonnes of iron ore since 1970. The town of Newman has a population of 6,200, about 30 percent of whom are employed at the mine.

A controversial monopoly

BHP began production in Newcastle in 1915, expanded in 1935, and following the Depression became established at Port Kembla. It first smelted iron and built ships at Whyalla in 1941. The continued growth and expansion of BHP has given rise to controversy over the control of such a basic industry by a monopoly, thought by some to be undesirable, especially with regard to potential exports, which are needed to help payment deficits. High levels of capital investment are needed for new equipment that may not be in use all the time, and might therefore prove unprofitable to BHP. The company is now based in Melbourne and has 21 subsidiaries that dominate the private sector in Australia.

Australian production of steel reached a peak in 1980, but demand and then production fell rapidly, a reflection of the depression that hit the industry worldwide. The world recession and overcapacity of steel caused serious problems for the Australian iron and steel industry. As profitability fell, so did capital investment, and in the early 1980s BHP was producing less than 30 percent of its steel by the continuous casting method that had been adopted by other oil-producing countries. Most of these countries were producing 70 percent of their steel in this way. Wages, state taxes and charges rose and in 1983 BHP's steel operations lost A$94 million. By this time the company had entered the more profitable energy sector and threatened to close down its three steel works.

In 1983 the government launched a five-year plan to establish a viable and internationally competitive steel industry that would also offer job security. New investment would reduce costs and improve quality. Productivity and work practices would be reviewed, and wages and state charges restricted. The government also helped to identify the factors that were limiting competitiveness and sought support for the self-help program from unions and industry.

The plan had limited success but output and sales grew at the expense of jobs, which fell by 15,000 between 1984 and 1991. However, in 1989 the biggest producer, BHP, turned its former loss into profits of A$451 million.

GLOSSARY

Added value
A higher price fetched by an article or RESOURCE after it has been processed. For example, CRUDE OIL has added value when it has been refined.
Aerospace industry
The industries involved in the manufacture or assembly of airplanes, rockets, satellites and other vehicles that operate in the atmosphere or the space beyond.
Alloy
A METAL such as brass which is made of a mixture of two or more metals. Other nonmetallic elements may also be included. They have properties very different from pure metals.
Alluvial
Materials such as silt and sand deposited by flowing water on river beds, flood plains and estuaries.
Aluminum
A METAL, refined from BAUXITE, which is very light and resists corrosion. Aluminum ALLOYS are used where light but strong metals are needed, for example, in airplane manufacture.
Anthracite
A very hard form of COAL which has a high energy value and burns slowly.
Arkwright and Crompton powered spindles
Sir Richard Arkwright (1732–92) invented machines to spin yarn previously done by hand. Samuel Crompton (1753–1827) improved on the designs and his machines were able to spin higher quality yarn at greater speeds.
Armaments industry
The industries manufacturing weapons and ammunition to be used in warfare.
Asbestos
A fibrous mineral that is resistant to fire and to some chemicals. If the fibers are inhaled they can cause severe inflammation in the lungs called asbestosis.
Assembly units
Factories that make products by assembling parts made by other factories.
Automated assembly line
The assembly of a product piece by piece by machines as it moves from one station to another.
Automated production
The manufacture of products by machines rather than people.
Automotive industry
The industries responsible for the manufacture of parts and the assembly of those parts into motor vehicles such as automobiles, trucks and buses.
Automotive products
Products manufactured for the AUTOMOTIVE INDUSTRY.
Avionics
The application of ELECTRONICS to the AEROSPACE industry. This includes the design and manufacture of radar and other instrumentation.

Barium
A metallic element that is used as a pigment in paint, in ALLOYS used to make bearings and by the medical profession in X-ray diagnoses.
Basic oxygen process (Linz-Donawitz process)
The process by which molten PIG IRON and other elements in a furnace are turned into steel. Oxygen is blown onto the molten METAL through a water-cooled lance. The impurities are oxidized and removed and the iron converted to STEEL.
Bauxite
The ORE that is smelted to make the METAL ALUMINUM.
Beryl
A valuable mineral found mainly in granite. Crystals can be up to 1 m (1.09 yd) in length. It is used on X-ray tubes and as a CERAMIC.
Bessemer converter
The long cylindrical vessel into which molten PIG IRON and FERROALLOYS are poured. Oxygen is blown through the molten METAL to oxidize the impurities and turn it into STEEL.
Biofuels
Fuel that is obtained from organic material decomposing. For example, methane gas formed in waste tips can be collected and used as a fuel.
Biotechnology
Technology that uses biological processes such as fermentation. Its scope has increased with the advent of genetic engineering.
"Biotic" resources
RESOURCES taken from the living world such as timber and food. They are renewable if the environment is not damaged by their exploitation.
Bituminous coal COAL that contains less carbon than ANTHRACITE and has a lower energy value.
Blast furnace
A tall furnace into which iron ore, limestone and coke are poured. Air is blown through to aid the combustion of the coke and the ore melts to release a METAL called PIG IRON drawn off at the bottom of the furnace.
Brewing
The process by which beer is made. It involves malting grain and fermenting it with hops to make it bitter. Fermentation uses yeast to convert the carbohydrates in the grain into alcohol.
Brown coal
A peat-like material, also known as LIGNITE, which is an immature form of COAL. It has a lower energy value than more mature forms of coal.

Calcium
One of the main elements found on Earth. It occurs in LIMESTONE and is used in manufacturing CEMENT, chloride and CARBIDE.
Canning
The process of preserving food by sealing it in airtight cans, heating them and rapidly cooling them. Bacteria are killed and the food does not deteriorate.
Capital goods
Manufactured goods that are not sold to consumers but used to make other products, for example robots made for automobile assembly.
Carbide
A compound of carbon and another substance, especially CALCIUM used in the production of the inflammable gas acetylene.
Capital-intensive industries
Industries that require a lot of financial capital to set them up. They include HEAVY industries such as iron and steel making and shipbuilding as well as those industries where expensive machines are needed such as automobile assembly.
Catalytic converter
The device fitted into the exhaust system of a petrol engine in an automobile to convert polluting gases into less harmful carbon dioxide and water before they are emitted into the air.
Catalytic cracking
The process by which CRUDE OIL is refined into PETROLEUM products. The oil is heated and put under pressure in a cracking tower in the presence of a catalyst. Different types of PETROLEUM-based substances condense at different levels in the tower.
Cement
A powdered mixture made by heating LIMESTONE with clay and then grinding it up. When mixed with water it is able to bind bricks together.
Ceramics
Nonmetallic inorganic materials that are baked and made into pots, insulators and other products. Some are able to withstand very high temperatures and are used for lining BLAST FURNACES.
Charcoal
Wood that has been heated to high temperatures but has not burnt because air was excluded. The resulting black carbon makes a good smokeless fuel. Before coke, charcoal was used to smelt IRON.
Chemicals industry
The industries responsible for the manufacture of chemicals that are used by other industries, for example sulfuric acid and ethylene. Many of the chemicals are extremely hazardous.
China clay
A fine white clay also known as kaolin. It is formed when feldspars in granite are weathered and is used in CERAMICS, medicines and for coating paper.
Chromium
A METAL used in STEEL ALLOYS for strengthening and to resist corrosion. It also takes a high polish and is used to plate steel.
Circuit boards
Boards onto which electronic components are soldered. The components are linked by metal strips to make electrical circuits.
Coal
Coal is formed in parts of the Earth's crust where partially decomposed vegetation has been subjected to intense pressure and heat. It is used as a fuel, to generate electricity, and as a raw material by the CHEMICAL INDUSTRY. There are three basic types: ANTHRACITE, the hardest form yielding high energy; BITUMINOUS COAL, the middle grade; and BROWN COAL or LIGNITE yielding low energy and creating significant atmospheric pollution as it burns.
Cobalt
A METAL that is mixed with STEEL to make an ALLOY. It is used in cutting steels and magnets.
Coking coal
COAL heated in a furnace in the absence of air so that gas and other chemicals are given off. It has properties similar to CHARCOAL and is used as a smokeless fuel and to smelt IRON in BLAST FURNACES.
"Combined-cycle" gas turbines
Jet engines that are used to drive TURBINES to generate electricity. The exhaust gases drive a generator and the heat from the gases is collected and used to make steam for a conventional steam-driven generator.
Commodities
RAW MATERIALS or products such as IRON or RUBBER that are traded on world markets.
Communications network
1. Individual roads and/or railroads that link up to form a system of routes. 2. A system of telephones, faxes, electronic mail, and written communications that convey information efficiently.
Components
The individual parts that can be put together to make a product such as a television or automobile.
Computer aided design (CAD)
The design of COMPONENTS, products, buildings, CIRCUIT BOARDS, etc using computer software. The product can be shown in three dimensions on the screen and instructions can sometimes be sent direct to the machine that will manufacture it.
Conditional reserves
Deposits of RESOURCES or fuels that are known to exist but not economic to extract using current technology.
Conglomerate
A large industrial CORPORATION made up of several companies involved in widely different activities.
Consortium
A group of companies working together for a common purpose.
Construction industry
The general term for the companies that build houses, offices, shopping malls etc.
Consumer goods
Goods that are sold to the general public rather than to other industries, for example, radios, televisions, automobiles, washing machines and furniture.
Continuous casting
An energy-saving process by which molten STEEL is cast directly into sheets rather than being cast into ingots, reheated and then rolled into sheets.
Conveyor-belt A flexible continuous belt that moves over rollers and is able to carry materials from one place to another. They are frequently used in factories to move the product through the various stages of manufacturing or to move MINERALS at a quarry.
Cooperative
Individuals who come together to benefit from the economies of scale derived from being able to buy raw materials or sell their goods in bulk.
Copra
The white flesh of the coconut after it has been dried. Oil is extracted from the copra to be used in soaps, synthetic rubber and edible oils.
Corporate restructuring
The changing of work patterns and reporting lines within a company to try to become more productive and therefore more competitive.
Corporation
In commerce, a large company with strong financial backing and often with interests in several countries.
Cottage industry
Strictly, the manufacture of products in private homes by individuals using their own equipment. However, the term is often used to describe very small factories producing HANDICRAFT products.

Crude oil
OIL in its raw state before it is refined.
Custom-made
Items which are made specially to suit a customer's requirements rather than MASS-PRODUCED.

Decentralization
The planned dispersal of commercial or government activities from a center of control to regional areas.
Defense industries
The industries that produce goods such as ARMAMENTS, military equipment, specialized vehicles, military aircraft or radar for the armed forces.
Deindustrialization
The decline of HEAVY ENGINEERING and other MANUFACTURING INDUSTRIES in an area and the corresponding rise of SERVICE INDUSTRIES.
Diamond
The hardest known naturally occurring MINERAL, formed from carbon exposed to extreme heat and pressure. Most diamonds are used in industry for cutting other materials but stones of exceptional quality, known as gem diamonds, are used in jewelry.
Direct reduction
The process of making STEEL direct from IRON-ORE pellets melted in a furnace missing out the stage when ORE is converted into PIG IRON in the BLAST FURNACE.
Distilling
The process of separating components in a liquid by heating it and then collecting and condensing the vapors as they are given off. It is used when refining CRUDE OIL and in the production of alcoholic spirits.
Drift mining
Underground mining of a near horizontal mineral seam. It normally outcrops at the surface and is accessible at that point.

Effluent
Waste liquid from a factory or sewage works.
Electric arc process
A furnace where STEEL is made using heat generated by electricity jumping from one electrode to another.
Electric hearth process
A large OPEN HEARTH furnace in which IRON pellets are heated by electric power and made directly into STEEL.
Electrical assembly (industry)
The assembly of electrical COMPONENTS to make electrical goods such as sound and recording equipment, radios and televisions.
Electrolysis
A way of breaking down or changing a substance by passing an electric current through it.
Electrosmelting
The use of electricity to provide the heat for SMELTING highgrade IRON ORE to make IRON.
Electronic (industry)
The industries that make electronic COMPONENTS such as MICROCHIPS, used in many modern electrical goods.
Energy exporting
Selling surplus electricity or fuels for energy to other states that generate less energy than they use.
Energy importing
Buying the surplus electricity or fuels of another state to supplement domestic supplies.
Energy-intensive industries
Industries that require large amounts of energy and therefore locate close to power supplies. For example the BAUXITE smelting industry generally establishes plants close to supplies of HYDROELECTRIC power.
Entrepreneurialism
A business practice in a capitalist economy whereby an individual or company risks capital in a new venture with the intention of making a profit.
Ethylene
Gas derived from PETROLEUM, used to make polythene.
European Community (EC)
An organization of West European states working together to become an economic unit.
Export-orientated industries
Manufacturing industries that aim to export a large proportion of their production. They are often located at the coast to make export easier.
Extraction industries
Industries that produce PRIMARY RESOURCES for other industries by mining or quarrying.

Fast-breeder reactor
A nuclear reactor that generates its own NUCLEAR FUEL, PLUTONIUM at the same time as generating electricity. Most are still at an experimental stage.
Ferroalloy metals
METALS blended with IRON in the manufacture of STEEL eg CHROMIUM, COBALT, MANGANESE, MOLYBDENUM, NICKEL, TITANIUM, TUNGSTEN, VANADIUM and zirconium. Most of these metals have additional industrial uses outside STEEL manufacture.
Fertilizers
Natural manure or manufactured chemicals, added to the soil to improve the yield of crops. They contain one or more of the main nutrients that plants need, including nitrogen, phosphate and POTASSIUM.
Fiber optics
Flexible glass fibers used to transmit light, which in turn can carry information. The fibers are only 1 mm thick and light travels through them with little loss of energy. The fibers can be used singly or in bunches.
"Flow" resources
Natural and renewable sources of food such as crops and fish, natural and renewable materials such as wood and cotton, and natural and renewable sources of energy such as wind, sunshine and running water.
Fluxes
Materials such as lime that are added to COKE and IRON ORE in BLAST FURNACES during STEEL-making to help remove impurities from the molten mixture.
Food processing
Preparing or altering raw food and packaging it as a new product. For example, fresh sardines are processed into canned sardines in oil.
Fossil fuels
COAL, PETROLEUM or NATURAL GAS, materials used as fuels that occur naturally, normally as the result of vegetation and organisms that died millions of years ago being compressed into a combustible substance. PEAT and LIGNITE are less compressed fuels and less efficient to burn.
Foundry
A factory where molten METALS are poured into molds and cast to make products such as engine blocks.
Free trade
The free exchange of goods and services between one country and another without any tariffs, quotas or other financial barriers.
Freight industries
Industries specialising in transporting goods. They grow up around ports and other transshipment points.

Gallium
A METAL that remains in a liquid state over a wide temperature range. It is used in high temperature thermometers and some ALLOYS.
Gasoline
A light, flammable liquid normally refined from CRUDE OIL. It is used to power automobiles, and other machines driven by internal combustion engines. It is also used in the CHEMICAL INDUSTRY as a solvent.
GATT
The General Agreement on Tariffs and Trade, a treaty that governs world imports and exports. Its aim is to promote FREE TRADE, but at the moment many countries impose TARIFF BARRIERS to favor their own industries and agricultural produce.
Gemstones
Precious and semiprecious stones used for decoration and in jewelry rather than for industrial purposes. Their value is based on beauty, hardness and rarity. Diamond, ruby, sapphire and emerald are precious stones. Others such as onyx are semiprecious.
Geothermal power
The heat of the Earth's interior harnessed as a source of energy. Geysers and hot springs are used to supply hot water and generate electricity in a few countries. Experimental projects are investigating other options.
Gold
A soft yellow PRECIOUS METAL. Used in ALLOYS it is a good conductor of heat and electricity, but its main uses are in jewelry and electrical contacts. It is also accepted as a currency standard.
Graphite
A form of pure carbon that is soft and flaky. It is used as a lubricant, in paint, pencil leads and batteries.

Gypsum
A mineral consisting of hydrated calcium sulphate formed when salty water evaporates. It is used in CEMENT, RUBBER, paper, plaster of Paris and writing chalk.

Handicrafts
Products that are handmade in small quantities and require a high degree of skill and dexterity.
Hardwoods
The wood from trees other than conifers (which produce softwoods). Hardwoods are generally stronger and less likely to rot.
Heavy engineering
The application of scientific knowledge to the design and construction of large pieces of engineering, machines and their main COMPONENTS, for example shipbuilding and the construction of bridges.
Heavy industry
Industries that are involved in manufacturing the products of HEAVY ENGINEERING.
Hematite
The principal ore of IRON varying in color from red to gray to black. The ore contains over 70 percent IRON.
Hi-tech (industries)
An abbreviation for high-technology industries, the research, development and application of advanced technologies such as miniature electronics, information technology and BIOTECHNOLOGY.
Holding company
A company that, through its share ownership, has control of one or more other companies.
Human resources
The workforce that is responsible for the production of goods and the provision of services, through its accumulated skills and labor.
Hydroelectric power
Electricity that is generated from a TURBINE which is spun by the energy of falling water.
Hypothetical resources
RESOURCES that are thought likely to exist, but for which there is still no evidence.

Immigrant labor
Workers in a country who are not native to that country but have been brought in to increase the LABOR FORCE or have come hoping to find work.
Import substitution industries
Industries that have been set up mainly in Third World countries to manufacture products that used to be imported. The industries are normally simple ones with an immediate local market such as cigarette, soap and textile manufacture. They are protected during their start-up phase by high TARIFFS on foreign rivals.
"In-bond" factories
Factories set up in Mexico by American companies to assemble parts. The factories make use of cheaper labor and do not have to pay high import TARIFFS.
Industrial chemicals
Chemicals manufactured by the CHEMICAL INDUSTRY.
Industrial park
A site where groups of factories, usually specializing in HI-TECH manufacturing are located. They are often established close to universities to benefit from their scientific and research expertise and are therefore sometimes known as science parks.
Industrial output
The total amount of goods and services produced by a country or region.
Industrialization
The process by which a country moves from an agricultural economy to an industrial one.
Integrated circuit
Electrical circuits that are etched onto a single SILICON WAFER. The end product is called a MICROCHIP. They are widely used in MICROPROCESSORS.
Internal combustion engine
An engine powered by fuel that is burnt inside the engine rather than in an outside furnace. Examples include automobile engines, jet engines and rockets.
Iron
One of the most common metals found in the Earth's crust in ORES such as HEMATITE. The ORE is smelted in a BLAST FURNACE to produce PIG IRON and then often processed further into STEEL.

Joint venture
Any of a number of industrial activities in which two or more companies are involved. It usually refers to cooperation across national boundaries.
Jute
Plants with straight stems processed to produce fibers used in cloth, ropes, hessian and tarpaulin.

Kaolin *see* **china clay**
Kay's flying shuttle
The shuttle invented by John Kay (1704–1764) that contributed to the mechanization of weaving. The invention allowed the shuttle carrying the weft thread to be passed between the warp threads automatically.
Kerosene
The liquid fuel for jet engines refined from CRUDE OIL. It is also known as paraffin.
Kiln
The large oven in which bricks and pots are baked during manufacture.
Kilowatt-hours
A unit of work or energy equal to that expended by 1,000 watts over one hour.

Labor force
The people that are employed by companies to carry out manufacturing or processing, or to provide services.
Labor union
An association of employees that exists to improve the terms and conditions of their employment by negotiation with the employers.
Labor-intensive industry
The industries where labor represents one of the major costs, or which require large numbers of workers.
Lead
A very heavy METAL frequently used in ALLOYS. It is used in domestic roofing, in accumulators, cable sheaths, in some paints and as a radiation shield.
Light engineering
The application of scientific knowledge to the design and construction of small objects, machines and their main COMPONENTS such as machine tools.
Light industry
The term used to describe the industries that are involved in LIGHT ENGINEERING.
Lignite *see* **brown coal**
Limestone
A sedimentary rock formed under the sea and consisting mainly of calcium carbonate. It is used as a building stone and in the manufacture of CEMENT.
Liquid nitrogen
Nitrogen gas cooled or compressed to become a low-temperature liquid.
Liquid propane gas (LPG)
A flammable gas found in CRUDE OIL which is stored under pressure in containers and used as a fuel. Under intense pressure the gas becomes a liquid.
Liquified natural gas
NATURAL GAS stored in tanks under pressure so that it becomes liquid. It is used as a fuel.
Lobbying
Seeking to influence the policy of government or any political organization to benefit the particular interests of a company, group or individual.
Locomotive
An engine driven by steam, electricity or diesel fuel, used to move railroad wagons and coaches.
Lumber
Logs or sawn timber used as fuel or as a RAW MATERIAL in other industries.

Magnesium
A METAL that burns with a bright flame. It is used in fireworks, flares and flash bulbs and in ALUMINUM ALLOYS for use in aircraft construction.
Manganese
A METAL used in STEEL ALLOYS to improve strength and hardness.
Manufacturing base
A place where a company establishes a key factory to make its product.
Manufacturing industry
The sector of industry responsible for making CAPITAL and CONSUMER GOODS.
Marble
LIMESTONE in which the crystals have been metamorphosed to form a rock that has attractive patterns and the ability to take a high polish.
Mass-produced
Goods that are produced on a large scale to a standard design using a high degree of mechanization.
Megawatts
One million watts. A watt is a unit of power equal to the work done by 1 joule per second.
Microchip
The collection of complex INTEGRATED CIRCUITS on one SILICON WAFER.
Microelectronics
The tiny electronic COMPONENTS such as MICROCHIPS used in many modern electrical goods.
Microprocessor
The central processing unit of a personal computer. It consists of very complex INTEGRATED CIRCUITS stored on one microchip.
Migrant workers
Part of the LABOR FORCE which has come from another country looking for temporary employment.
Mineral resources
The Earth's NONRENEWABLE, inorganic substances used in manufacturing or construction industries.
Molybdenum
A METAL used in METAL ALLOYS to improve strength.
Multinational
A large business CORPORATION with a head office in one country but with subsidiaries in one or more other countries.

Natural gas
A FOSSIL FUEL in the form of a flammable gas that occurs naturally in the Earth. It is often found in association with deposits of PETROLEUM.
Natural resources
The general term for agricultural resources, MINERAL deposits, FOSSIL FUELS and other natural substances that occur naturally and are used in FOOD PROCESSING, MANUFACTURING and other industries.
Newcomen engine The steam engine invented by Thomas Newcomen (1663–1729). It was mainly used for pumping water out of coal mines, but the principle helped to develop steam power for other industries.
Nickel
A METAL mainly used in STEEL ALLOYS such as armor plate and stainless STEEL.
Nonmetallic minerals
MINERALS, including DIAMOND, GRAPHITE, LIMESTONE and SILICON, that are not one of the known METALS.
Nonrenewable resources
Naturally occurring substances that are found in finite deposit and cannot be replaced, for example COAL.
Nuclear fuel
The fuel (usually enriched uranium) used to power NUCLEAR REACTORS. Energy is released by splitting the atoms from which it is formed.
Nuclear power
The term used to describe the electricity generated using the energy derived from NUCLEAR FUELS.
Nuclear reactor
The part of a NUCLEAR POWER station where the nuclear reaction takes place and generates heat.

Offshore production of oil
The extraction of OIL from under the sea bed using wells drilled by rigs at sea.
Oil
The general term for all kinds of CRUDE OIL.
Oil palm
A palm tree whose fleshy fruits contain a kernel from which edible oil is extracted. It is used in soaps, margarine and lubricants.
Oil refinery
The processing plant where CRUDE OIL is refined into different products such as GASOLINE and KEROSENE.
Onshore production of oil
The extraction of OIL from wells drilled on land.
Opal
A semiprecious GEMSTONE.
OPEC
The Organisation of Petroleum Exporting Countries, a cartel that represents the interests of the 11 chief petroleum exporting nations. It is able to exercise a degree of control over the price of their product.
Open-cast mining
Mining minerals from excavations on the surface.
Open hearth furnace
A process for making STEEL from PIG IRON, scrap and IRON ORE using gas as a fuel.
Ores
A deposit from which one or more useful minerals can be obtained economically after SILICON and other waste materials have been discarded, for example IRON ORE.
Outcrop mining
Mining a mineral vein (or bed) of COAL that projects out at the surface.

Palm oil
The edible oil extracted from the kernel of the fleshy fruit of the OIL PALM.
Paper mill
A factory where wood pulp is processed into paper, often of several different quality grades.
Paperboard
A thick cardboard made from compressed layers of wood PULP.
Particleboard
Wooden boards made from a mixture of bits of wood and glue compressed together.
Petrochemical complex
A large plant using OIL or NATURAL GAS to produce a variety of chemicals used by industry.
Petroleum
The thick green mineral oil that occurs in porous underground rocks and is derived from the remains of living organisms deposited millions of years ago.
Petroleum complex
A large plant where OIL is refined into a number of different fuels and products.
Pharmaceuticals
Drugs used for medicinal purposes.
Pig iron
The IRON that is produced in a BLAST FURNACE. It has to be processed further to make it into STEEL.
Plant
A factory or works usually manufacturing goods. Also the equipment used in manufacturing.
Plastics
The general term for a large number of synthetic organic materials that can be molded when they are soft but set to form a tough resistant fabric.
Platinum
A PRECIOUS METAL that is found in NICKEL ORE. It is used as a catalyst in CATALYTIC CONVERTERS for automobile exhausts and when making sulfuric acid.
Plutonium
A highly toxic METAL that occurs in tiny quantities in URANIUM and is produced in a FAST-BREEDER NUCLEAR REACTOR. It can be used as a NUCLEAR FUEL in other reactors or to make nuclear weapons.
Plywood
A kind of wooden board used in building made up of an odd number of layers of wood glued together with the grain running at right angles to each other.
Potash
A compound of POTASSIUM harvested from the sea, and used in the manufacture of fertilizer.
Potassium
A substance that when made into an ALLOY with sodium is used as a coolant in NUCLEAR REACTORS. As a compound it is widely used in fertilizers because it one of the elements that growing plants need.
Precious metal
A METAL of high economic value, used in jewelry as well as industry – usually GOLD, SILVER, PLATINUM or copper.
Pressurized Water Reactor (PWR)
A type of NUCLEAR REACTOR in which the NUCLEAR FUEL heats water which turns into steam, driving TURBINES to generate electricity.
Primary fuel
A naturally occurring fuel such as OIL or COAL that releases its energy as heat. PRIMARY FUELS are often used to generate secondary fuels such as electricity.
Primary industry
An industry responsible for producing PRIMARY PRODUCTS – often MINING, or drilling for minerals.

Primary products
A naturally occurring mineral or the product of agriculture mined, quarried, or harvested in its raw state.
Privatization
The process of transferring companies from state to private ownership.
Processing industries
Industries that take RAW MATERIALS and process them into added-value products with a higher price.
Production-line method
A way of assembling a product piece by piece as it moves from one station to another.
Proven reserves
RESOURCES that are known to exist but have not yet been exploited.
Public works program
A major development project undertaken by the government, often as a way of providing employment.
Puddling process
A process rarely used today by which PIG IRON is turned into pure IRON. The pig iron is melted in a furnace and powdered iron oxide stirred in to drive off the slag. Eventually the iron forms a pasty mass and is hammered or rolled into the required shape.
Pulp
The moist mixture of fibers obtained from woodchips when they are shredded to make paper.

Quaternary industries
The SERVICE INDUSTRIES that supply manufacturers and businesses with legal and financial advice, education and training, information and quality control, catering, and a range of other support activities.

Radioactivity
1. The spontaneous emission of a particle from the atomic nucleus of elements such as URANIUM and PLUTONIUM. 2. The rays emitted – exposure can be dangerous to living things.
Raw materials
Metals, minerals, fuels or foodstuffs in their natural state, used as the basic ingredients during industrial processing or manufacture.
Reactor *see* **nuclear reactor**
Renewable energy
Sources of energy, such as the Sun, wind and waves, that are not found in finite deposits and which cannot be exhausted by exploitation.
Research and development
The exploration of new products or new production methods, aimed at protecting the future of an industry as products grow outdated or old methods become uneconomic.
Resource-based industry
An industry that is directly involved in processing a RAW MATERIAL, often so that it can be sold on to a manufacturing or another processing industry. For example, refining OIL or METALS or harvesting timber.
Resource
RAW MATERIALS, information or qualities in the workforce that are of economic value to industry and business including raw materials, land, labor, capital and less tangible things such as market research, specialist knowledge, and professional skills.
Rolling-stock
All the moving vehicles used on a railroad including LOCOMOTIVES, freight containers and carriages.
Rolling process
The process carried out at a STEEL-rolling mill. Steel slabs are squeezed as they pass through rollers to make sheet steel of varying thickness.
Rubber
A flexible but durable organic substance used in tires, children's toys and other products. Natural rubber is made from the sap of the rubber tree, but industrial needs can be fulfilled by synthetic rubber.
Rutile
A mineral found in plutonic and metamorphic rocks which is the source of TITANIUM and titanium dioxide.

Science park *see* **industrial park**
Semiconductor devices
Devices used in MICROELECTRONICS as tiny switches usually made of SILICON.
Semiskilled employee
An employee who is partly skilled but not sufficiently trained or experienced to undertake very specialized work.
Service industries
Industries that supply services to customers rather than manufacturing goods. They include banking, transport, insurance, education, retailing and distribution.
Shale
A sedimentary rock formed by the compression of clay. Some deposits contain OIL which can be DISTILLED from the shale and used as fuel.
Shipbreaking
The process of dismantling ships, recycling some of the parts to be used again and selling the rest as scrap to be melted down. Few parts are wasted.
Silicon
The second most common element in the Earth's crust after oxygen. Pure silicon is used in making SEMICONDUCTORS by the ELECTRONICS INDUSTRY.
Silicon Valley
The Santa Clara valley in California where many computer and other HI-TECH industries using MICROCHIP technology have grown up close to a concentration of research-based institutions.
Silicon wafer
A thin slice of SILICON on which minute complex INTEGRATED CIRCUITS are built.
Silver
One of the major PRECIOUS METALS. It has the highest electrical and thermal conductivity among known METALS and is sometimes used in printed circuits. It is also used in solder and silver salts are used in photography because they are sensitive to light.
Skilled employee
An employee with sufficient training and skills to take on specialized tasks.
Smelting
The extraction of METAL from an ORE by heating the RAW MATERIAL and separating out impurities.
Smoke-stack industries
Industrial plants, particularly iron and steel works, which use large amounts of solid fuels emitting a great deal of smoke and atmospheric pollution.
Solar energy
The energy produced by the Sun which powers all the natural processes on the Earth. It can be harnessed to produce electricity and to heat solar furnaces.
Speculative resources
RESOURCES which are not PROVEN or HYPOTHETICAL but may exist in areas as yet unexplored.
State-owned industries
Industries that are owned and run by the state for the benefit of the community.
State subsidies
Financial help or incentives, given by the government, to ensure that a commercial activity remains viable and to protect the national economy.
Steam engine
An engine that uses steam to drive a TURBINE or pistons to produce mechanical work.
Steel
Any one of a number of ALLOYS, consisting of a high percentage of IRON, with other METALS added for strength, resistance to corrosion or other qualities.
Steel-rolling mill
A factory where sheet STEEL is produced using the ROLLING PROCESS.
"Stock" resources
NONRENEWABLE RESOURCES of FOSSIL FUELS and MINERALS used by industry to support modern life including COAL, OIL and IRON ORE.
Subassemblies
A collection of parts assembled ready for another product. For example, an automobile engine is assembled separately and fitted into the body.
Sub-bituminous coal
COAL which contains less carbon and more impurities than ANTHRACITE or BITUMINOUS coal and therefore has a lower energy value. See BROWN COAL.
Sulfur
A yellow nonmetallic element used in the manufacture of sulfuric acid and fungicides and in treating RUBBER to give it elasticity. It is found in FOSSIL FUELS and released into the atmosphere as sulfur dioxide.
Synthetic fibers
Artificial fibers such as nylon and polyester, used in the textile industry.
Synthetic oil
A liquid fuel manufactured from solid fuels such as COAL or LIGNITE.

Taconite
A low-grade IRON ORE used where high-grade HEMATITE is not available.
Tariff
A government tax levied on goods from overseas.
Tariff barrier
The obstacle to trade that imported goods have to overcome when a TARIFF is imposed.
Telecommunications
The movement of information from one place to another using wires, fiber optics or radio waves.
Teletext
An information service that transmits messages to computers or television screens.
Tertiary industry
A subdivision of the SERVICE INDUSTRIES which are sometimes divided into tertiary (services supplied directly to consumers) and QUATERNARY (services such as auditing or marketing supplied to producers).
Textiles
Woven fabrics and cloth used in garment manufacture and house furnishings.
Thermal reactors
A type of NUCLEAR REACTOR in which a coolant, often water, absorbs heat from the nuclear reaction and uses it to make steam, which in turn provides the energy to drive a generator to make electricity. PRESSURIZED WATER REACTORS are the most common type.
Tin
A silvery-white METAL commonly used in tin-plate coating on steel cans in the FOOD PROCESSING industry. Mixed with copper it forms the ALLOY bronze.
Titanium
A light strong METAL which is stronger and lighter than STEEL. It is used in ALLOYS for the aircraft industry.
Tourist industry
A general term covering all the services, manufacturing and construction required to cater for tourists.
Trade union see **labor union**
Transistor
A SEMICONDUCTOR device that is combined with other COMPONENTS in an INTEGRATED CIRCUIT to amplify the current. It can be extremely small and cheap to make.
Transportation industries
Road, rail, sea and air transportation, catering for passenger and business travel as well as the movement of goods in bulk. *See also* freight industries.
Tungsten
A brittle METAL with the highest melting point of any metallic element. It is used as a filament in electric lights and in hard ALLOYS used to make cutting tools.
Turbine
A machine that converts the kinetic energy of flowing water or steam into mechanical energy. The most simple example is the water wheel.

Underground mining
Obtaining RESOURCES such as COAL from beneath the surface by digging a shaft to where the resources are found and extracting material to the surface.
Unskilled employee
An employee of a company who has few skills and little training and therefore is unable to do specialized jobs.
Uranium
The RADIOACTIVE METAL used as a fuel in NUCLEAR REACTORS.
Uranium enrichment
The addition of URANIUM-235 or PLUTONIUM-239 to natural uranium so that it can sustain a chain reaction in a NUCLEAR REACTOR.

Vanadium
A METAL which is used in STEEL manufacture to make it stronger and resistant to rust.

Woodchip
The offcuts of wood left over when a log has been sawn into timber, used to make PARTICLE BOARD or PULP.

Further reading

Bateman, M., *Office Development: A Geographical Analysis*, St. Martins Press, NY, 1985

Cohen, S. S., and Zysman, J., *Manufacturing Matters: The Myth of the Post-Industrial Economy*, Basic Books, NY, 1987

Daniels, P. W., *Service Industries: A Geographical Appraisal*, Methuen, London, 1986.

Forester, T., *High-Tech Society: The Story of the Information Technology Revolution*, Blackwell, Oxford, 1987.

Forester, T., *The Materials Revolution*, Blackwell, Oxford, 1990.

Giscard d'Estaing, Valérie-Anne, *The Book of Innovations and Discoveries 1992*, MacDonald, London, 1991.

Hamilton, A., *The Financial Revolution*, Penguin, Harmondsworth, 1986.

Hamilton, J. (ed.), *They Made Our World: Five Centuries of Great Scientists and Inventors*, BBC Books, London, 1991.

Harvey-Jones, J., *Making It Happen: Reflections on Leadership*, Collins, London, 1988.

Hawkes, N., *Structures: Man-made Wonders of the World*, Readers' Digest Association Ltd., London, 1990.

Hobsbawn, E., *Industry and Empire*, Pelican, Harmondsworth, 1982.

Horsley, W., and Buckley, R., *Nippon New Superpower: Japan since 1945*, BBC Books, London, 1990.

Kerrod, R., *Collins Guide to Modern Technology*, Collins, London, 1983.

Messadié, G., *Great Modern Inventions*, Chambers, London, 1991.

Messadié, G., *Great Inventions through History*, Chambers, London, 1991.

Negrine, R., and Papathanassopoulos, S., *The Internationalization of Television*, Pinter, London, 1990.

Olins, W., *Corporate Identity*, Thames & Hudson, London, 1990.

Palfreman, J., and Swade, D., *The Dream Machine: Exploring the Computer Age*, BBC Books, London, 1992.

Porritt, J., *Where On Earth Are We Going?*, BBC Books, London, 1990.

Powell, D., *Counter Revolution: The Texco Story*, Grafton, London, 1991.

Wallace, I., *The Global Economic System*, Unwin Hyman, London, 1990.

Sources used for Data Panels and Energy Balance artwork

Data on Industrial Output is from: The Economist, *Book of Vital World Statistics: A Complete Guide to the World in Figures*, London: Hutchinson, 1990.

Data on "Energy Balance" (Diagram) and "Sources of Energy Output" (in Data Panel) from the same source.

Data on "Industrial Workers" is drawn from the: *1991 Britannica Book of the Year*, Encyclopedia Britannica Inc., Chicago (from both National data section and the Employment and Labour section).

Data on Output of Energy Resources (Coal, Oil, Natural Gas) is drawn from: British Petroleum, *Statistical Review of World Energy* 1990, 1991 (London: British Petroleum).

Data on Production of Energy Resources, Metals, Minerals and Manufactures is drawn from: *1991 Britannica Book of the Year* (but mainly): United Nations, *Industrial Statistics Yearbooks: Commodity Production Statistics* (annually), New York: United Nations.

Acknowledgments

Picture credits

Key to abbreviations: ABCL Allcomm Business Communication Limited, Switzerland; **AD** Art Directors, London, UK; **ANT** Australasian Nature Transparencies, Victoria, Australia; **C** Colorific!, London, UK; **E** Explorer, Paris, France; **EU** Eye Ubiquitous, Sussex, UK; **FSP** Frank Spooner Pictures, London, UK; **HDC** The Hulton Deutsch Collection, London, UK; **HL** The Hutchison Library, London, UK; **IB** The Image Bank, London, UK; **IP** Impact Photos, London, UK; **M** Magnum Photos Limited, London, UK; **MEPL** Mary Evans Picture Library, London, UK; **PEP** Planet Earth Pictures, London, UK; **PF** Popperfoto, Northampton, UK; **PP** Panos Pictures, London, UK; **PR** Photo Researchers, New York, USA; **RHPL** Robert Harding Picture Library, London, UK; **SP** Select Photos, London, UK; **SAP** South American Pictures, Suffolk, UK; **SGA** Susan Griggs Agency, London, UK; **SPL** Science Photo Library, London, UK; **Z** Zefa Picture Library, London, UK

t=top; c=center; b=bottom; l=left; r=right

1 FSP/P Maitre **2** E/Gerard Boutin **3** FSP/Marc Deville **4** FSP/Matt/Liaison **6–7** SGA/William Strode **8–9** E/Francois Gohier **10–11** C/Tom Price **11t** E/A Des Essartons **12** SGA/Michael Maur Sheil **13** IB/Michael Coyne **14** Z/T Horowitz **16** Christine Osborne **17** E/Francois Gohier **18b** Robert Estall **18–19t** IB/Flavio del Mese **20t** FSP/Fornaciari **20–21b** Helene Rogers/TRIP **21t** M/Fred Mayer **22t** MEPL **22–23b** BAL/Science Museum, London **23c** MEPL **24t** MEPL **24–25c** Bildarchiv Preussischer Kulturbesitz **24–25b** BAL/National Railway Museum **26** E D Photos **27** BAL **28–29t** E/Nicholas Thibault **28–29b** PF 30t SPL/David Parker **30–31b** B & C Alexander **31b** SPL/Hank Morgan **32** C/Peter Charlesworth **34b** Reuters **34–35** PR/Spencer Grant **35** M/Rene Burri **36** HL **37** M/Micha Bar-Am **38b** Christine Osborne **38–39** M/James Nachtwey **40b** E D Photos **40–41** FSP/P Aventurier **44–45** AD/Les Hannah **46** IB/Werner Bokelberg **47** AD/George Hunter **48t** M/Paul Fusco 48b AD **49** B & C Alexander **50b** Mach 2 Stock Exchange **50–51** Mach 2 Stock Exchange **52–53** Z **54** B & C Alexander **55** E/Francois Gohier **56b** IB/Jay Freis **56–57** Z/K Goebel **58–59** PR/Fred McConnaughey **60–61** IB/Harald Sund **61t** PR/Dick Luria **62** IB/Michael Melford **63** C/Jim Pickerell **64–65t** PR/David Frazier **65t** C/Jim Pickerell **65b** PR/Art Stein **66** HDC **66–67b** M/Eli Reed **67t** PR/Will McIntyre **68** C/Dan McCoy **69** IB/Ted Kawalerski **70–71** IB/Barry Rokeach **73** M/Alex Webb **74t** FSP/Liaison/Burrows **74b** M/Thomas Sennett **76–77t** C/Randa Bishop **76b** RHPL/Robert Cundy **78b** RHPL/Joe Clark **78–79t** RHPL/Joe Clark **80** E/Maja Koene **82–83t** SAP/Nicholas Bright **83t** David Phillips **84–85t** Z/Kurt Goebel **84b** M/Sebastian Salgado **85t** SAP/Tony Morrison **86–87** SAP/Tony Morrison **88–89** E/Samuel Costa **90** Knudsens Fotosenter **92–93t** Z/B Croxford **93b** Knudsens Fotosenter **94–95t** Knudsens Fotosenter **95b** Knudsens Fotosenter **96** Volvo, Marlow **96–97t** Volvo Media, Sweden **98–99** LEGO **100–101** Rolls Royce Cars Limited **102t** IB/Arthur D'Arazien **102b** B & C Alexander **103b** Nimbus Records **104–105t** IP/Homer Sykes **104–105b** AD **106bl** M/Ian Berry **106br** B & C Alexander **108–109** M/Jean Gaumy **110–111** M/Patrick Zachmann **112b** M/Harry Gruyaert **112–113t** E/DPA **113t** M/Gilles Peress **114t** M/Gilles Peress **114b** M/Bruno Barbey **115** M/Richard Kalvar **116** RHPL/Y Arthus Bertrand **118–119b** IB **120–121t** Z/E Streichan **120b** FSP/Marc Deville **122t** M/Dennis Stock **122–123b** Sygma/ E Prean **123** FSP/Marc Deville **124t** E D Photos **124–125** M/Harry Gruyaert **126–127** SGA/Adam Woolfitt **128–129** M/Fred Mayer **130** IB/Luis Castaneda **131** M/Fred Mayer **132t** M/Fred Mayer **132–133t** M/Fred Mayer **132b** M/Fred Mayer **134bl** SGA/Robert Frerck **134br** DSP/Patrick Landmann **134–135t** SGA/Anthony Howarth **136–137** M/Fred Mayer **138–139** Z/P Freytag **141t** Granata Press Service **141b** FSP/Fornaciari **142b** SP/Robert Koch **142–143** M/Richard Kalvar **143** SGA/Julian Nieman **144** M/Marc Riboud **145t** Aspect/Peter Carmichael **145b** Aspect/Peter Carmichael **146–147t** Z/Streichan **148** Z/Havlicek **149bl** Z/Boleslaw Edelhajt **149br** IB/Gladstone **150** RHPL **151** IB/Peter Grumann **152t** ABCL/Ciba-Geigy **152b** ABCL/Ciba-Geigy **152–153** ABCL/Ciba-Geigy **154** SP/Liba Taylor **156t** SP/Liba Taylor **156–157** M/Bruno Barbey **158t** SP/Serge Attal **158b** SP/Serge Attal **159** SP/Liba Taylor **162–163** M/James Nachtwey **164** M/Fred Mayer **166** M/Abbas **167** M/Fred Mayer **168–169t** FSP/V Shone **168–169b** FSP/Ph Novosti **169t** M/Abbas **170–171** FSP/Novosti **172–173** M/Abbas **174** Christine Osborne **176–177** SGA/Adam Woolfitt **176b** EU/Rhoda Nottridge **177b** C/Israel Talby **178t** M/Abbas **178b** Christine Osborne **178–179t** IB/Kay Chernush **180** Shell Photo Service **182–183** E/J F Gerard **184t** Peter Sanders **184–185t** E/Gerard Boutin **186–187t** M/Bruno Barbey **186b** HL/Liba Taylor **187t** Peter Sanders **188–189** E/Gerard Boutin **188b** Christine Osborne **190–191** Z **192–193t** HL/Bernard Regent **192b** HL **193t** HL **194–195t** Tropix/R Cansdale **195** HL **196t** M/Jean Gaumy **196–197** M/Ian Berry **199** IB/Guido Alberto Rossi **200–201t** SGA/Marc and Evelyne Bernheim **201b** M/Chris Steele-Perkins **202t** PP/Bruce Paton **202b** M/Gideon Mendel **203** HL **204–205** M/Fred Mayer **205t** M/Fred Mayer **206** M/Raghu Rai **208** RHPL/J H C Wilson **209t** M/Bruno Barbey **209c** Christine Osborne **210–211t** Christine Osborne **211b** Christine Osborne **212t** PP/Shahidul Alam **212–213** PP/B Klass **214** Z/K Goebel **216t** IP/Alain le Garsmeur **216–217** IP/Alain le Garsmeur **216b** M/Hiroji Kubota **218t** IP/Alain le Garsmeur **218c** HL/Felix Greene **220t** C/Black Star/Joseph Rupp **220–221t** C/Black Star/Charles Moore **222** Z/Damm **224–225t** RHPL/RKK/Luca Invernizi Tettoni Photobank **225t** Tien-Wah Press **226c** HL/J G Fuller **226b** PP/Paul Harrison **227** HL/Ian Lloyd **228c** PEP/Richard Matthews **228–229** PP/Ron Giling **230** IB/Grant V Faint **232t** M/Richard Kalvar **232–233t** M/Hiroji Kubota **232b** M/Philip Jones-Griffiths **234–235** HL/Michael MacIntyre **235t** AD/Jim Brandenburg **236** PR/Tadanori Saito **238–239** HL/Michael MacIntyre **240** ANT/Bill Bachman **242–243t** Z/Thiele **242b** B & C Alexander **244t** SGA/Robert McFarlane **244–245b** ANT/ L Zann **245** C

Editorial, research and administrative assistance
Neil Alexander, Nick Allen, Brad Bates, Lauren Bourque, Joanna Chisholm, Penny Commerford, Reina Foster-de Wit, Professor Modest Goossens, Robin Kerrod, Matthew Kneale, Hilary McGlynn, Jo Rapley, Dr Alisdair Rogers, Lin Thomas

Artists
The Maltings Partnership, Derby, England

Cartography
Maps drafted by Euromap, Pangbourne, England

Index and Glossary
Barbara James and John Bains

Production
Clive Sparling

Typesetting
Brian Blackmore, Niki Whale

Color origination
Scantrans pte Ltd, Singapore

INDEX

Page numbers in **bold** refer to extended treatment of the topic; in *italic* to illustrations or maps

D

E

F

G

H

I

J

K

L

M

N

O

P

Q